Climate and Ancient Societies

Climate and Ancient Societies

Edited by Susanne Kerner, Rachael J. Dann
and Pernille Bangsgaard

Museum Tusculanum Press
University of Copenhagen

Climate and Ancient Societies
Edited by Susanne Kerner, Rachael J. Dann and Pernille Bangsgaard

© 2015 Museum Tusculanum Press and the authors
Composition and cover design: Erling Lynder
Set with Arno Pro and printed by Tarm Bogtryk
ISBN 978 87 635 4199 2

Cover illustrations: See pp. 116, 117, 122 and 320.

This volume is published with financial support from
The Danish Resarch Council, The Department of Cross-Cultural Studies,
University of Copenhagen, E. Lerager Larsens Fond

Museum Tusculanum Press
Birketinget 6
DK-2300 Copenhagen S
Denmark

www.mtp.dk

Contents

Acknowledgements 9

Foreword: Stine Rossel (1975–2007): An Appreciation
Richard H. Meadow 11

Introduction: Can Archaeology Save the World?
Rachael J. Dann 19

Holocene Climate Reconstruction

Holocene Climate Change and Archaeological Implications, with Particular Reference to the East Mediterranean Region
Neil Roberts 27

Hunter-Gatherers Living in a Flooded World: The Change of Climate, Landscapes and Settlement Patterns during the Late Palaeolithic and Mesolithic on Bornholm, Denmark
Lasse Sørensen and Claudio Casati 41

Complex Society's Responses to Climatic Variation

Urban Adaptations to Climate Change in Northern Mesopotamia
Jason Ur 69

Cultural Transformation and the 8.2 ka Event in Upper Mesopotamia
Peter M. M. G. Akkermans, Johannes van der Plicht, Olivier P. Nieuwenhuyse, Anna Russell and Akemi Kaneda 97

Climate and Social Change during the Transition between the Late Neolithic and Early Chalcolithic in Central Anatolia
Peter F. Biehl 113

A Narrow Place Can Contain a Thousand Friends: Irrigation as a
Response to Climate in the Zerqa Triangle, Jordan
Maurits Ertsen and Eva Kapteijn 137

The Late Bronze Age Collapse and the Early Iron Age in the Levant:
The Role of Climate in Cultural Disruption
*David Kaniewski, Elise Van Campo, Karel Van Lerberghe, Tom Boiy,
Greta Jans, Joachim Bretschneider* 157

Long Term or Short Term? Climate Change and the Demise of the
Old Kingdom
Miroslav Bárta 177

Archaeological Evidence for Pollution and its Ecological Implications

New Data on Animal Exploitation from the Mesolithic to the Neolithic
periods in Northern Sudan
Louis Chaix and Matthieu Honegger 197

Large Game Depression and the Process of Animal Domestication in
the Near East
Benjamin S. Arbuckle 215

Living in a Marginal Environment: Climate Instability and Possible
Lathyrism in the Syrian Neolithic
Deborah C. Merrett and Christopher Meiklejohn 245

Perceptions of Pasture: The Role of Skill and Networks in Maintaining
Stable Pastoral Nomadic Systems in Inner Asia
Joshua Wright and Cheryl Makarewicz 267

Stable Isotope Analysis in the Middle East

Understanding the Reasons for Non-Sustainability in Past Agricultural Systems
Simone Riehl 291

AMS ^{14}C-dated Plants as a Tool for Investigating Palaeoclimate: New Data for Analysing Social Complexity in Ebla and Qatna (Northwestern Syria) in the Light of 3rd Millennium BC Climate Change
Girolamo Fiorentino and Valentina Caracuta 313

Provenance Studies of Ancient Textiles: A New Method Based on The Strontium Isotope System
Karin Margarita Frei 335

Contributors 349

Stine Rossel

Acknowledgements

The chapters in this volume were originally presented at a workshop that took place in April 2009, held in memory of Stine Rossel, who was for far too short a time a colleague at the Institute for Cross-Cultural and Regional Studies.

The workshop and subsequent publication were aided financially by Danish Research Council, the Institute for Cross-Cultural and Regional Studies, the Faculty of Humanities of the University of Copenhagen and E. Lerager Larsens Fond. We are enormously grateful for this support.

The editors would like to express their gratitude to the institute director, Ingolf Thuesen, for his support and the Geological Institute for housing us during the workshop. We would also like to thank Stine Rossel's parents for their help during the entire process from workshop to book. We would like to thank the following students who volunteered their help during the workshop: Nora Abdel-Hak, Kasper Anderson, Marc Fenchel, Mia Høgh, Ditte Kvist, Maria Mawla, Hanne Nyman, Rune Rattenborg, Jonas Steffensen, Elise Thing, Hülja and Marius. Mikkel Thykier put finishing touches to the manuscript during his internship and is thanked for his invaluable work.

We would also like to thank our anonymous readers, who did an excellent job. We also wish to thank Tony Wilkinson for his advice. Sadly, he passed away before he could see the completion of this volume.

Susanne Kerner, Rachael J. Dann and Pernille Bangsgaard

Foreword

Stine Rossel (1975–2007): An Appreciation

Richard H. Meadow

The present volume, and the conference that spawned it, were conceived as a memorial to Dr. Stine Rossel, a young Danish zooarchaeologist who met her death in a freak accident on October 20th, 2007, while hiking with her husband in the White Mountains of New Hampshire (USA). In the ten months before being taken from us, Stine had attained three important milestones in her young life. In January, having just turned 32, she began her professional career as an Adjunkt in the Department of Cross-Cultural and Regional Studies, University of Copenhagen. On August 18th, she was married at Rungstedlund, Karen Blixen's old home north of Copenhagen, to Brian M. Wood, a biological anthropologist whom she met while studying archaeology at Harvard University. And in late September, she submitted her dissertation, *The Development of Productive Subsistence Economies in the Nile Valley: Zooarchaeological Analysis at El-Mahâsna and South Abydos, Upper Egypt*, to Harvard's Faculty of Arts and Sciences for the PhD in Anthropology.

Stine began her archaeological fieldwork in Denmark (at Ballum) in 1996, followed in 1997 by excavations in Syria (Tell Jurn Kebir) and Turkey (Gavur Kalesi and Kerkenes Dag). Her introduction to zooarchaeological research was at Tell Brak, Syria, in 1998, where, once her excavation responsibilities were completed for the day, she enthusiastically served as laboratory assistant to Jill Weber. She decided that the study of animal remains was an interesting and important way to gain insight into past human activities and especially those related to the development of increasing complex socio-economic systems. Thus, at Jill Weber's suggestion, she contacted me about pursuing a higher degree in archaeology with a specialisation in zooarchaeology. This resulted in her application and subsequent acceptance to Harvard's Department of Anthropology where, in September 1999, she began her PhD studies after earlier in the year having received her BA from the Carsten Niebuhr Institute, University of Copenhagen, with courses in Near Eastern archaeology, social anthropology and biological anthropology. As testimony of her precocious archaeological abilities and pedagogical skills, before she had completed her undergraduate degree, Stine had already helped teach field methods to first-year students of the Carsten Niebuhr Institute on site at Gavur Kalesi in Turkey (summers 1997, 1998). Subsequently, in 1999, she joined the University of Pennsylvania/Yale/Institute of Fine Arts of New York University Expedition to Abydos, Egypt, where she spent the field season before beginning her studies at Harvard. Stine worked on the faunal remains from midden deposits outside of the mortuary temple of the fifth monarch of the Twelfth Dynasty of the Middle Kingdom Senwosret III at South Abydos that was being excavated by a team led by Josef Wegner (University of Pennsylvania). A preliminary analysis of material from these contexts was the subject of her first publication (Rossel 2000), with the completed study later to form a significant part of her PhD dissertation.

Once at Harvard, Stine began graduate level study of world archaeology together with archaeological method and theory and was not to return to Abydos until 2001 after she had successfully completed her PhD qualifying examinations in January of that year. Even so, during the summer of 2000, she had found an occasion to broaden her field experience by assisting at the joint Harvard–Georgian excavations of Ortvale Klde, a late Middle–Early Upper Palaeolithic rockshelter in the western Georgian Republic. This was followed in the fall by her participation in my "boot-camp" course on osteoarchaeology, during which she systematised her analytical technique and recording protocols and honed her already considerable ability to reliably discriminate the usually fragmentary remains of domestic animals. Then in the late winter, I took Stine and her classmate Benjamin Arbuckle to Pakistan to excavate at the Indus Civilisation site of Harappa and – after hours – to analyse some of the recently excavated faunal remains from the aceramic Neolithic levels at Mehrgarh that had been transferred to Harappa for storage. This was my first experience working with and observing Stine in the field, and I found not only that she was an excellent excavator, but that she took the best field notes of anyone I had ever worked with. Particularly remarkable, however, was her ability to quickly bond with her crew of seasoned Pakistani excavators. Without knowing any Punjabi when she arrived, she soon picked up enough to communicate effectively with them about the excavations and to learn about aspects of their lives and families.

Following the field season at Harappa, Stine returned to Egypt to continue her analysis of the South Abydos fauna that had been excavated that season. Then in August she travelled to Turkey to participate in what turned out to be the last of three advanced workshops on zooarchaeological methods that I taught in 1994, 1997 and 2001 at the Japanese Institute of Anatolian Archaeology, in Çağırkan village, Kaman, Kırşehir, Turkey (sponsored by the Middle Eastern Culture Center in Japan). Stine thrived on the more than eighty hours of practical training and intense discussion spread over ten days, happily offering up her developing expertise, experience and love of teaching to help five somewhat bewildered Japanese students to comprehend the intricacies of developing a faunal reporting protocol, dealing with the trade-offs inherent in problem-based analyses, using the resulting database to characterise the studied assemblage, and interpreting the results.

In order to better carry out her analyses, Stine took the initiative to obtain additional practical training of particular relevance to the diversity of Egyptian faunal remains that she was encountering in her research. Thus her time at my 2001 workshop in Turkey was followed in 2002 by study with Wim van Neer at the Royal Museum for Central Africa in Tervuren-Brussels on the comparative osteology of freshwater fishes. Then in 2003, she went to Munich to study the comparative osteology of African freshwater fishes with Angela von den Driesch, and again in 2004 to Munich to work with Joris Peters on the comparative osteology of birds – on both occasions at the Institut für Paläoanatomie, Domestikationsforschung und Geschichte der Haustiere.

The next five years saw Stine spending considerable amounts of time carrying out faunal analyses in the field and assisting with excavations. During 2003 and 2004, not only did she continue her work in Egypt on midden assemblages from the Middle Kingdom South Abydos temple and elite residences A and E for her dissertation, but she added another dimension to her project by studying faunal material from the nearby Predynastic settlement site of el-Mahâsna, excavated in 2000 by a team led by David Anderson (University of Pittsburg). This provided her with data nearly 1500 years earlier than that from South Abydos, thus permitting her to begin to look at continuities and transformations in animal exploitation practices over time in the same limited area of Upper Egypt.

Also at Abydos, early in the new millennium, excavations by a team under the leadership of David O'Conner (Institute of Archaeology of New York University) came upon the skeletons of twelve donkeys, which had been intentionally buried in three contiguous grave chambers subsidiary to an initial Early Dynasty (ca. 3000 cal BC) cult enclosure of one of the earliest kings to initiate construction activities in the area of what is today termed "North Abydos". Stine's work on these donkeys began in 2002 and culminated in 2005, when Joris Peters (Ludwig Maximilian University, Munich) and Fiona Marshall (Washington University in St. Louis) joined her at Abydos. Their investigations led to an important article in *PNAS* providing some of the earliest evidence for incipient morphological changes in the donkey skeleton that can be associated with domestication, including stress-related pathologies and modifications in skeletal proportions (Rossel et al. 2008). Stine's encounters with donkeys – both ancient and modern – that were used for draft purposes and as beasts of burden focused her attention on the importance of the use of animals in trade, transport, communication and conflict, particularly as societies became increasingly dependent on distant resources and spread their spheres of influence ever more widely.

Stine's research was not confined to Egypt. In 2003 and 2004, she joined an Oriental Institute, University of Chicago, team at Bronze Age Tell Atchana (ancient Alalakh) in the Amuq region of southern Turkey. Here at the very northern margin of Egypt's influence she began to investigate how the social and economic organisation of an urban polity as reflected in its faunal remains differed from that of contemporary sites in ancient Egypt. When sponsorship of the Atchana project changed after the 2004 season, Stine was prepared to continue working at the Oriental Institute's Iron Age site of Zincirli somewhat farther north, but decided not to do so until after her dissertation was completed. Similarly in 2006, near the southern margin of Egypt's influence, in Sudan, Stine began to work with Louis Chaix and the Kerma prehistoric project of the Université de Neuchâtel at El-Barga and Wadi el-Arab in order to investigate how early experimentation with pastoral production became manifest under changing environmental conditions. Her untimely death robbed both projects of her enthusiasm, dedication and expertise, and her Atchana work remains unpublished.

On May 22nd, 2006, Stine successfully defended a draft of her Ph.D. dissertation with Nanna Noe-Nygaard coming from Copenhagen for the occasion. Late in the summer she was notified that she had been awarded a Postdoctoral Fellowship from the Danish Council for Research on Culture and Communication for the fall of 2006. A Young Elite Researcher's Award from the Danish Councils for Independent Research soon followed, and she began her Adjunkt duties in early 2007 by teaching courses on Ancient Egyptian Archaeology and on Zooarchaeology and supervising BA papers at the University of Copenhagen.

When Stine came to America in early October of 2007, she had just submitted a reworked dissertation for her PhD degree. Over the fifteen months between her oral defense and submission she had completely rethought the significance of her work and had come to some rather provocative conclusions. These are best summarised in her own words, as the work from which they come represents her primary academic legacy to posterity:

> The results of the analysis [presented in the dissertation] are not in agreement with the traditional substantivist economic model of ancient Egypt, which states that the primary mode of resource circulation was by redistribution. Instead the results support a household-based *(oikos)* model in which resources are distributed only partially through redistribution from large households, such as temple estates, to smaller affiliated households. A second result of the analysis suggests that many of the strategies employed in the use of animal resources were similar during the Predynastic and the Dynastic periods. Hence the establishment of the first ruling dynasties ca. 3000 BC, did not lead to a *break* in former subsistence practices. Instead previously established traditions *continued* but went through restructuring to meet the demands of a society characterized by a more integrated economy. A final result suggests that the Predynastic subsistence economy appears to have been closely adapted to the environmental conditions of the fourth millennium. During the Pharaonic periods, although some adaptations to increasingly arid conditions are reflected in the animal economy, basic features of the Predynastic economy continued despite environmentally deteriorating conditions. I argue

that the core features of the Pharaonic subsistence economy were established under different environmental conditions and do not represent the ideal adaptation to prevailing conditions during the Dynastic period. This, I propose, is connected to production strategies of Pharaonic society that were deeply rooted in Predynastic lifeways. (Rossel 2007: iii–iv)

When Stine and I spoke just days before her death, it was evident that the results of her work on the South Abydos and el-Mahâsna materials together with that on the donkeys from North Abydos were now leading her in new directions. As alluded to above, she had become interested in the role that donkeys – but also horses, cattle, and (later) camels – played in moving goods and people across and integrating vast areas of northern Africa. To her, key features were landscapes and the routes through them and how people used both. She was getting involved with the Theban Desert Road Survey led by John and Debby Darnell (Yale University) in Egypt's Western Desert and was in contact with environmental geographers at the Massachusetts Institute of Technology about developing a project that would include a south–north transect across Niger to monitor the pre-, early and more recent history of the Sahel and Sahara. She was terribly excited about her opportunities in Copenhagen and the promise that a new generation of scholars was beginning to hold out for a more nuanced understanding of the development of societies – ancient and more recent – in North Africa.

Stine's passing was a heart-wrenching loss to her husband Brian, to her parents Dorrit and Ole, and to her legion of friends, many of whom were also her academic peers and colleagues. To me, as her PhD supervisor, losing Stine was to lose one of my intellectual offspring. I saw her grow from an enthusiastic beginning researcher to an equally enthusiastic accomplished young scholar with plans and dreams for the future of archaeological research in Africa and for the teaching of ever younger generations in Denmark and elsewhere in the world – wherever her peripatetic ways might take her. She was a joy to work with and a joy to be with. We miss her greatly.

Publications of Stine Rossel

Rossel, Stine, Fiona Marshall, Joris Peters, Tom Pilgram, Matthew D. Adams and David O'Connor. 2008. "Domestication of the donkey: Timing, processes, and indicators." *Proceedings of the National Academy of Sciences (PNAS)* 105(10): 3715–3720.

Rossel, Stine. 2007. *The Development of Productive Subsistence Economies in the Nile Valley: Zooarchaeological Analysis at El-Mahâsna and South Abydos, Upper Egypt*. PhD Dissertation, Department of Anthropology, Harvard University, Cambridge, Massachusetts. ProQuest document ID: 1464110981; ISBN: 9780549278788.

Rossel, Stine. 2006. "A tale of the bones: Animal use at South Abydos." *Expedition* 48(2): 41–43.

Rossel, Stine. 2006. Review of *Modeling Socioeconomic Evolution and Continuity in Ancient Egypt* (C. Yokell, 2004). *Environmental Archaeology* 11: 284–286.

Rossel, Stine. 2004. "Food for the dead, the priest, and the mayor: Looking for markers of status at South Abydos, Egypt." In S. J. O'Day, W. Van Neer and A. Ervynck (eds.), *Behaviour Behind Bones: The Zooarchaeology of Ritual, Religion, Status and Identity*. Oxbow Books. Oxford.

Rossel, Stine. 2000. "Preliminary report on the faunal remains." In J. Wegner (eds.), The Organization of the Temple NFR-K3 of Senwosret III at South Abydos. *Ägypten und Levante* 10: 83–125.

Acknowledgements

Thanks to Jill Weber and Brian M. Wood for helping write the obituary on which this appreciation is partly based. Thanks to Dorrit and Ole Rossel for their wonderful hospitality in Denmark even at the most trying of times. Thanks to Susanne Kerner for organising the conference, editing the book, inviting me to write this Foreword, and for being very, very patient.

Introduction: Can Archaeology Save the World?

Rachael J. Dann

At first glance, the subject of climate and climate change may not appear to be core aspects to archaeological research. Based on the study of material culture, archaeologists might be expected to access information about climate in an oblique manner, for example via types of clothing for particular environments. Yet, alongside its enduring focus on material culture, archaeology and archaeologists have always wrestled with grander themes such as migration, trade or state formation. It is as one of these encompassing themes that climate, and questions concerning human causation of and human adaptations to climate change, can be positioned.

Both environment and climate (two inherently interrelated "objects") have featured in archaeological studies from the very infancy of the discipline. During the 18th and 19th centuries the relationship between the developing discipline of archaeology and other sciences including geology, palaeontology and botany was an imbricated one. The supposedly catastrophic biblical flood episode marked an historical horizon after which human history truly began. It was Frere's recognition and documentation of environmental processes producing deep stratigraphic layers at Hoxne in England that enabled him to posit that the Acheulean hand axes found in the lower layers "may tempt us to refer them to a very remote period indeed" (Frere 1800: 205). Fundamental changes in the understanding of the depth of human antiquity were informed by advances in the recognition of site formation processes at both geological and archaeological timescales and the association of material culture with those contexts (Lyell 1863).

It is perhaps with the advent of processual archaeology that studies of climate and environment in the past really found a home. Studies of ancient climate in archaeology (but also in history and geography) were influenced by cultural ecology – the study of how humans adapted to their environment (and for archaeologists this was given added impetus with the development of new scientific techniques). This adaptation was to be figured in different ways, from Steward's (1949, 1955) strong emphasis on the environment as a core factor influencing cultural development (Fagan 1996: 155–156), to more subtly interwoven explanations of the two.

Rowley-Conwy contends (1999: 511) that the 1939 and then 1953 publication by Grahame Clark of a flow chart linking subsistence, habitat and biome was the first to explicitly connect environmental systems and human culture as part of an interrelated system. Clark's many publications throughout the 1950s reiterate this important reflexive relationship (see particularly *Prehistoric Europe: The Economic Basis*, 1953; *Excavations at Star Carr*, 1954), his theories complementing his practical work. Methodologically, Clark undertook interdisciplinary collaborations in his excavation work with, for example, palynologists and animal bone specialists, using their expertise to build theories concerning the imbricated interaction between humans in the environment in the British Mesolithic and Neolithic. Although Clark never appeared as a leading light of the Processual movement, ideas that prefigure and were taken up by that movement figure strongly in his work (Rowley-Conwy 1999: 524).

It is also in the emergence of processualism that the understanding of human activity as an adaptive mechanism to an external environment was explicitly developed (see Johnson 2010: 23–27 for other distinctive aspects of processual thought). The dissatisfaction with the limitations of historical archaeology in term of its methodological and inferential basis, and in terms of its failure to attempt to grasp and explain the past (see Binford 1968), had left environment, climate, and very often landscape as a whole, absent as an object of study. However, as Kushner observed (1970: 125) the processual emphasis on culture as adaptive meant that culture was viewed as response to an environment (itself positioned as non-culture[al]). Whilst bringing the environment, climate and ecology back into historical explanation, it was simultaneously relegated to a causal factor in cultural change, in what Kushner describes as "a curious blend of Radcliffe-Brownian structural-functionalism, Whitean evolutionism and culturology, and Stewardian multilinear evolution" (ibid).

Yet a co-developing appreciation of the possibilities for explaining variability in the past via reference to ethnographic studies and the coincidental progression of ethnoarchaeological work, steered archaeology along a humanistic route, and opened up new possibilities. Evans-Pritchard's now classic studies of the Nuer of the Sudan were published over a number of years

(1940, 1956). Evans-Pritchard's wide-ranging work has remained influential in many fields, but particularly in archaeology due to his imbricated analyses of human practices, the uses of material culture, and symbolic thought and action. In the present context, one can point to Evans-Pritchard's analyses of Nuer pastoral-nomadism, the cattle complex and the entwining interaction between humans, animals and environment which present an attempt at an holistic description which elides a simple nature-culture divide. It is in this sense that anthropologists helped archaeologists to see the potential in their various objects of study.

In the present theoretical climate, and particularly with a decline in interest in grand narratives, one might have expected "environment" and "climate" to have slipped off the archaeological radar. The shape of past environments can be addressed rather obliquely by archaeologists employing technologies such as GIS, whilst the feel of those environments are also being explored from phenomenological perspectives. Yet studies of ancient climate change are increasingly germane in the widest socio-economic and socio-political contexts which we now occupy as individuals, and as a discipline.

For archaeologists working in the heritage sector, there are a series of practical problems which are currently considered likely to emerge as our climate changes. English Heritage led-research considering climate change predictors in the UK over the next 100 years (with relative degrees of confidence) suggests particular areas of concern. Whilst pointing out that many standing buildings and monuments have survived the last two periods of global warming (c. 1500–1200 BC and c. AD 800–1200), Cassar and Pender (2005: 610) outline renewed or developing threats. Whilst torrential rainfall and subsistence are likely to affect standing monuments and buildings, changes in soil moisture and soil chemistry may lead to losses of archaeological heritage below ground. Rising sea levels may also destroy low lying heritage sites either directly by submersion, or indirectly via erosion. Heritage sites may also have to contend with the invasion of new species, or the loss of others, resulting in a visual alteration to the landscape context of a given site (ibid.: 613–614).

For archaeologists working in a university context, van der Leeuw and Redman (2002) argue that collaborative work on climate change could prove to be a dynamic avenue for research whilst offering a means by which to "reinvigorate archaeology" (597). The academic focus on abrupt episodes of climate change as the end of the 20th century approached (Cleuziou 2009: 727) had perhaps a rather millennial character. Disaster events such as volcanic eruptions or large-scale flooding may punctuate the archaeological record, and may entail major environmental or cultural change (see comparative discussions in Haberle and Lusty 2000), but are perhaps most informative when viewed comparatively across time and space, especially as in that interpretative scenario, archaeology has much to offer. The fundamental ability of the archaeological discipline to provide insight into long-term development, coupled with its understanding that the environment is neither outside nor beyond culture, make it uniquely placed to contribute to any analysis or evaluation of climate change and human causes and responses to it. By virtue of their subjects of study, social science or arts disciplines such as history, economics or geography are unable to work with both extended scales or periods of activity *and* the traces of nuanced human relations.

Whilst demonstrating the wealth of relevant expertise that archaeologists have to offer in interdisciplinary or multidisciplinary collaborations, van der Leeuw and Redman note that archaeologists are not seen as natural collaborators for earth scientists or natural scientists (2002: 599). Whilst certain cross-disciplinary boundaries can be identified such as the specificities of disciplinary practices, differences in vocabulary (ibid.: 597), and chronological differences of the BC and BP systems between the hard sciences and archaeology (Cleuziou 2009: 727), these obstacles do not seem to be so fundamental that they are insurmountable. The English Heritage study mentioned above, for example, combined the expertise of conservators, archaeologists, a climate modeller and a building physicist (Cassar and Pender 2005). The long history of multidisciplinary work which archaeologists have undertaken, and the explicit encouragement of interdisciplinarity within university working environments and from funding bodies, makes collaborative approaches advantageous in various ways (van der Leeuw and Redman 2002: 597).

Perhaps the key underlying aspect in this discussion is the place and relevance of archaeology in a (post)modern, late capitalist, highly technologised world. Building a bridge between the past and the present, and communicating archaeological discoveries with a general public who have an appetite for knowledge of the ancient world, are central responsibilities of our discipline. With the professionalisation of the museum and heritage fields, such issues have been brought into sharper focus. Varied programmes of engagement in the museum space, outreach programmes and diversified use of museum and heritage sites are commonplace. Schemes via which professional archaeologists work with local non-professionals such as community groups, schools or other groups who may not normally be drawn into a museum setting are increasingly viewed as an effective tool for engagement with a general public, often in a reflexive manner. It is in these ways that community archaeology or public archaeology seek to make the archaeological past relevant. Undoubtedly these approaches can be highly successful, enjoyable and thought provoking ventures (but see Dawdy 2009, however, for a critique). Yet, as van der Leeuw and Redman put it, "society maintains academic research and rewards university departments for a variety of reasons. At the top of the list is that research may help solve present day problems" (2002: 598). Van der Leeuw and Redman's surmise prefigures that of Dawdy in her explicit call for "a louder conversation about how archaeology can be socially useful" (2009: 140). A turn away from a well-embedded belief that our disciplinary relevance and importance lies largely in the management of cultural resources and our ability to effectively and diversely engage and educate a public, in order to address current (socio-economic-cultural-environmental) concerns is the challenge these authors pose. No one can expect *all* archaeologists to turn their attention in a new direction, nor to cede ground in domains in which we are already explicitly involved, and are the credible experts. It will be difficult to walk a line that attempts to remain appealing to and engaged with the public, relevant to and credible to policy makers with respect to solving present day issues, whilst also being rigorous and specific to our academic peers.

Van der Leeuw, Redman and Dawdy's perspectives demand a debate on a number of questions concerning our subjects of study, our methodologies, and more challengingly, our identity as a discipline (what should, or could, archaeology be? What should it try to do?). One answer that the authors agree on is a greater disciplinary concern for the study of environment, climate and human interactions therein. In December 2009, the United Nations Climate Change Conference (COP15) was hosted in Copenhagen, just a few months after the conference on which this volume is based. Two weeks of high level international talks ended in the so-called "Copenhagen Accord", an acknowledgement that keeping global temperature rises to less than two degrees centigrade was scientifically beneficial, but without any legally binding treaty, and with major uncertainty concerning the number of countries (out of a total of 192) who would agree to it.

One may consider the range of consequences that climate change could cause to environments (drought, desertification, sea level rise, dramatic weather events) and the humans and animals that live in them (changing population levels and profiles, food shortages, loss of species, colonisation by new species, epidemiological changes). These are all long-term, broadly ranging developments, which will be different according to particular conditions in specific, local and regional settings and which will have visible consequences socially, culturally, economically, and so on. These issues are exactly those with which archaeology has long engaged. Dawdy's most provocative question (2009: 139) is that which forms the title of this introduction: "Can archaeology save the world?" and its repetition here is not intended to be merely flippant. Uniquely placed, due to well-developed methodological and interpretative strategies, and with our *longue durée* perspective, archaeologists clearly have much to bring to the table in forming strategies to anticipate and solve current and future climate and environment based problems. It remains for individuals to meet that challenge.

Bibliography

Clark, G. 1952. *Prehistoric Europe: The Economic Basis*. Methuen. London.

Clark, G. 1954. *Excavations at Star Carr*. Cambridge University Press. Cambridge.

Cassar, M., and R. Pender. 2005. *The Impact of Climate Change on Cultural Heritage: Evidence and Response*. 14th Triennial Meeting. The Hague Reprints. James & James. London.

Cleuziou, S. 2009. "Extracting Wealth from a Land of Starvation by Creating Social Complexity: A dialogue between Archaeology and Climate?" *Comptes Rendus Geoscience* 341(8–9): 726–738.

Dawdy, S. L. 2009. "Millennial Archaeology: Locating the discipline in the age of insecurity." *Archaeological Dialogues* 16(2): 131–142.

Evans-Pritchard, E. E. 1940. *The Nuer: Political Institutions of a Nilotic People*. Clarendon Press. Oxford.

Evans-Pritchard, E. E. 1956. *Nuer Religion*. Oxford University Press. Oxford.

Frere, J. 1800. "On the Flint Weapons of Hoxne in Suffolk." *Archaeologia* 13: 204–205.

Haberle, S. G., and A. C. Lusty. 2000. "Can Climate Influence Cultural Development? A View Through Time." *Environment and History* 6: 349–369.

Johnson, M. 2010. *Archaeological Theory: An Introduction*. Wiley-Blackwell. Oxford.

Leeuw, S. van der, and C. L. Redman. 2002. "Placing Archaeology at the Center of Socio-Natural Studies." *American Antiquity* 67(4): 597–605.

Lyell, C. 1863. *The Geological Evidences of the Antiquity of Man, with Remarks on Theories of the Origin of Species by Variation*. Murray. London.

Rowley-Conwy, P. 1999. "Sir Graham Clark, 1907–1995." In T. Murray (ed.), *Encyclopedia of Archaeology: The Great Archaeologists*, Vol. II, ABC-CLIO, 507–529. Santa Barbara, California.

Schrag, D. P., and R. B. Alley, 2004. "Ancient lessons for our future climate." *Science* 24. 821–822.

Wylie, A. 2000. "Questions of Evidence, Legitimacy and the (Dis)Union of Science." *American Antiquity* 65: 227–237.

Holocene Climate Reconstruction

Holocene Climate Change and Archaeological Implications, with Particular Reference to the East Mediterranean Region

Neil Roberts

Abstract

Global and regional climatic variations during the Holocene are recorded in a wide array of proxy data. Within these, primary palaeoclimate proxies (e.g. stable isotopes from cave speleothems) may be distinguished from those which behave as response variables (e.g. pollen data or alluvial histories). The latter group are influenced not only by climate, but also by human impact along with ecological or geomorphological dynamics. This distinction is particularly significant in regions such as the Mediterranean which have a long-standing human "footprint". Although a few proxy data have the time resolution needed to identify inter-annual climate variability, most natural archives only reveal multi-decadal to multi-millennial climate trends. In addition there are very important limits to absolute dating precision (e.g. calibrated ^{14}C ages), which reduces our ability to correlate reliably short-lived climate events (e.g. drought periods) and archaeological records (e.g. cultural collapse). This problem may be circumvented via the multi-proxy approach and by the use of time-synchronous marker horizons, such as volcanic tephra layers. These approaches are illustrated with reference to coupled lake isotope and pollen records from the eastern Mediterranean region at the end of the Late Holocene Beyşehir Occupation cultural landscape phase.

Introduction: Holocene climatic variability

One of the characteristic features of the Holocene climate has been its relative stability. Nonetheless, Holocene climate has varied on a hierarchy of timescales and for a variety of different causes (Wanner et al. 2008). Long-term (i.e. multi-millennial) climatic changes can be linked to variations in the Earth's orbit around the sun, the so-called Milanković cycles. Although orbital changes in solar radiation took place gradually, the Earth's climate responded much less smoothly, due to complex interactions between factors such as greenhouse gas concentrations and ice-ocean albedo effects. Shorter-term "Sub-Milanković" climatic variations have also occurred as a result of secular variations in the receipt of solar radiation (e.g. due to sunspot cycles), the effects of large volcanic eruptions, and other factors. Sometimes these have been abrupt in character; for example, around 8200 years ago, when there was a 100 to 150 year long cooling across much of the Northern Hemisphere (Alley and Ágústsdóttir 2005). Its cause was a pulse of glacial meltwater into the North Atlantic ocean from the final breakup of the Laurentide (Canadian) ice sheet over Hudson Bay. This event caused major drought in the Old World tropical drylands, as indicated by an abrupt fall in African lake levels (Gasse 2000). The "8.2 ka" event probably also affected southwest Asia, although independent palaeoclimatic records of it in this region are not common.

Other centennial scale, abrupt climatic events have been identified for later periods of the Holocene, but it is unclear how widespread these were. One such drought event has been recognised around 4200 BP in southwest Asia with links postulated to the collapse of the Akkadian Empire and the Egyptian Old Kingdom (Cullen et al. 2000; Drysdale et al. 2006; Stanley et al. 2003; Weiss et al. 1993). Its spatial extent may have extended to the Indian Ocean monsoon weather system, since the Blue Nile flood is controlled by summer rainfall over Ethiopia; however, its ultimate cause is unknown. Late Holocene climate changes include those associated with the Little Ice Age (~AD 1400–1850) and the preceding mediaeval climate anomaly (~AD 800–1400). While these periods are well marked in the circum-North Atlantic region, their spatial extent and manifestation elsewhere on the planet are much less certain.

There were, thus, multiple timescales of climate variation during the Holocene. Some, but not all, changes were abrupt; some, but not all, were quasi-cyclical in character. Shorter-term fluctuations were often superimposed on longer-term trends, and most, if not all, Holocene climate changes were regional rather than global in extent.

Climate archives and culture change in the eastern Mediterranean

Global and regional climatic variations during the Holocene are recorded in a wide array of proxy data. There are numerous potential palaeoclimatic data sources, but some are more useful than others if we want to relate them to long-term societal change. Ideal palaeoclimate indicators should be long, continuous and well-dated, they should be able to register sub-Milanković as well as long-term climate changes, and they should not respond to non-climatic factors (e.g. the cultural filter that can bias evidence such as faunal assemblages from archaeological sites). Finally, from an archaeological point of view they should not be geographically remote from areas of human occupation and there should be robust ways to correlate palaeoclimatic and archaeological sequences, not just by wiggle matching curves.

Using these criteria, some of the most useful palaeoclimate proxies in the eastern Mediterranean region are

- cave speleothems (e.g. Bar-Matthews et al. 1998; Fleitmann et al. 2007; Verheyden et al. 2008)
- deep sea sediment cores; although it should be noted that these were also subject to exogamous influences, such as changing Nile discharge in relation to early Holocene sapropel formation (Calvert and Fontugne 2001)
- lake levels, salinity and isotope hydrology (e.g. Roberts et al. 2008); while normally free of non-climatic influences, human impact on catchments can alter lake water balances.

These proxies may be distinguished from secondary and/or response variables, which can be influenced not only by climate but also by human impact along with ecological or geomorphological dynamics. This distinction is particularly significant in regions, such as the eastern Mediterranean, which have a longstanding human "footprint" and where there was no baseline environment prior to the impact of Neolithic farming societies, unlike, for example, in northern Europe. These proxies include pollen evidence (vegetation and land use; see Schab et al. 2004), microcharcoals (wildfire history; see Turner et al. 2008), onsite archaeological evidence (Fiorentino et al. 2008) and geomorphological data (e.g. alluvial histories; see Dusar et al. 2011; Wilkinson 1999).

During the first half of the Holocene, increased receipt of summer radiation in the Northern Hemisphere led to a strengthened and expanded African-Asian monsoon circulation. This caused tropical rains to extend further northwards; for example, into what is today the Sahara desert. This "African Humid Period" came to an end around 6000–5000 BP (Gasse 2000). Composite $\delta^{18}O$ records from the eastern Mediterranean also show a long-term drying trend between 8000 and 3000 BP (Roberts et al. 2011; Finné et al. 2011). This desiccation was oscillatory, with dry phases occurring in the later 6th, 5th and 4th millennia BP, and coinciding with major breaks in the cultural record at the end of the Chalcolithic, the Early Bronze Age and the Late Bronze Age, respectively. The decline in rainfall was therefore not gradual, but punctuated, and the 4200 BP drought event may therefore have been one of several negative rainfall fluctuations during a period of long-term climatic desiccation.

The decline in rainfall in many areas of the Old World tropics and sub-tropics during the mid-Holocene was also the period of major cultural transformation that saw the rise of the first complex Near Eastern civilisations in regions such as Mesopotamia. In northern Africa, as summer monsoon rains weakened and retreated, so the modern day Saharan and Arabian deserts came into existence. Cattle-raising pastoralists were pushed southward into sub-Saharan Africa for the first time. Other human groups responded to mid-Holocene climatic desiccation by relocating from the eastern Sahara into well-watered "oases", notably along the river Nile where flood irrigation developed to provide the economic basis for pre-Dynastic Nilotic civilisation (Kuper and Kröpelin 2006).

Testing cause and effect relationships

Comparing and combining these various archives poses non-trivial methodological challenges, for instance in terms of reliably correlating records. This is often best attempted via integrated regional multidisciplinary field projects. Accurate dating plays a key role in inter-archive comparison, because past causal relations are commonly inferred from serial relations; that is, from the timing between events. As it is not possible for societal consequences to have preceded a purported environmental (or other) cause, we can use this logic to falsify or verify hypothesised relationships.

For most of prehistoric and early historic times, ^{14}C dating provides the main basis for sequence chronologies, both palaeoclimatic and archaeological. After allowing for standard statistical errors and the time-varying additional uncertainty introduced by the need for ^{14}C calibration, the precision attributed to any individual date almost always rises to more than ±50 years. This is longer than the sub-centennial duration of many Holocene drought events recorded in continuous high resolution palaeoclimatic sequences. Chronological uncertainties of non-annually resolved climate records are too large to answer decadal timescale questions, including some of most direct societal relevance. Instead there is a real danger that the problem of "suck-in and smear" can lead to erroneous – or at least not proven – correlations, with obvious consequences in terms of correctly or wrongly inferred causal relationships (Baillie 1991). For example, the 4200 BP drought event has been linked to ^{14}C-dated sediment cores from the Arabian-Persian Gulf, archaeological site abandonment in Upper Mesopotamia, a dust layer in the Kilimanjaro ice cap, and historical records of failing Nile floods in Egypt. But does the dating precision allow us to be sure that these are all recording the same event?

One way to improve correlation between palaeoclimatic and archaeological sequences is by the use of time-synchronous marker horizons, such as volcanic tephra layers (Zanchetta et al. 2011). Another way to test the synchroneity, or otherwise, of climatic and cultural changes is by employing multiple analytical techniques on the same stratigraphic sequence; the so-called multi-proxy approach. In this case, pollen or sedimentological parameters can provide an index of human land use change, while stable isotope or other geochemical/biological indicators offer a proxy for past climate. Because these can be derived from the same sediment records, there is no possibility of mis-correlation between them, even if their absolute age is not known precisely. For example, an annually varved late Holocene sedimentary sequence from Nar crater lake in central Anatolia shows pollen evidence for the latter stages of the Beyşehir Occupation cultural land use phase (England et al. 2008). This phase came to an end during early Byzantine times and was followed by apparent landscape abandonment. The inferred age of its termination varies between ~400 and 700 AD in different individual pollen diagrams based on ^{14}C dating (Eastwood et al. 1998). Did the inferred societal "collapse" in this region coincide with a period of drought conditions? The Nar record has also been analysed for stable isotopes which show that markedly dry climatic conditions prevailed between AD 400 and 500 (Jones et al. 2006). With a ±150 year dating uncertainty for the end of the Beyşehir Occupation phase, it would be tempting to align inter-site chronologies so that drought and landscape abandonment were synchronised. In fact, cultural pollen indicators in the Nar record do not decline until ~670 AD based on varve count ages; that is, two centuries later than the period of greatest aridity in the same record (England et al. 2008). Examination of climatic and cultural history from the same sedimentary archive provides a rigorous test of the hypothesis that the late Antique societal crisis was prompted or inflamed by climatic stress, and in this case it has been falsified.

Even when climate archives and archaeological histories can be correlated with confidence – via tephrostratigraphy, by using the multi-proxy approach or by other means – a key issue remains their commensurability in space and time. For example, we can examine the societal consequences of contemporary (i.e. 20th/21st century) climatic events, such as the Sahel drought of the 1970s and 1980s, using social anthropological studies coupled to observational data from weather stations and satellites. Similarly we can examine the effects of the climatic deterioration of the Little Ice Age on Norse settlement in Greenland using historical archive sources, such as the Icelandic sea ice cover chronicles. Finally we can hope to compare long-term archaeological trajectories recording the development of complex urban societies in the Near East at the start of the Bronze Age with proxy-climate records from east Mediterranean caves and lakes registering an oscillatory decline in rainfall during the mid-Holocene. However, the timescale of change in each of these three cases was very different – from decadal to centennial to millennial – and therefore so, too, were the processes involved. In the case of a single event, even a very severe one such as the Sahel drought, people would normally expect both the weather and their livelihoods to return to pre-drought conditions, unless other factors intervened. However, human responses change once climate perturbations are repeated, and droughts, storms or cold winters became the norm rather than the exception, as they did in the North Atlantic after AD 1400 (Dawson et al. 2003). The Greenland Norse perceived this adverse shift in the climatic baseline as a function of cosmological disorder, and their response was to build ever more impressive churches, in order to appease "Almighty God". In the absence of technological or economic adaptations, population declined and Norse settlement in Greenland was abandoned in the mid-15th century AD (Buckland et al. 1996; Diamond 2005, ch. 6; Dugmore et al. 2007, 2012).

In the eastern Mediterranean region, the 6th millennium BP trend towards a drier climate is paralleled in the archaeological record by the Late Chalcolithic-EBA cultural transition. But rather than having a "negative" effect, this synchronism points, if anything, to a "positive" relationship because it corresponds to the start of EBA city states and flourishing urbanisation and trade (Roberts et al. 2011). The relationship between climatic stimulus and cultural change is particularly clear in Egypt, where previously dispersed cattle cultures were faced with a range of alternative choices in the face of climatic desiccation. The alternatives included migration southwards into sub-Saharan lands, the adoption of desert nomadism, or nucleation and agricultural intensification within the Nile valley (Kuper and Kröpelin 2006). Different strategies were adopted by different population groups, and the last of them provided the foundation of Dynastic Egypt at the start of the 5th millennium BP. Elsewhere in the Near East, the centralisation of power in emerging states may also have been a response to the need to protect growing EBA societies from the possibility of climatically induced instability and risk (Roberts et al. 2011).

Thus, the dynamics of the climate-culture relationship has varied with different timescales of change; and, from a research perspective, it is important that we do not attempt to mix them. For example, the observed responses of pastoral or agrarian societies in dryland regions to short periods of drought are likely to involve adaptive strategies such as food storage and redistribution. However, this will be insufficient to explain human responses to repeated mega-droughts of multi-decadal duration, such as those which affected the Pueblo cultures of the American southwest between AD 990 and 1300, and is even less relevant to understanding the responses of the Cattle Cultures of the northeast Sahara to 6th millennium BP climatic desiccation. What determines household responses to a single drought year is not the same as the way an urban polity reacts to five decades of declining average rainfall and crop yield. Dependent and independent variables consequently change at different temporal and

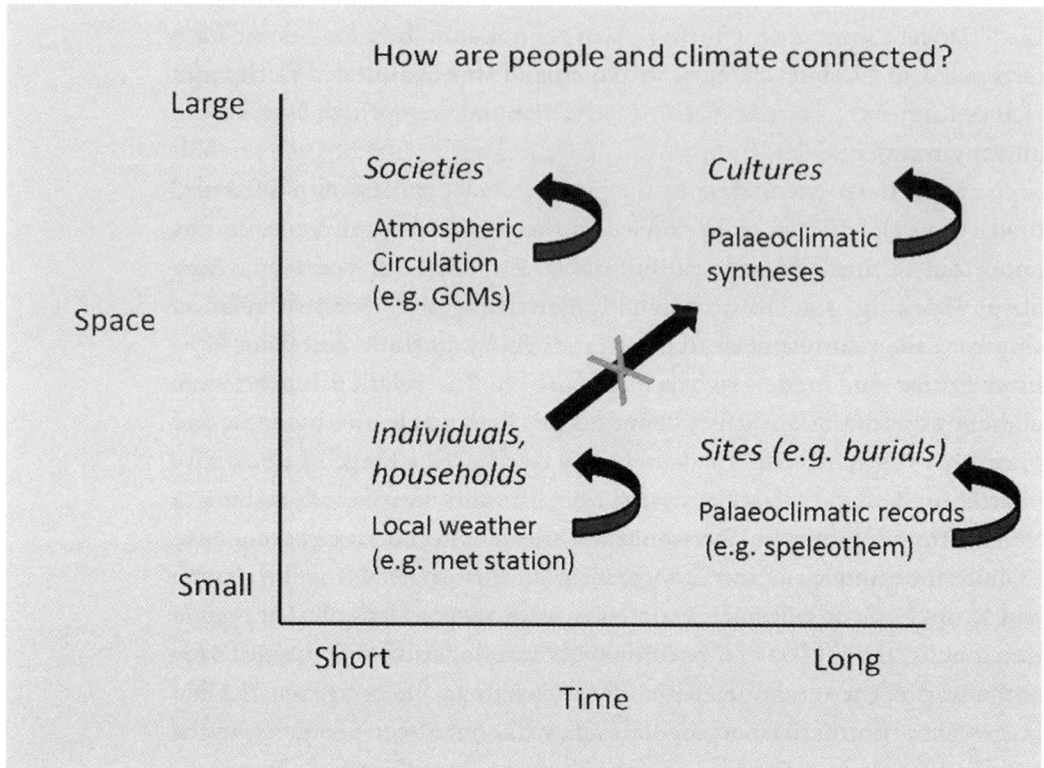

FIGURE 1. *Connecting people and climate: A schematic framework over different temporal and spatial scales.*

spatial scales. The issue of commensurability of scale in understanding the relationship between climate and cultural change is shown schematically in figure 1. As this diagram attempts to show, we can hope to compare regional scale palaeoclimatic synthesis with long-term reconstructions of cultural change, but not with individual or household responses to variations in weather. If we can avoid mixing "apples and oranges", then debates over cultural responses to climate change are likely to be less sterile and more productive.

Conclusion

The Holocene provides valuable insights into how human beings have responded to climate changes in the past. When cultural and climatic changes appear to coincide in time, it can become temptingly easy to slip into environmental determinism and use climatic fluctuations as a catchall explanation for socio-economic changes (cf. deMenocal 2001; Weiss and Bradley 2001), when in reality non-environmental factors were at least as important (Crumley 2000; Rosen 2007). Societies were perfectly capable of "bucking" the climatic trend under appropriate circumstances, as shown by the example given above from early Byzantine Anatolia or from those Bronze Age societies in the Near East which survived through the climatic stresses of the late 5th millennium BP (Rosen 2007; Kuzucuoğlu and Marro 2007). Alternatively, it would be equally naïve to deny the possibility of a role for climatic variations in past human affairs. Past human responses to climate change are not just about salutory lessons and societal collapse. Equally informative and more useful to us are those societies in the past who coped successfully with climatic stress and did not collapse. What is clear is that societal responses to environmental stresses and opportunities have not been predetermined in the past, but have been chosen from within a range of constraints, based on belief systems, economic dependency, social organisation and governance (Diamond 2005; Rosen 2007). Understanding different societal choices in the past can help inform contemporary responses to present or future environmental risk, such as predicted human induced climate change (Roberts 2011).

Bibliography

Alley, R. B., and A. M. Ágústsdóttir. 2005. "The 8k event: Cause and consequences of a major Holocene abrupt climate change." *Quaternary Science Reviews* 24: 1123–1149.

Baillie, M. G. L. 1991. "Suck-in and smear: Two related chronological problems for the 90s." *Journal of Theoretical Archaeology* 2: 12–16.

Bar-Matthews, M., A. Ayalon and A. Kaufman. 1998. "Middle to Late Holocene (6,500 Yr. Period) paleoclimate in the Eastern Mediterranean region from stable isotopic composition of speleothems from Soreq Cave, Israel." In A. S. Issar and N. Brown (eds.), *Water, Environment and Society in Times of Climatic Change*, Kluwer, 204–214. Dordrecht.

Buckland, P. C., T. Amorosi, L. K. Barlow, A. J. Dugmore, P. A. Mayewski, T. H. McGovern, A. E. J. Ogilvie, J. P. Sadler and P. Skidmore. 1996. "Bioarchaeological and climatological evidence for the fate of Norse farmers in medieval Greenland." *Antiquity* 70: 88–96.

Calvert, S. E., and M. R. Fontugne. 2001. "On the late Pleistocene-Holocene sapropel record of climatic and oceanographic variability in the eastern Mediterranean." *Paleoceanography* 16: 78–94.

Crumley, C. L. 2000. "From Garden to Globe: Linking Time and Space with Meaning and Memory." In R. J. McIntosh, J. A. Tainter and S. K. McIntosh (eds.), *The Way the Wind Blows: Climate, History, and Human Action*, Columbia University Press, 193–208. New York.

Cullen, H. M., P. B. deMenocal, S. Hemming, G. Hemming, F. H. Brown, T. Guilderson and F. Sirocko. 2000. "Climate change and the collapse of the Akkadian empire: Evidence from the deep sea." *Geology* 28: 379–382.

Dawson, A. G., L. Elliott, P. Mayewski, P. Lockett, S. Noone, K. Hickey, T. Holt, P. Wadhams and I. Foster. 2003. "Late-Holocene North Atlantic climate 'seesaws', storminess changes and Greenland ice sheet (GISP2) palaeoclimates." *The Holocene* 13: 381–392.

deMenocal, P. B. 2001. "Cultural Responses to Climate Change During the Late Holocene." *Science* 292: 667–673.

Diamond, J. 2005. *Collapse: How Societies Choose to Fail or Succeed*. Allen Lane. London.

Drysdale, R., G. Zanchetta, J. Hellstrom, R. Maas, A. Fallick, I. Cartwright, L. Piccini and M. Pickett. 2006. "Late Holocene drought responsible for the collapse of Old World civilisations is recorded in an Italian cave flowstone." *Geology* 34: 101–104.

Dugmore, A. J., C. Keller and T. H. McGovern. 2007 "Norse Greenland settlement: Reflections on climate change, trade, and the contrasting fates of human settlements in the North Atlantic Islands". *Arctic Anthropology* 44: 12–36.

Dugmore, A. J., T. H. McGovern, O. Vésteinsson, J. Arneborg, R. Streeter and C. Keller. 2012. "Cultural adaptation, compounding vulnerabilities and conjunctures in Norse Greenland". *Proceedings, National Academy of Sciences USA* 109: 3658–3663.

Dusar, B., G. Verstraeten, B. Notebaert and J. Bakker. 2011. "Holocene environmental change and its impact on sediment dynamics in the Eastern Mediterranean." *Earth-Science Reviews* 108: 137–157.

Eastwood, W. J., N. Roberts and H. F. Lamb. 1998. "Palaeoecological and archaeological evidence for human occupance in southwest Turkey: The Beyşehir Occupation Phase." *Anatolian Studies* 48: 69–86.

England, A., W. J. Eastwood, C. N. Roberts, R. Turner and J. F. Haldon. 2008. "Historical landscape change in Cappadocia (central Turkey): A palaeoecological investigation of annually-laminated sediments from Nar lake." *The Holocene* 18: 1229–1245.

Finné, M., K. Holmgren, H. S. Sundqvist, E. Weiberg and M. Lindblom 2011. "Climate in the eastern Mediterranean, and adjacent regions, during the past 6000 years – a review." *Journal of Archaeological Science* 38: 3153–3173.

Fiorentino, G., V. Caracuta, L. Calcagnile, M. D'Elia, P. Matthiae, F. Mavelli and G. Quarta. 2008. "Third millennium B.C. climate change in Syria highlighted by Carbon stable isotope analysis of ^{14}C-AMS dated plant remains from Ebla." *Palaeogeography, Palaeoclimatology, Palaeoecology* 266(1–2): 51–58.

Fleitmann, D., S. J. Burns, A. Mangini, M. Mudelsee, J. Kramers, I. Villa, U. Neff, A. A. Al-Subbary, A. Buettner, D. Hippler and A. Matter. 2007. "Holocene ITCZ and Indian monsoon dynamics recorded in stalagmites from Oman and Yemen (Socotra)." *Quaternary Science Reviews* 26: 170–188.

Gasse, F. 2000. "Hydrological Changes in the African Tropics since the Last Glacial Maximum." *Quaternary Science Reviews* 19: 189–211.

Jones, M. D., N. Roberts, M. J. Leng and M. Türkeş. 2006. "A high-resolution Late Holocene Lake Isotope Record from Turkey and Links to North Atlantic and Monsoon Climate." *Geology* 34(5): 361–364.

Kuper, R., and S. Kröpelin. 2006. "Climate-Controlled Holocene Occupation in the Sahara: Motor of Africa's Evolution." *Science* 313: 803–807.

Kuzucuoğlu, C., and C. Marro (eds). 2007. *Human societies and climate change at the end of the Third Millennium: Did a crisis take place in Upper Mesopotamia? (Sociétés humaines et changement climatique à la fin du Troisième Millénaire: une crise a-t-elle eu lieu en Haute Mésopotamie?)*, Varia Anatolica XIX, IFEA (Istanbul) and de Boccard (Paris).

Roberts, N. 2011. "'Living with a moving target': Long-term climatic variability and environmental risk in dryland regions." In N. F. Miller, K. M. Moore and K. Ryan (eds.), *Sustainable Lifeways. Cultural Persistence in an Ever-changing Environment.* University of Pennsylvania Press. Philadelphia, pp. 13–38

Roberts, N., W. J. Eastwood, C. Kuzucuoğlu, G. Fiorentino and V. Caracuta. 2011. "Climatic, vegetation and cultural change in the eastern Mediterranean during the mid-Holocene environmental transition." *The Holocene* 21: 147–162.

Roberts, N., M. D. Jones, A. Benkaddour, W. J. Eastwood, M. L. Filippi, M. R. Frogley, H. F. Lamb, M. J. Leng, J. M. Reed, M. Stein, L. Stevens, B. Valero-Garcés and G. Zanchetta. 2008. "Stable isotope records of Late Quaternary climate and hydrology from Mediterranean lakes: The ISOMED synthesis." *Quaternary Science Reviews* 27: 2426–2441.

Rosen, A. M. 2007. *Civilizing Climate: Social Responses to Climate Change in the Ancient Near East.* Altamira Press. Lanham, Maryland.

Schwab, M. J., F. Neumann, T. Litt, J. Negendank and M. Stein. 2004. "Holocene palaeoecology of the Golan Heights (Near East): Investigation of lacustrine sediments from Birket Ram crater lake." *Quaternary Science Reviews* 16–17, 1723–1732.

Stanley, J.-D., M. D. Krom, R. A. Cliff and J. C. Woodward. 2003. "Nile flow failure at the end of the Old Kingdom, Egypt: Strontium isotopic and petrologic evidence." *Geoarchaeology* 18: 395–402.

Turner, R., N. Roberts and M. D. Jones. 2008. "Climatic pacing of Mediterranean fire histories from lake sedimentary microcharcoal." *Global and Planetary Change* 63: 317–324.

Verheyden, S., F. H. Nader, H. J. Cheng, L. R. Edwards and R. Swennen. 2008. "Paleoclimate reconstruction in the Levant region from the geochemistry of a Holocene stalagmite from the Jeita cave, Lebanon." *Quaternary Research* 70: 368–381.

Wanner, H. and 17 co-authors. 2008. "Mid- to Late Holocene Climate Change – An Overview." *Quaternary Science Reviews* 27: 1791–1828.

Weiss, H., and R. S. Bradley. 2001. "What Drives Societal Collapse?" *Science* 291: 609–610.

Weiss, H., M.-A. Courty, W. Wetterstrom, F. Guichard, L. Senior, R. Meadow and A. Curnow. 1993. "The Genesis and Collapse of Third Millennium North Mesopotamian Civilization." *Science* 261: 995–1004.

Wilkinson, T. J. 1999. "Holocene valley fills of Southern Turkey and Northwestern Syria: Recent geoarchaeological contributions." *Quaternary Science Reviews* 18: 555–571.

Zanchetta, G., R. Sulpizio, N. Roberts, R. Cioni, W. J. Eastwood, G. Siani, B. Caron, M. Paterne and R. Santacroce. 2011. "Tephrostratigraphy, chronology and climatic events of the Mediterranean basin during the Holocene: an overview." *The Holocene* 21: 33–52.

Hunter-gatherers Living in a Flooded World: The Change of Climate, Landscapes and Settlement Patterns during the Late Palaeolithic and Mesolithic on Bornholm, Denmark

Lasse Sørensen and Claudio Casati

Abstract

The change towards a warmer climate and flooded landscapes during the Early Holocene is investigated on the basis of a regional study on Bornholm. The environmental consequences are discussed together with the impact all these changes had on the hunter-gatherers' ability to colonise new regions, when living in a flooded world. Bornholm was, despite being a peninsula, a marginal region during the Late Palaeolithic although visits of pioneering hunter-gatherers became more frequent during the Preboreal. Improved environmental conditions during the following Boreal stage were characterized by an expansion and consolidation of animals and humans in the region. During the following Middle and Late Boreal, the land bridge was flooded and Bornholm became an island. The flooded landscape changed the way hunter-gatherers interacted and communicated, thus creating new social systems, where Bornholm again became a marginal zone. The island was not recolonised before the Late Ertebølle (4500–4000 cal BC). The hunter-gatherers' most important faculties, when living in this flooded world, were the ability to adapt to the changing landscape and maintaining cultural as well as social contacts with other synchronic societies in the Baltic region. When social interaction became too difficult some hard priorities had to be made and some regions were abandoned.

Introduction

During the Late Palaeolithic and Mesolithic the northern European Plain was characterized by a gradual flooding of huge landmasses, which was caused by a warmer climate change and the melting of the Northern American ice core. Few archaeological studies have discussed what consequences these changes of climate and landscape would have for hunter-gatherers, who lived in a flooded world. We will discuss if the change of climate and landscapes had an effect on the settlement patterns on Bornholm, Denmark, during the Maglemose Culture (9400–6500 cal BC). Bornholm was alternately connected to modern Germany-Poland or an island during certain stages of the Late Palaeolithic (12500–9400 cal BC) and Early Mesolithic (9400–6000 cal BC), making it an interesting area for studying processes of colonisation. In particular, the impact of the changing sea level and changing environmental situation makes it especially fascinating to investigate how these hunter-gatherers adapted to the changed conditions in the landscape. Finally, there are some interesting local perspectives which emerge in studying the site densities from the Late Palaeolithic and Early Mesolithic, as they can reveal issues of mobility and shed light on when Bornholm was colonised, or possibly isolated.

Exotic raw materials reveal the pioneering habitation

The raw materials in the southern Baltic are all procured from secondary deposits, brought to the region by glaciers in the quaternary period (Becker 1990). The local hunter-gatherers had detailed knowledge of where to find the different raw materials in the Mesolithic landscape. Contact between the Baltic regions and Bornholm can be identified, because Senonian flint does not occur naturally on Bornholm (Casati and Sørensen 2006: 10–11). Our investigations have shown that Late Palaeolithic and Mesolithic hunter-gatherers did have some contact with the flint producing areas in the western Baltic area, based on the imported Senonian artefacts found on several sites (see below). It is difficult to understand how this exchange worked. It is hardly a question of trade, but rather a result of a mobility pattern with sporadic contact with the western Baltic area. The possibilities of maintaining contact with other hunter-gatherers were also dependent on the geographical development in the Baltic region.

The paleogeography of Bornholm and the southern Baltic

The changes in land and sea that take place in the development of the Baltic Sea during the Late Glacial and Early Holocene had a great influence on the environmental conditions for both the humans and animals that inhabited Bornholm and the southern Baltic region (fig. 1). The Late- and Postglacial landscape on the Northern European Plain is influenced by some major changes in sea levels (*eustacy*) and uplift of land (*isotacy*). At the end of the last Ice Age, large quantities of water were stored in the icecaps of both the North and South Poles and in Scandinavia. This resulted in a sea level much lower than at the present, where the North Sea was an area of settlement during the Late Palaeolithic and Early Mesolithic (Coles 1998: 45ff).

The development of the Baltic Basin during the Late- and Postglacial is complex due to differences in the glacio-isostatic uplift, which had a major impact on the geographical development of Bornholm, and the land bridges that connected Bornholm with Germany and Poland (Björck 1995: 19ff; Jensen et al. 1999: 437ff, 2002: 5ff). This is one of the main reasons why the paleogeography plays such an important role in understanding when humans and different types of animals migrated to Bornholm during the Late Palaeolithic and Early Mesolithic.

During the Baltic Ice Lake stage, 12,500–9500 cal BC, the Yoldia Sea stage, 9500–8500 cal BC, and the Ancylus Lake stage, 8500–7000 cal BC, Bornholm was either the northern part of a peninsula, or an island with a substantial land bridge towards Vorpommeren (fig. 2). This land bridge created a bay on either side of Bornholm. The western side was orientated towards the Arkona Basin, and the eastern side was located where the Oder River had its outflow. The land bridge consists of two different ridges called Rønne Bank and Adler Ground, which stretch in a southwestern direction with a width of 15–17 km (fig. 3A and 3B). Today the shoals lie 12–20 m underwater, as confirmed by the detailed bathymetric investigations made by Nielsen et al. (2004: 87ff) and Uścinowicz (2006). Furthermore, the shoals are separated by a northwest to southeast depression a few kilometres wide, with water depths of 20–25 m. The chart also shows that between Adler Ground and the Bay of Pomerania, there is a stretch of 10 km where the shoals drop away and the depth is as much as 30 m (Bennike and Jensen 1998: 30f; Jensen et al. 1999: 439–440).

Drowned tree stumps of pine (*Pinus Sylvestris*) have been fished out

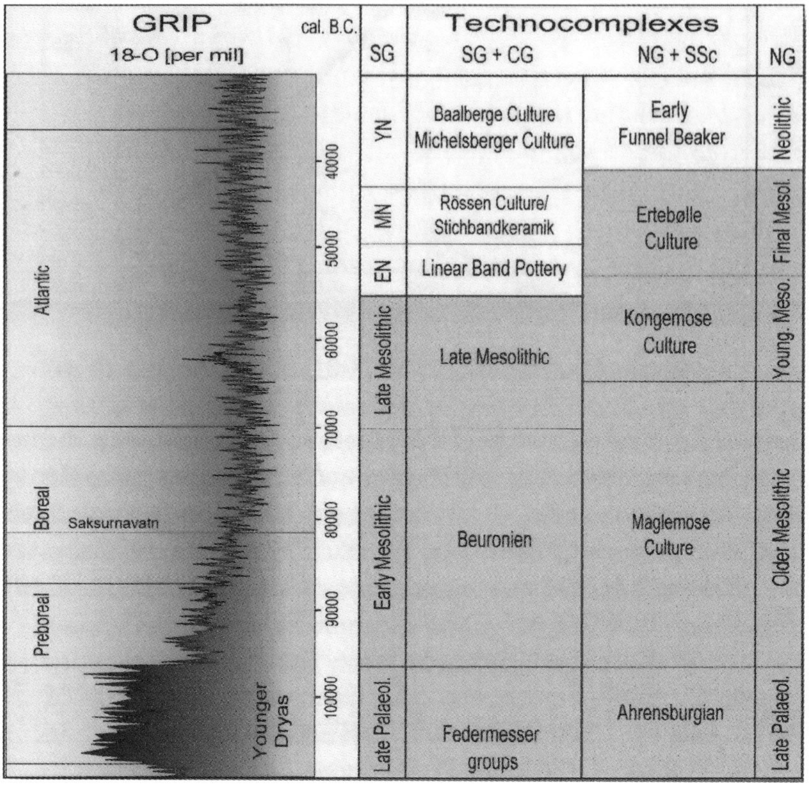

FIGURE 1. *Chronostratigraphy of the Mesolithic. SG – Southern Germany. CG – Central Europe. NG – Northern Germany. SSc – Southern Scandinavia (after Terberger 2006: 113).*

previously by fishermen on Rønne Bank and Adler Ground, indicating that these shoals were once part of a land bridge (fig. 3; Nielsen 1986a). However, none of these tree stumps have been ^{14}C dated, and until the beginning of the BALKAT project,[1] a lack of data limited any understanding of the shore level curves in this part of the Baltic Sea (Jensen et al. 2002: 2ff). The land bridges were permanently flooded during the Littorina Sea transgressions (7000–3000 cal BC). Between 7000–6000 cal BC, the transgressions flooded the land bridge gradually, creating several smaller islands including Rønne Bank, Adler Ground, Bank of Oder and Słupsk Shoal (fig. 3C). It is currently not clear when these smaller shoals were flooded, but sometime

---

1 The purpose of the BALKAT project is to investigate the geographical development since the last ice age in the southwestern parts of the Baltic Sea. The results give new insights within climate history, environmental history and cultural history of the Baltic Sea.

FIGURE 2. *The development of the Baltic Sea.*
1: The first stage of the Baltic Ice Lake (12,500 cal BC).
2: The second stage of the Baltic Ice Lake before the breakthrough at Mount Billingen (11,000 cal BC).
3: The drainage at Mt. Billingen in the south-central part of Sweden lowered the sea level to about 25 m below the current sea level (10,300 cal BC).
4: The Yoldia Sea with the open strait in the south-central part of Sweden (9500–9300 cal BC).
5: The closing of the strait in the south-central part of Sweden created the Ancylus Lake with drainage through the Göta Älv and the Otteid-Stenselva system. The drainage lowered the water level to 10–15 m below the current sea level (8600–8400 cal BC).
6: A regression followed around 8200–8000 cal BC, which recreated the land bridge towards Bornholm with water levels of approximately 25–35 m below the current sea level. The proposed drainage through the Dana River is questioned (after Björck 1995: 19ff; Lemke 2004: 46; Jensen et al. 2002).

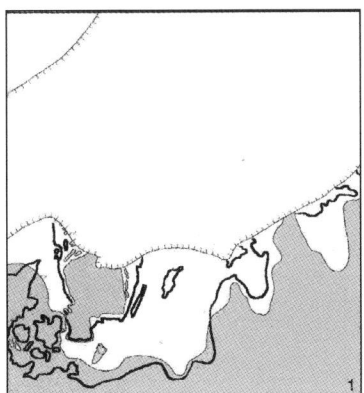
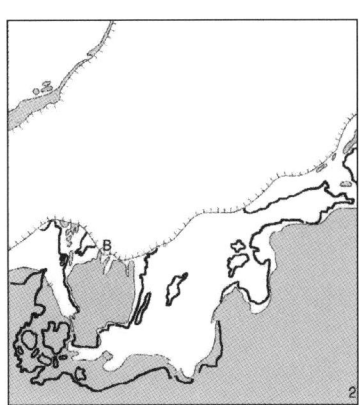
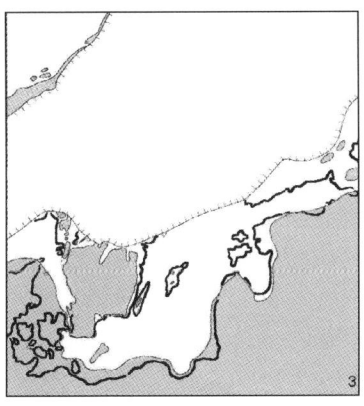
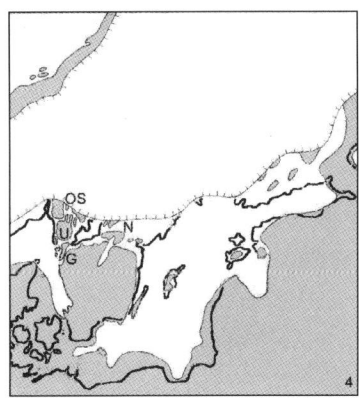
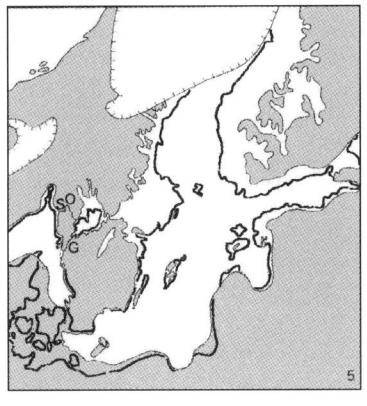
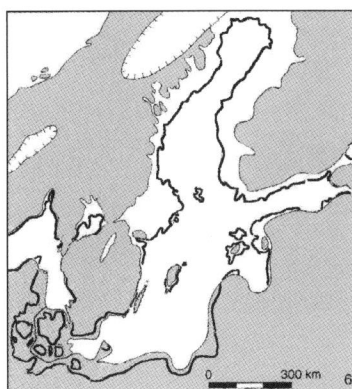

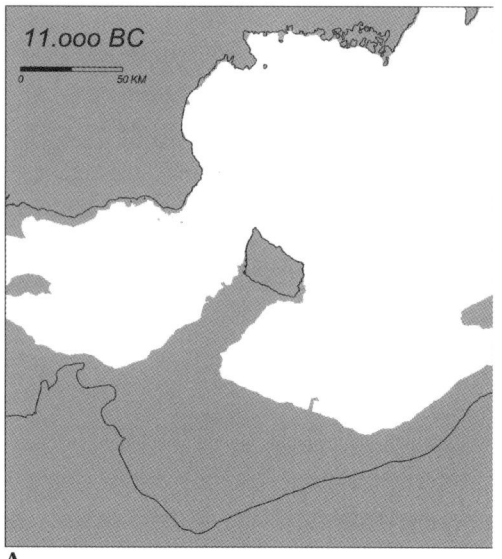
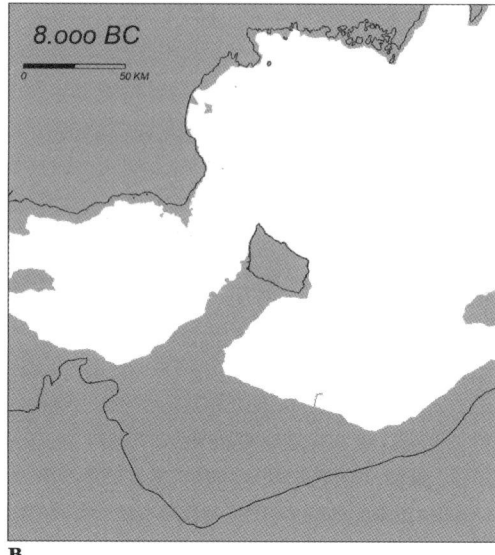
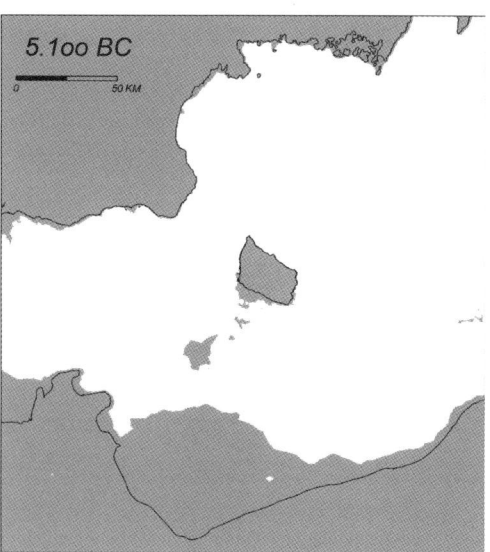

FIGURE 3. Paleogeographic map of the southern part of the Baltic Sea.
A: Around 11,000 cal BC, at the time of the Baltic Ice Sea.
B: Around 8000 cal BC, when Bornholm was the northern part of a peninsula, due to the constant low water level in the Ancylus Lake and the strong regression in the area.
C: Around 5100 cal BC. Bornholm is an island and towards the south some smaller islands on the Adlers Ground and the Banks of Rønne and Oder are visible (after Casati and Sørensen 2006: 13).

after 6000 cal BC seems probable. The land bridge was more or less permanent from 12,000 to 7000 cal BC, inviting animals and humans to migrate into this part of the Baltic region.

Flora and fauna on Bornholm during the Late Glacial

The environmental conditions on Bornholm are of cardinal importance in order to understand what awaited the first humans, who potentially could walk across the land bridge during the Late Glacial and Early Mesolithic. The pollen diagram from the Vallensgård peat bog (Iversen 1954: 87ff; Usinger 1977: 5ff) offers vital information on the Late Glacial and Preboreal vegetation history on Bornholm. The flora that spread across the land bridge after Bornholm became ice free during Bølling (zone Ib) and Older Dryas (Dryas II) (zone Ia) was a treeless tundra vegetation. The flora of this tundra consisted of the hardy white dryas (*Dryas octopetala*), which could resist long periods of hard winter.

Currently, there are no data to indicate any migration of either animals or humans during Bølling (12,500–12,000 cal BC) and Older Dryas (12,000–11,800 cal BC). Reindeer (*Rangifir tarandus*) is the only larger mammal registered on Bornholm, as proven by three early ^{14}C dates to the following Allerød period (11,800–11,000 cal BC) from Pellegård, Dyndeby and Klemensker (Aaris-Sørensen et al. 2007: 917ff) (fig. 4). According to Aaris-Sørensen (1998: 21), it is probable that also elk (*Alces alces*) and beaver (*Castor fiber*) were present on Bornholm during the Allerød.

According to Usinger (1977), the Allerød layer in Vallensgård Mose during the following period shows a threefold division, with a cooler phase in the middle. Both before and after this phase, higher values of junipers (*Juniperus*) seem to indicate two climatic optima. These interpretations correspond to the pollen diagram made by Iversen (1954), which concludes that Bornholm had a warmer climate than Jutland during the Allerød (11,800–11,000 cal BC), thus explaining the presence of reindeers.

The main support for stating a warmer Allerød climate on Bornholm is the higher number of pollen from pine and birch (*Betula*) in the Vallensgård Mose, compared with other pollen diagrams from Jutland. The approximately 40% of pine and birch indicate that a light open forest was present on Bornholm during Allerød (Usinger 1977). However, many of these bogs may reveal more about the local history of flora than about general trends,

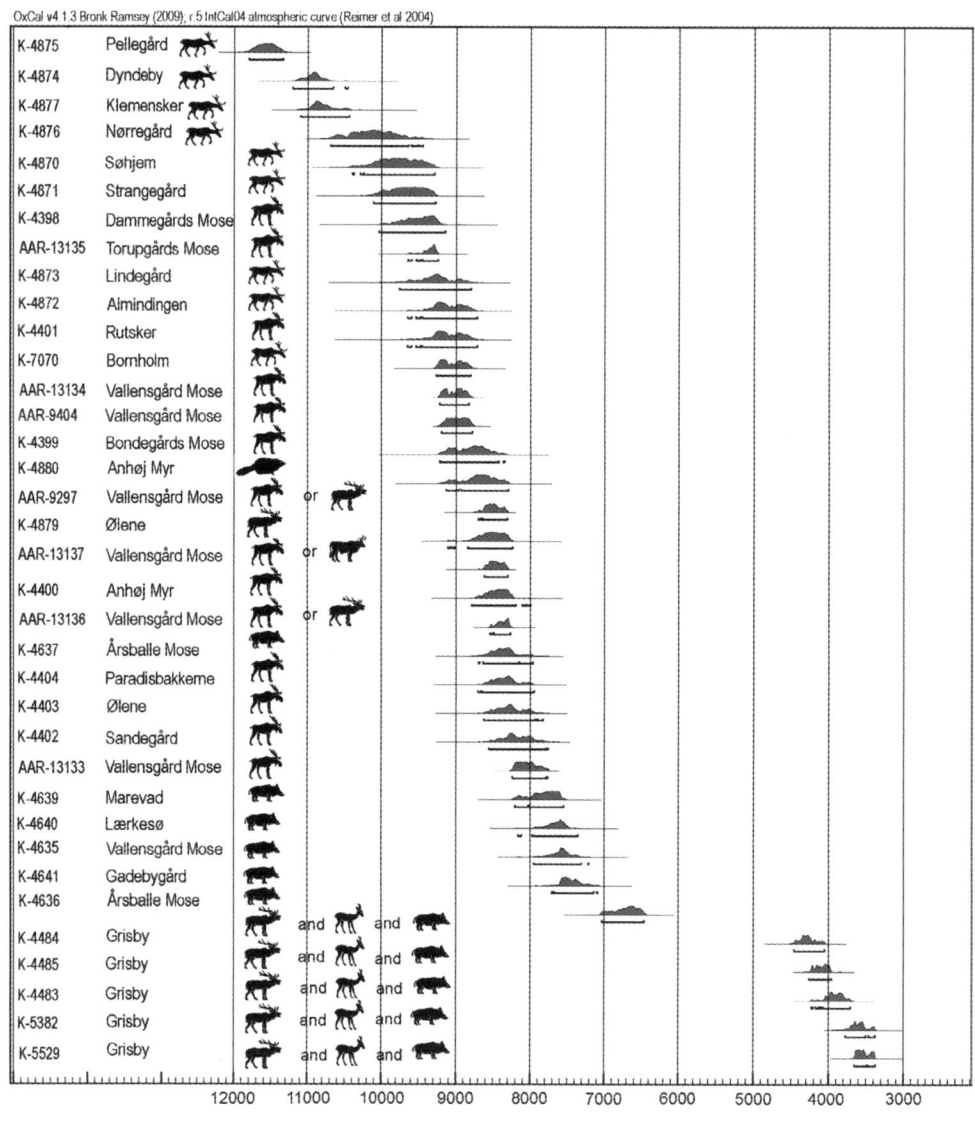

FIGURE 4. ^{14}C dates of animals on Bornholm, showing the faunal development during the Late Palaeolithic and Mesolithic (data compiled from Aaris-Sørensen 1998; Vang Petersen 2001: 161ff; Aaris-Sørensen et al. 2007: 914ff).

and Vallensgård Mose is no exception, because the bog is more or less sheltered by protruding bedrock to the north (fig. 6A). As such, the bog may have provided a "sanctuary" for some forms of vegetation in both harsher and milder climates, and this could in turn mean that the pollen diagram does not reflect more general conditions (Casati et al. 2004: 113ff). Thus, it must be concluded that Bornholm conforms to the general picture of the Allerød climate and flora in southern Denmark. Within the Allerød layer in Vallensgård Mose, Usinger found some microtephra associated with the volcanism of the Laacher See in the Eifel Mountains, dated to 10,932 cal BC, thus placing the layer containing the tephra immediately after the cooler middle part of the Allerød period, which corresponds with the Late Allerød layer in the pollen diagram (Usinger 1977; Street 1986; Baales et al. 1999, 2002; Schmincke et al. 1999; Riede 2008: 591ff; Sørensen 2010: 276ff).

When was Bornholm settled for the first time?

The lithic assemblage from the Vallensgård site, which can be associated with the Bromme Culture, suggests that Bornholm was settled for the first time sometime during the Allerød period. The Vallensgård site is located on a small sandy hilltop on the edge of a bog at the centre of the island near Rytterknægten (fig. 6A). Stray finds from three neighbouring Palaeolithic sites have been registered, as well as some Neolithic settlements overlying these sites (Casati and Sørensen 2006: 14ff). The Vallensgård site was excavated in 1986 and has only been published briefly (Nielsen 1986b: 77). The assemblage consists of worked flint found in a layer that was cryoturbated during either the permafrost in Dryas III (10,650–9500 cal BC) or during the Preboreal Oscillation (9300–9150 cal BC). Nielsen (1986b: 77) argues for a typological dating of the lithic material to the Bromme Culture.

The lithic material consists of flakes, blades, cores, two end scrapers and one tanged point. The main part of the assemblage originates from the surface, including the tanged point which was lying stratigraphically above the cryoturbated layer. Recent surveys of the Vallensgård sites have confirmed that it is still possible to find Palaeolithic artefacts on the surface. The bluish patina, shape and technology are especially characteristic of the Palaeolithic artefacts, making them different from the Neolithic material. The Late Palaeolithic material consists of Senonian flint of the highest quality. Some flakes from the cryoturbated layer had a thick chalky cortex, indicating that the raw material at Vallensgård Mose was imported and probably procured in the western part of the Baltic region, although it cannot be excluded that it was possible to find Senonian flint within the Late Glacial landscape on the land bridge towards Bornholm.

The hunter-gatherers at the Vallensgård site had limited knowledge of exploiting the local raw nodules. Furthermore, the lithic technology is characterised by an expedient production, thus indicating that we are dealing with pioneering hunter-gatherers settling this region during the Allerød. Many of the Late Palaeolithic sites have been associated with possible reindeer routes. However, recent analysis of seasonal markers proves that southern Scandinavia acted as a calving area throughout the Late Glacial and Early Holocene, thus refuting the north–southern route and supporting the east–western route. The reality of these pioneering hunter-gatherers is therefore much more complex than previously thought, and encouraged them to supplement reindeer and elk hunting with marine resources from the shores of the Baltic Ice Lake and Yoldia Sea. Bornholm was a marginal area during the Late Glacial, proving that environmental advantages did not always attract hunter-gatherers during the Late Palaeolithic.

Flora and fauna during the Late Glacial and Early Holocene

The following Younger Dryas (Dryas III) was a colder period, as shown in the Vallensgård pollen diagram. The vegetation consisted mainly of birch (*Betula*) and willow (*Salix sp.*) (20–30%), pine (*Pinus*) (20–30%) and wind resistant, herbaceous plants and heather (approx. 50%; Usinger 1977). These percentages are roughly comparable with the rest of southern Denmark, although Bornholm seems to have a slightly higher percentage of pine pollen and thus a larger amount of pine trees. However, once again the sheltered surroundings of Vallensgård Mose could explain these differences. The colder climate, and the fact that Bornholm was an island at high sea levels and had a substantial land bridge with shallow seas, did not stop the reindeer from migrating to the area during Dryas III. Seventy of the 280 finds of reindeer from Danish prehistory have been made on Bornholm, and nine have been ^{14}C dated (fig. 4). Three of these date back to the Dryas III, where the results from Nørregård, Søhjem and Strangegård belong to the middle and later part of the Dryas III. The dates prove that reindeer were present throughout the Dryas III. A study by Aaris-Sørensen et al. (2007: 917) also proves that reindeer moved around in the vicinity of the Baltic Ice Lake, as shown by ¹⁴C dates from Køge Bugt. Additionally, an elk bone from Veddelev Havn dated to 10,540 ± 110 BP, K-3493 (10,703–10,150 cal BC) indicates the possibility of elk living near the Baltic Ice Lake (Aaris-Sørensen 1998: 105). Furthermore, the recently published site of Hässleberga, a group of kettle holes in southwestern Scania, has also yielded ¹⁴C dates of reindeer from the Dryas III chronozone (Larsson et al. 2002: 61ff). Currently, the reindeer remains found on Bornholm display no certain signs of human activity such as cut marks and marrow fracturing. Nevertheless, reindeer were present on Bornholm during Dryas III; only the question of their numbers remains to be answered.

Lack of settlements from the Ahrensburgian culture during the Dryas III

The lack of settlements from the Ahrensburg Culture is remarkable because the island was connected to the mainland, giving a safe passage for reindeer during the entire period of Dryas III. A climatic explanation involving colder climate during Dryas III has been suggested for the lack of Ahrensburg material on Bornholm. However, the same climate is observed in the areas surrounding Bornholm, where Ahrensburgian sites occur. The identification of Ahrensburgian sites becomes even more complicated if these settlements consist of lithic assemblages from the Early Mesolithic, as illustrated by the Epi-Ahrensburgian inventories, which is defined by the presence of microliths and the absence of tanged points (Brinch Petersen 2009: 110ff). Future studies are needed to confirm the existence of the transitional Epi-Ahrensburgian group. However, the fact remains that while purposeful reconnaissance is responsible for finding a number of Maglemose settlements, no Ahrensburgian sites have so far been discovered on Bornholm.

Flora and fauna during the Preboreal

The warmer climate during the Preboreal (9700–8300 cal BC) changed the vegetation to a more dense pine and birch forest, as confirmed by different pollen diagrams in southern Scandinavia and by the deposited snow from the Greenland ice cores which show a fast rise in temperature (Iversen 1954; Noe-Nygaard et al. 2006: 303ff). The faunal picture from the beginning of the Preboreal – containing reindeer, elk and beaver – proves that Bornholm had a wide range of animals during this period (fig. 4). The Preboreal Oscillation (9300–9150 cal BC), a colder phase lasting approximately 100 to 150 years, could be one of the reasons why reindeer have been found on Bornholm and southern Scandinavia during the Early Preboreal (Aaris-Sørensen 2007: 918).

Elk were also present in the Preboreal and are known from forty bog finds on Bornholm. Seven of these have been ^{14}C dated, indicating a regular and stable population of elk (fig. 4). Another important animal from the Preboreal, the beaver, has been found on Anhøj Myr and dated to 9380 ± 130 BP, K-4880 (9137–8304 cal BC). Finally, towards the end of the Preboreal, the fauna becomes even more diversified, with the earliest evidence of fox (*Vulpes vulpes*), pine marten (*Martes martes*), polecat (*Mustela putorius*), roe deer (*Capreolus capreolus*), red deer (*Cervus elaphus*) and wild boar (*Sus scrofa*). It is, though, only the red deer from Ølene (9270 ± 130 BP, K-4879, 9114–8244 cal BC) and the wild boar from Årsballe Mose (9120 ± 120 BP, K-4637, 8695–7966 cal BC) which confirm the presence of these animals during the later stages of the Preboreal (Aaris-Sørensen 1998: 128). Another important event during the Preboreal was the open passage to the North Sea through the Yoldia Sea (9500–8500 cal BC), which attracted a wide range of marine animals (fig. 2). Around 8500 cal BC, the glasio-isostatic uplift closed the south central Swedish connection from the Baltic Basin. This development had a major impact on the ring seal (*Phoca hispida*), which became a relict population (Aaris-Sørensen 1998: 94; Fredén 1988). The same thing happened to the Baltic salmon (Karlsson and Karlström 1994: 62f). The detailed knowledge of the flora and fauna suggests some environmental advantages, which could be exploited by migrating hunter-gatherers. The sparse finds from these pioneers are presented below.

The early Maglemose on Bornholm during the Preboreal

The Lundebro material is the only lithic assemblage from the Early Preboreal and consists of four blades, four lanceolates and some microburins (fig. 6A). These are made of Senonian flint, and are three times larger than artefacts made from local raw materials (Nielsen 2001: 91ff). The local raw material is dominated by the so-called Kugleflint, a Maastrichtian flint type commonly found as small round nodules of approximately 5–8 cm in diameter. The artefacts produced from these nodules are, by default, small. The thick chalky cortex observed on one of the blades of Senonian flint from Lundebro indicates that we are dealing with a unique assemblage which includes imported raw materials. The raw nodules used in this assemblage were probably procured from a primary source in the western Baltic area or on the land bridge towards Bornholm. The site was located near the coast where 58 m² were excavated in 1994–95. The site could benefit from further excavation in order to document the stratigraphy and collect a larger lithic assemblage and materials for ^{14}C dates. The small assemblage of Senonian flint was found in the lower layers of the site, whereas the upper layers were dominated by lanceolates and triangles made from local raw materials (Nielsen 2001: 93ff).

Organic finds from the Early Maglemose culture during the Preboreal

A few stray organic finds from Bornholm can also be associated with the Early Maglemose Culture. These finds consisted of harpoons, leisters and a perforated tooth from a wild boar. Their exact position within the bog is unknown, but they were all found at the bottom of the calcareous marl, indicating that these objects were used by some of the earliest hunter-gatherers during the Dryas III and Preboreal. However, all artefacts were ^{14}C dated to the Preboreal, except for one bone point that was dated to the Early Boreal (fig. 5; Becker 1952: 167ff). The large-barbed harpoons have previously been interpreted as fishing implements (Andersen 1976: 17; Heidelk-Schacht 1984: 11). But no "fish corrosion" has been observed on any of the harpoons at hand (Sarauw 1903: 258). An alternative interpretation of the harpoons from Vallensgård Mose and other inland lakes on the Northern European Plain proposed by Vang Petersen (2009) is that they indicate a hunting method where the elk or reindeer were driven into the water, possibly with the help of dogs, and then harpooned from a canoe. The method is also known from ethnographic records (Charnley 1983; Grønnow et al. 1983: 29ff). It is clear that the large-barbed harpoons should detach from the shaft when the animal was harpooned (Andersen 1972: 74), thus explaining why so many harpoons have been found near inland lakes (Andersen and Vang

Petersen 2009: 8ff). The wide geographical and chronological distribution of the double-rowed harpoons and the large-barbed harpoons suggests that the proposed hunting strategy was important already from the Late Palaeolithic and onwards into the Early Maglemose Culture. During the following Middle and Late Boreal period, this hunting strategy became impossible to carry out, because many inland lakes became overgrown (figs. 6B and 6C).

Another important stray find from Bornholm was an elk antler adze from the small bog called Torupgaard near Klemensker. The artifact was found in 1928 at the bottom of the calcareous marl, indicating a similar date to the finds from Vallensgård. The adze had a length of 20.5 cm and a width of 13 cm. It had a perforation for the shaft at the proximal end measuring approximately 3.5 in diameter. At the distal end, the adze had a 9.5 cm deep oval shaped groove made for an inset or blade (fig. 5). The adze was ^{14}C dated to 9885 ± 55 BP, AAR-13135 (9460–9249 cal BC), making it the earliest artifact on Bornholm. The distribution of these adzes is concentrated around the Baltic Sea (Indreko 1948: 170). But two adzes belonging to the Butovo Culture and situated in the Upper Volga area were found in Russia at Stanovoye 4, indicating a wide mobility pattern of these hunter-gatherers during the Preboreal period (Zhilin 2007: 97ff; Zhilin and Matiskainen 2003: 699).

The Lundebro assemblage and the organic stray finds can thus be interpreted as belonging to a second pioneering group that colonised Bornholm during a favourable period, a period synchronic with a stable land bridge and the migration of reindeer, elk and beaver.

Flora and fauna during the Boreal

During the following Boreal phase (8300–7000 cal BC), a denser forest emerged and the hazel tree (*Corylus*) became a dominant species, although pine could still survive on sandy ground. The earliest ^{14}C results from burnt hazelnut shells dates them to 8900 BP (8300 cal BC) (Casati and Sørensen 2009: 248ff). The warmer temperature lowered the ground water level and the lakes such as Vallensgård Mose became overgrown and filled with peat during the earliest parts of the Boreal phase. The ^{14}C dates from the harpoons and leisters from Vallensgård Mose support this hypothesis (fig. 5). Other flat-bottomed lakes on Zealand such as Barmose, Lundby, Sværdborg and Holmegård also became overgrown at the same time. The same phenomenon is observed in the larger lakes in northern Poland (Alexandrowicz

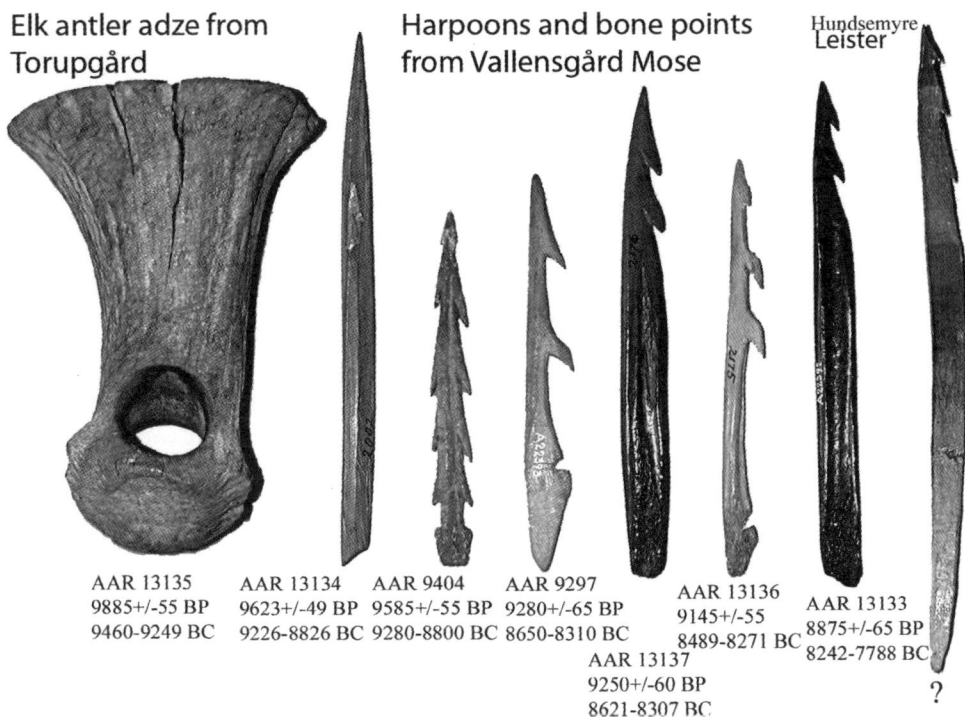

FIGURE 5. ^{14}C dates of the elk antler adze from Torupgård and harpoons and bone points from Vallensgård Mose and Hundsemyre – still undated.

1999: 67). The fauna is unchanged from the Late Preboreal into the Boreal (fig. 4; Aaris-Sørensen 1998: 128). In particular, severely burned bones from wild boar have been observed at Melsted (Becker 1952: 100f), Kobbebro and Ålyst. The only mammal missing so far in the Boreal fauna on Bornholm is the aurochs (*Bos primigenius*). The reindeer became extinct during the Late Preboreal, followed by beaver and elk during the Early Boreal (fig. 5). There could be several reasons for the extinction of the elk, including disease, the emergence of a dense forest and perhaps increased hunting pressure.

Site density during the Maglemose culture on Bornholm

The sparse finds from the early and middle part of the Preboreal indicates that Bornholm was a marginal zone. However, during the Early Boreal, the

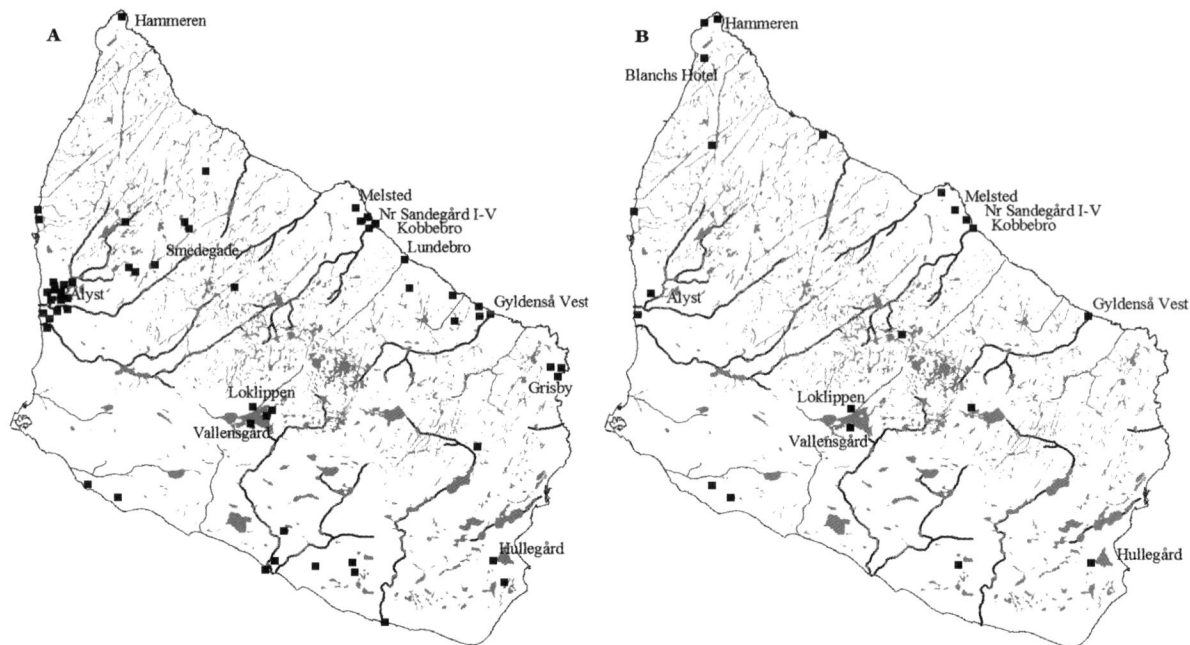

FIGURE 6.
A: Site distribution from the Early Boreal part of the Maglemose Culture.
B: Site distribution from the Middle Boreal part of the Maglemose Culture.

typical Boreal fauna migrated into Bornholm together with the hunter-gatherers, resulting in a high number of Maglemosian sites. These hunter-gatherers exploited the local raw materials, thus revealing a more detailed knowledge of the resources and more permanent habitation within this region. The settlement density during the Boreal on Bornholm can therefore shed some new light on the colonisation process.

Today, more than 125 Maglemosian settlements are known, all associated with the Boreal, a quadrupling of the number published by Becker in 1952 (1952: 152). But most of them are a result of surface collections (figs. 6A, 6B, 6C). The specific location of the different types of settlements on Bornholm gives us a unique opportunity to study the topographical characteristics in our examination. Becker had already demonstrated that the

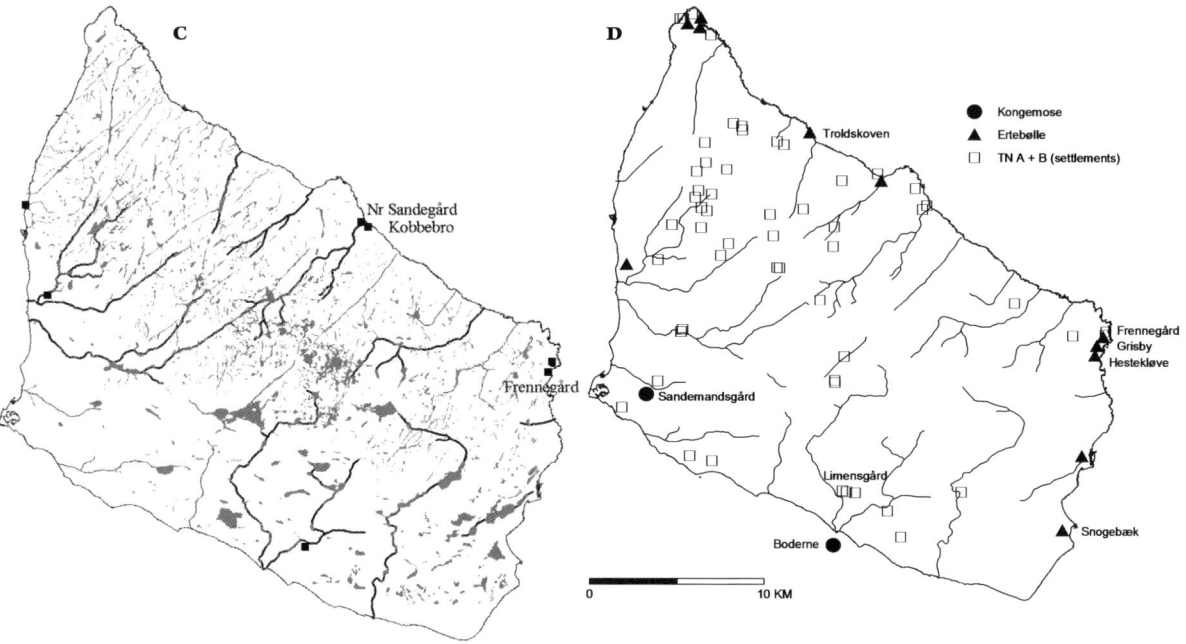

C: Site distribution from the Late Boreal part of the Maglemose Culture.
D: Site distribution in the Atlantic period during the Kongemose (dot), Ertebølle (triangle) and Early Funnel Beaker (square) Cultures.

Maglemose sites were often found on Late Glacial shorelines, on deposits close to the mouths of larger streams, or on sandy ground alongside these streams (Becker 1952: 152ff). Recent investigations initiated by Nielsen (2001: 89ff) have demonstrated that many settlements are linked to springs or are located on elevated ground with a broad outlook. We have also initiated a survey focusing on sites from inland areas, situated unusually far from any water resources. However, by virtually raising the ground water level on modern maps using GIS, it was possible to recreate the size of the former inland lakes and thereby reconstruct the former landscape and lakes (Casati and Sørensen 2006: 9ff, 2009: 248ff). The majority of the Maglemose sites were then located exactly on the edge of these former lakes (figs. 6A and 6B).

The settlement pattern during the Maglemose culture

The interpretation of the settlement pattern from Bornholm is based on typological studies of the microliths recovered, combined with both technological and ^{14}C analyses of various sites, making it possible to detect long-term changes (Casati and Sørensen 2006: 18ff). The settlement patterns include both coastal, inland and transit camps during the Early to Middle Boreal, suggesting that the habitants lived on the peninsula for a longer period (figs. 6A and 6B). However, it is difficult to determine the seasonality because of poor preservation of the organic materials (Nielsen 2001: 87). The sites are located in areas where it was possible to exploit many resources connected to fishing, hunting and gathering. The majority of excavated sites show a higher density of occupation especially within the Early and Middle Boreal, indicating a repeated settlement pattern closely connected to the larger creeks.

During the Late Boreal there appears to have been a rejection of the interior of the peninsula as a habitation zone (fig. 6C). Instead, sites are clearly concentrated near the coastal zones. Furthermore, the smaller size and limited variability in the tool assemblages, and at some flint concentrations such as Nørre Sandegård II and Ålyst, indicate a shorter duration and habitation within this region (Becker 1990; Casati and Sørensen 2006: 21ff). Why, then, do we see a marked reduction in the population density during the Late Boreal and Early Atlantic?

One of the reasons may be that all the larger inland lakes on Bornholm, including the Vallensgård Mose, became overgrown during the Middle and Late Boreal, resulting in a lost resource and leading to an abandonment of the inland zone. A similar phenomenon is observed in other flat-bottomed lakes on Zealand, where Late Boreal and especially Early Atlantic sites are rare. Another reason could be that the peninsula was flooded during the Late Boreal, creating smaller islands at the banks of Rønne, Adler Ground, Oder bank and Słupsk Shoal (fig. 3C). ^{14}C dates from Adler Ground prove that it was still an island during the Late Boreal and Early Atlantic. The earliest marine ^{14}C date demonstrates that the island was flooded sometime during the Middle Atlantic (Nielsen et al. 2004: 87ff).

The Atlantic flora and fauna

When the land bridge gradually disappeared, something caused by the Littorina transgressions, this did not have any immediate effect on the larger mammals such as red deer, roe deer and wild boar (Aaris-Sørensen 1998: 128). This is confirmed by the faunal remains found at the Late Ertebølle site Grisby, where charcoal from the find-bearing layers has been dated from 5450 ± 90 BP, K-4484 (4456–4051 cal BC) to 4750 ± 70 BP, K-5529 (3651–3371 cal BC) (fig. 4; Vang Petersen 2001: 168). Vang Petersen concludes that the deer population especially was probably much reduced after Bornholm became an island. Arguably, the zoological material presented by Aaris-Sørensen and Vensild (1996: 89ff) indicates a continuous population of red and roe deer on Bornholm, as documented by several Subboreal finds. Only the wild boar seems to have maintained a large population during the Atlantic period, having been found at several zoological locations and at Maglemose (Melsted, Kobbebro and Ålyst) and Late Ertebølle sites (Frennemark and Grisby). Surprisingly, the larger animals in the faunal assemblage thus appear to have survived the isolation of Bornholm, despite the small size of the island (587 km^2). The limited amount of resources could have had an effect on the hunter-gatherer population, as discussed in the following.

The settlement pattern during the Late Boreal and Early Atlantic

During the Late Boreal and Early Atlantic (7000–6000 cal BC), it became impossible to walk to Bornholm, although it is uncertain how low or high the water level was in the Ancylus Lake (8500–7000 cal BC) during these stages. The water level in the Ancylus Lake was probably regulated by the amount of rainfall together with the gradual melting of the North American ice cores. In general, the Atlantic phase has been described as a warm and wet period, but if rainfall was limited during the summer months it could have serious consequences for the trout population, resulting in a smaller population and thus a smaller resource. This could be one of the reasons why we have identified few sites near the streams during the Late Boreal (fig. 6C). The fact that Bornholm became an island does not seem to have resulted in the extinction of the larger mammals (fig. 4). Red and roe deer and the wild boar continued to be part of the prehistoric fauna on Bornholm, indicating that there were economic reasons for visiting the island. Could the lack of sites be explained by a collapse of the social system within the eastern parts of the Maglemose Culture?

According to Nico Arts (1989: 291), the collapses of social systems from the Late Palaeolithic or Mesolithic are evidenced by:

- A marked reduction in population density
- A decrease in the volume of exchange
- An increase in the proportion of smaller settlements

Our research has already documented a marked reduction in population density and an increase in the number of smaller settlements, resulting in a decrease in the size of individual residence groups and a fragmentation of the social organisation. It is more difficult to identify a decrease in the volume of exchange. The Late Boreal settlements on Bornholm do not contain the same amount of Senonian flint as sites in the Early Boreal phase. Arguably, we need to investigate more sites in order to verify or reject this hypothesis, although it is apparent that there is a continuous decrease in the number of sites from the beginning of the Middle Boreal. Most collapses are caused by persistent stressors, which in the case of Bornholm could be a combination of natural disaster and reduced resources, followed by an emigration away from the region. The natural disaster is documented by the continuous rising of the sea level caused by a rapid climate change, which gradually changed the mobility pattern of the hunter-gatherers. Bornholm was located in the central part of the Maglemose Culture and could have acted as a region of communication between Polish Pomerania and the eastern part of Scania. This could be one of the reasons why Bornholm was visited so many times during the Early and Middle Boreal. During the following phases, the overgrown lakes and warmer summer months, together with an increased hunting pressure on larger mammals when Bornholm became an island, could have reduced the faunal resources. These changes could have resulted in a changed mobility pattern, making Bornholm a marginal area during the Late Boreal and Early Atlantic, as shown by the lower settlement density (figs. 6C and 6D).

Was Bornholm totally abandoned during the Kongemose and Early Ertebølle culture?

Bornholm continued to be a marginal zone, as illustrated by the lack of finds from the Kongemose Culture (6500–5500 cal BC). The exception being a recent find from Sandemarksgård (fig. 6D), which, based on typological grounds, has been dated to the Blak phase (6500–6000 cal BC) (Sørensen 1996). Currently, we have not found a single site from the later phases of the Kongemose and Early Ertebølle Culture (6000–5000 cal BC). A limited occupation during the Early and Middle Atlantic (7000–6000 cal BC) has also been documented on Gotland in St. Förvar (Lindqvist and Possnert 1999: 66ff). However, some submerged sites have been found at Boderne and Sose Odde, proving that the area was not totally abandoned during the early and middle part of the Atlantic (fig. 6D; Nielsen 1986a; Casati and Sørensen 2006: 40). Several submerged sites from the Kongemose and Ertebølle Cultures documented at Lübeck, Wismar Bay and Rügen (Lübke 2004: 85, 2009: 556ff) also add weight to this theory. The use of this submerged landscape and its impact on the settlement pattern is uncertain, but it demonstrates that Bornholm was never out of sight or out of mind for the hunter-gatherers of the Kongemose and Early Ertebølle Cultures. During the following Late Ertebølle Culture (4500–4000 cal BC), many coastal sites were observed on Bornholm. The cylinder-shaped base of the pottery is a typical feature within the sites in Scania and Bornholm, indicating a well-established network across the Baltic Sea during the Late Ertebølle Culture (figs. 1 and 6D). Bornholm was once again an important region located right in the centre of the Baltic network.

Concluding remarks

The investigation of the Late Palaeolithic and Mesolithic sites has given us a detailed knowledge of the colonisation process and settlement pattern on Bornholm. The region was settled for the first time during the Bromme Culture, traditionally connected to the Allerød, which was a warm period. In the following Dryas III there is a lack of finds connected to the Ahrensburgian Culture on Bornholm, despite the fact that there were many reindeer in the region. During the following Preboreal, visits to the area are more frequent, but Bornholm was still a marginal area, only occasionally visited by pioneering hunter-gatherers. This stage was followed by an expansion and consolidation phase during the Early Boreal, caused by improved environmental conditions, shown by a high concentration of sites where local raw materials were exploited and a more residential habitation can be identified. The repeated exploitation of the landscape created some "hotspots" containing a complex of several settlements during the Maglemose Culture.

During the following Middle and Late Boreal, the region became gradually marginalised when Bornholm became an island and lost some of its environmental resources. Additionally, the flooded landscape could have created a collapse of the social system during the Late Boreal. The distances between the mainland and Bornholm that the flooded landscape created gradually reduced the island to a marginal zone, only occasionally visited by hunter-gatherer groups. It is not until the Late Ertebølle (4500–4000 cal BC) that we have been able to observe an actual recolonisation of the island.

We have investigated how climate and geographical changes challenged the Stone Age hunter-gatherers on Bornholm. The region is thus an example of how a global climate change during the Early Holocene can create a totally different landscape, which on a smaller regional scale had some major consequences for the hunter-gatherers living in this flooded world. The hunter-gatherers' most important faculties were the ability to exploit and maintain cultural as well as social contacts with other Late Palaeolithic and Mesolithic societies in the Baltic region. But the changed landscape, isolating Bornholm as an island, also changed the way hunter-gatherers interacted and communicated. Some regions like Doggerland had to be permanently abandoned, while other regions like Bornholm became marginalised, thus showing the hard priorities the hunter-gatherers had to make, as the abandoned landscape had a history within these people's minds.

Bibliography

Aaris-Sørensen, K. 1998. *Danmarks forhistoriske Dyreverden*. 3rd ed. Gyldendal. Copenhagen.

Aaris-Sørensen, K., and H. Vensild. 1996. "Kronhjortene på Bornholm og især i Aaker sogn." *Bornholmske Samlinger*. III. Række, bind 10, 89–96. Rønne.

Aaris-Sørensen, K., R. Mühldorff and E. Brinch Petersen. 2007. "The Scandinavian reindeer (*Rangifer tarandus L.*) after the last glacial maximum: Time, seasonality and human exploitation." *Journal of Archaeological Science* 34: 914–923.

Alexandrowicz, W. 1999. "Evolution of the Malacological assemblages in North Poland during the Late Glacial and Early Holocene." In M. Kobusiewicz and J. Kozlowski (eds.), *Post-Pleniglacial Re-Colonisation of the Great European Lowland. Folia Quaternaria* vol. 70, 39–69. Krakow.

Andersen, S. H. 1972. "Ertebøllekulturens harpuner." *Kuml* 1971, 73–125. Aarhus.

Andersen, S. H. 1976. "Nye harpunfund." *Kuml* 1975: 11–28. Aarhus.

Andersen, S. H., and P. Vang Petersen. 2009. "Maglemosekulturens stortandede harpuner." *Aarbøger for Nordisk Oldkyndighed og Historie* 2005, 7–41. Copenhagen.

Arts, N. 1989. "Archaeology, Environment and the Social Evolution of Later Band Societies in a Lowland Area." In C. Bonsall (ed.), *The Mesolithic in Europe. Papers presented at the third international symposium in Edinburgh 1985*, 291–312. Edinburgh University Press. Edinburgh.

Baales, M., F. Bittmann and B. Kromer. 1999. "Verkohlte Bäume im Traß der Laacher See-Tephra bei Kruft (Neuwieder Becken). Ein Beitrag zur Datierung des Laacher See-Ereignisses und zur Vegetation der Allerød-Zeit am Mittelrhein." *Archäologisches Korrespondenzblatt* 28: 191–204.

Baales, M., O. Jöris, M. Street, F. Bittmann, B. Weininger and J. Wiethold. 2002. "Impact of the Late Glacial Eruption of the Laacher See Volcano, Central Rheineland, Germany." *Quaternary Research* 58: 273–288.

Becker, C. J. 1952. "Maglemosekultur på Bornholm." *Aarbøger for nordisk Oldkyndighed og Historie*: 96–177. Copenhagen.

Becker, C. J. 1990. *Nørre Sandegård. Arkæologiske undersøgelser på Bornholm 1948–1952*. Historisk-filosofiske Skrifter 13. Copenhagen.

Bennike, O. and J. B. Jensen. 1998. "Late- and postglacial shore level changes in the southwestern Baltic Sea." *Bulletin of the Geological Society of Denmark* 45: 27–38.

Björck, S. 1995. "A review of the history of the Baltic Sea, 13.0-8.0 ka BP." *Quaternary International* 27: 19–40.

Brinch Petersen, E. 2009. "The Human settlement of Southern Scandinavia 12500–8700 CAL BC." In M. Street, N. Barton and T. Terberger (eds.), *Humans, environment and chronology of the glacial of the North European Plain. Proceedings of Workshop 14. Commission XXXII, The Final Palaeolithic of the Great European Plain of the 15th U.I.S.P.P. Congress, Lisbon, September 2006*, 89–129. Römisch-Germanisches Zentralmuseum, Band 6. Mainz.

Casati, C., and L. Sørensen. 2006. "Bornholm i ældre stenalder. Status over kulturel udvikling og kontakter." *Kuml* 2006: 9–58. Aarhus.

Casati, C. and L. Sørensen. 2009. "The settlement patterns of the Maglemose Culture on Bornholm, Denmark." In S. B. McCartan, R. Schulting, G. Warren and P. Woodman (eds.), *Mesolithic Horizons. Volume I*, 248–254. Oxbow Books. Oxford.

Casati, C., L. Sørensen and M. Vennersdorf. 2004. "Current research of the Early Mesolithic on Bornholm, Denmark." In T. Terberger and B. V. Eriksen (eds.), *Hunters in a Changing World*, 113–132. Verlag Marie Leidorf. Rahden, Westfahlen.

Charnley, S. 1983. *Moose hunting in two central Kuskokwim communities: Chuathbaluk and Sleetmute*. Alaska Department of Fish and Game, Technical Paper No. 76.

Coles, B. J. 1998. "Doggerland: A speculative survey." *Proceedings of the Prehistoric Society* 64: 45–81.

Fredén, C. 1988. *Marine life and Deglaciation Chronology of the Vänern Basin, Southwestern Sweden*. Sveriges Geologiska Undersökning 71. Uppsala.

Grønnow, B., M. Meldgaard and J. B. Nielsen. 1983. *Aasivissuit – The Great Summer Camp: Archaeological, ethnographical and zoo-archaeological studies of a caribou-hunting site in West Greenland*. Meddelelser om Grønland: Man and Society 5.

Heidelk-Schacht, S. 1984. "Knochen- und Geweihgeräte des Spätpaläolithikums und Mesolithikums aus Mecklenburg." *Bodendenkmalpflege in Mecklenburg, Jahrbuch 1983*, 7–82. Berlin.

Indreko, R. 1948. *Die mittlere Steinzeit in Estland*. Kungl. Vitterhets Hist. Och Antikvitets Akad. Handlingar 66. Stockholm.

Iversen, J. 1954. "The late-glacial flora of Denmark and its relation to climate and soil." *Danmarks Geologiske Undersøgelser*. Række 2. Nr. 80, 87–119. Copenhagen.

Jensen, J. B., O. Bennike, A. Witkowski, W. Lemke and A. Kuijpers. 1999. "Early Holocene history of the southwestern Baltic Sea: The Ancylus Lake stage." *Boreas* 28: 437–453. Oslo.

Jensen, J. B., A. Kuipers, O. Bennike and W. Lemke. 2002. "BALKAT. Østersøen uden grænser." *Nyt fra GEUS* 4: December 2002, 2–19. Copenhagen.

Karlsson, L., and K. Karlström. 1994. "The Baltic salmon (*Salmo salar* L.): Its history, present situation and future." *Dana* 10: 61–85.

Larsson, L., R. Liljegren, O. Magnell and J. Ekström. 2002. "Archaeo-faunal aspects of bog finds from Hässleberga, southern Scania, Sweden." In B. V. Eriksen and B. Bratlund (eds.), *Recent Studies in the Final Palaeolithic of the European Plain. Proceedings of a U.I.S.P.P. Symposium, Stockholm 1999*. Jutland Archaeological Society Publications 39, 61–74. Aarhus.

Lemke, W. 2004. "Die kurze und wechselvolle Entwicklungsgeschichte der Ostsee – Aktuelle meeresgeologische Forschungen zum Verlauf der Litorina-Transgression." *Bodendenkmalpflege in Mecklenburg-Vorpommeren*. Jahrbuch 2004, Band 52, 43–54.

Lindqvist, C., and G. Possnert. 1999. "The First Seal Hunter Families on Gotland. On the Mesolithic Occupation in the Stora Förvar Cave." *Current Swedish Archaeology* 7: 65–88.

Lübke, H. 2009. "Hunters and fishers in a changing world. Investigations on submerged Stone Age sites off the Baltic coast of Mecklenburg-Vorpommern, Germany." In S. B. McCartan, R. Schulting, G. Warren and P. Woodman (eds.), *Mesolithic Horizons, Vol. 2*, 556–563. Oxbow books. Oxford.

Lübke, H. 2004. "Spät- und endmesolithische Küstensiedlungsplätze in der Wismarbucht. Neue Grabungsergebnisse zur Chronologie und Siedlungsweise." *Bodendenkmalpflege in Mecklenburg-Vorpommeren, Jahrbuch 2004*. 83–110. Lübsdorf.

Nielsen, F. O. 1986a. *Registrering og vurdering af fortidsmindeinteresser på havbunden omkring Bornholm*. Unpublished report to Fredningsstyrelsen 21.04.1986.

Nielsen, F. O. 1986b. *Arkæologiske udgravninger i Danmark 1986*. Det Arkæologiske Nævn, 77. Copenhagen.

Nielsen, F. O. 2001. "Nyt om Maglemosekultur på Bornholm." In O. Lass

Jensen, S. A. Sørensen and K. M. Hansen (eds.), *Danmarks Jægerstenalder – status og perspektiver*, 85–99. Hørsholm.

Nielsen, P. E., J. B. Jensen, M. Binderup, S. Lomholt and A. Kuijpers. 2004. "Marine aggregates in the Danish sector of the Baltic Sea: Geological setting, exploitation potential and environmental assessment." *Zeitschrift für Angewandte Geologie*, Sonderheft 2, 2004, 7–9.

Noe-Nygaard, N., K. L. Knudsen and M. Houmark-Nielsen. 2006. "Fra Istid til og med Jægerstenalderen." In K. Sand-Jensen (ed.), *Naturen i Danmark*, vol 1., 303–332.

Riede, F. 2008. "The Laacher See-eruption (12,920 BP) and material culture change at the end of the Allerød in Northern Europe." *Journal of Archaeological Science* 35: 591–599.

Sarauw, G. F. L. 1903. "En stenalders boplads i Maglemose ved Mullerup, sammenholdt med beslægtede fund." *Aarbøger for Nordisk Oldkyndighed og Historie*, 148–315. Copenhagen.

Schmincke, H.-U., C. Park and E. Harms. 1999. "Evolution and environmental impacts of the eruption of Laacher See volcano (Germany) 12,900 a BP." *Quaternary International* 61: 61–72.

Sørensen, L. 2010. "The Laacher See volcanic eruption: Challenging the idea of cultural disruption." *Acta Archaeologica* 81: 276–287.

Sørensen, S. A. 1996. *Kongemosekulturen i Sydskandinavien*. Egnsmuseet Færgegården.

Street, M. 1986. "Ein Wald der Allerödzeit bei Miesenheim, Stadt Andernach (Neuwieder Becken)." *Archäologisches Korrespondenzblatt* 16: 13–22.

Terberger, T. 2006. "The Mesolithic Hunter-Fisher-Gatherers on the Northern German Plain." In K. B. Pedersen and K. M. Hansen (eds.), *Across the Western Baltic*, 111–184. Sydsjællands Museum. Vordingborg.

Uścinowicz, S. 2006. *How the Baltic Sea was Changing*. Online report on the Polish Geological Institute webpage: www.pgi.gov.pl

Usinger, H. 1977. "Bölling-Interstadial und Laacher Bimstuff in einem neuen Spätglacial-Profil aus dem Vallensgård Mose/Bornholm. Mit pollengrössenstatistischer Trennung der Birken." *Danmarks geologiske undersøgelser, årbog 1977*, 5–29. Copenhagen.

Vang Petersen, P. 2001. "Grisby – en fangstboplads fra Ertebølletid på Bornholm." In O. Lass Jensen, S. A. Sørensen and K. M. Hansen (eds.), *Danmarks jægerstenalder – status og perspektiver*, 161–174. Hørsholm.

Vang Petersen, P. 2009. "Stortandede harpuner – og jagt på hjortevildt til vands." *Aarbøger for nordisk Oldkyndighed og Historie*, 43–54. Copenhagen.

Zhilin, M. 2007. "The Early Mesolithic of the Upper Volga: Selected Problems." In M. Masojć., T. Płonka, B. Ginter and S. K. Kozłowski (eds.), *Contributions to the Central European Stone Age*, 89–103. University of Wrocław Institute of Archaeology. Wrocław.

Zhilin, M., and H. Matiskainen. 2003. "Excavations at the Mesolithic sites Stanovoje 4 and Sakhtysh 14, Upper Volga region." In L. Larsson, H. Kindgren, K. Knutsson, D. Loeffler and A. Åkerlund (eds.), *Mesolithic on the Move. Papers presented at the Sixth International Conference on the Mesolithic in Europe, Stockholm 2000*, 694–702. Oxbow books. Oxford.

Complex Society's Responses to Climatic Variation

Urban Adaptations to Climate Change in Northern Mesopotamia

Jason Ur

Abstract

Only in rare circumstances does climate change force a uniform response from human communities; adaptations can vary dramatically even between neighbouring settlements. This chapter considers variation in response to proposed climate changes in third millennium BC northern Mesopotamia. Using spatial data from surface collection, satellite remote sensing and landscape archaeology, it considers how urban-based farmers and pastoralists responded at three of the largest Early Bronze Age cities in northern Mesopotamia: Hamoukar, Tell Brak, and Tell Leilan. Despite their close proximity, the different adaptive responses at these three places resulted in three different settlement histories.

Introduction

Complex societies pose a particular problem to the study of climate and human responses, since their complexity often involves the development of cooperative technologies and social institutions designed to compensate for variable conditions. On the other hand, it may be that in the development of technologies and institutions, these societies overextend themselves, or develop to the point of losing resilience in the face of such variable conditions. In the process, they can impact their environments in ways that mimic, or are indistinguishable from, the effects of climate change. Ongoing debates about the causes of deforestation and soil erosion in the late Holocene attest to these difficulties (Roberts et al. 2011a; Roberts et al. 2011b).

The focus of this study is squarely on this second aspect: the responses of human communities to environmental change. In general, models for human-environment interaction have fallen along a continuum. On one end, environment has been seen as determining aspects of social change, including some of the most dramatic events of human history. At the most extreme, a sudden infilling of the Black Sea basin has been hypothesised (Ryan and Pitman 1998). According to this account, in a span of a few weeks, this event displaced members of a society that had emerged around a freshwater lake at the centre of the basin. They migrated out in all directions, eventually settling in Europe, the Eurasian Steppes and Mesopotamia, where they gave birth to the Linearbandkeramik culture, the Indo-Europeans and the Sumerians, respectively. This last civilisation was kind enough to preserve this event in the form of the Flood myth.

Equally implausible is the idea that human societies exist somehow independently of their environments, and all social change is the endogenous result of shifting relationships between individuals and groups. Such interpretations rarely make these assumptions explicit, but the social actors in such models exist in an abstract world without physical environments or cultural landscapes.

The intermediate position places humans within an ecosystemic context. The environment is a real factor in the decision making processes of individuals, whose actions are constrained by it, but whose adaptability can take advantage of certain elements (McIntosh, Tainter and McIntosh 2000b; Rosen and Rosen 2001; Rosen 2007). It is important to recognise that the human-environmental relationship is complex and non-linear, and that different societies, and indeed different individuals in a society, can respond to identical environmental changes in variable ways (Coombes and Barber 2005; Rosen and Rosen 2001). In only the most extreme cases is a single human response the only possible option.

In considering the human-environmental dynamic, and particularly the social response, there are several issues that are worth explicit discussion as we evaluate competing reconstructions. Of particular importance is the nature of what it is that responds. "Society" is a convenient expression for an amazingly complex, internally diverse system of individuals, institutions and the relationships between them. To say that a society "responds" to some event is to speak of an aggregate of individual actions, but never a uniformity. Societies, settlements and institutions should not be reified into decision making agents.

Closely related to this issue is the chronological scale of events. Are "abrupt" events those that occur in a day, a week, over a decade, or a century? Past societies could persevere for centuries or longer, but the decision making individuals of which they were comprised lived in a human timescale. A century-long process of aridification might seem abrupt in an ice core, but four or five generations would have witnessed this "event", if we can use that term. Human response is closely related to human perception and social memory (McIntosh, Tainter and McIntosh 2000a: 16–17). Our reconstructions of social responses must consider how individual members of past societies would have perceived environmental change, and how it fit within the understood normal range of climatic variation.

A frequently reconstructed human response to climate change, particularly to abrupt events, has been "collapse", but all too frequently there is no explicit discussion of what this term means. What is it that collapses? Do we mean a political collapse, where centralised administrative institutions lose their legitimacy, and society breaks down into smaller autonomous units? Do we mean economic collapse, where trade networks are disrupted, and agricultural or pastoral systems become non-viable? Or do we mean a demographic collapse, where conditions are so drastically altered that communities are reduced in number or must physically relocate themselves? A "collapse" could mean any or all of these things; they are related but not necessarily synchronous (Yoffee 1988; Schwartz 2007: 46–47).

Archaeologists frequently assume that social responses took the form of a shift from a high state of social complexity to a lower state. These shifts include the political fragmentation of states and empires, and the reduction or disappearance of settlement nucleation in the form of urbanism. It is important to recognise that in such cases, climatic or environmental variations may be stressing societies that have developed in particular ways which have reduced their resiliency (Wilkinson 1997). Such societies may not need dramatic events to impact them; something as simple as high floods, a long winter, or a run of dry years might be sufficient to set off a non-linear chain of responses whose ultimate result might be radical social transformation (Coombes and Barber 2005). In such cases, the ultimate causes are as much social as they are environmental.

Climate and society in the Early Bronze Age (EBA) Northern Fertile Crescent

One of the most discussed of the proposed abrupt climatic events is placed at 4.2 thousand years ago, or 2200 BC. Initially this event was used to explain settlement history at Tell Leilan in northeastern Syria but was expanded to address hypothesised regional settlement abandonment in northern Mesopotamia (Weiss and Courty 1993; Weiss et al. 1993) and later throughout the Old World at the end of the 3rd millennium BC (Weiss 1997; Weiss 2000b). The initial publications in 1993 were provocative, proposing previously unrecognised regional abandonments and mass population movements at the hands of environmental events. In the succeeding 20 years, much research energy has gone into identifying social responses to this proposed event elsewhere in northern Mesopotamia and throughout the ancient Near East (see, most recently, Kuzucuoğlu and Marro 2007; Rosen 2007; Wossink 2010; Danti 2010). The current state of research makes it an excellent case study in the complex relationship between climate and society, the difficulties of integrating paleoclimatic data, archaeology and ancient texts, and the variability of human responses and adaptations. Contrary to the expectations of the 4.2 kya abrupt climate event model, a close consideration of the settlement trajectories of three large cities reveals variable adaptations resulting in three different histories, rather than a single simultaneous response.

Early Bronze Age urbanism

Early Bronze Age cities emerged between 2600–2500 BC in an arc along the southern edge of the Taurus Mountains (fig. 1). Today, this is a Mediterranean climate, with cold wet winters and hot dry summers. Most of this region receives between 300 and 600 mm of rainfall per year, almost all of it between November and March. At the upper end of this range, rainfall is sufficient for successful dry farming, and those regions have witnessed nearly continuous sedentary agricultural communities since the Neolithic. At the lower end, toward the steppes of central Syria, settlement occurred only in certain periods, generally times of political stability under states and territorial empires (Geyer and Calvet 2001; Wilkinson et al. 2005), but earlier in the Holocene possibly because of more favourable rainfall conditions (Hole 1998; Bar-Matthews and Ayalon 2011).

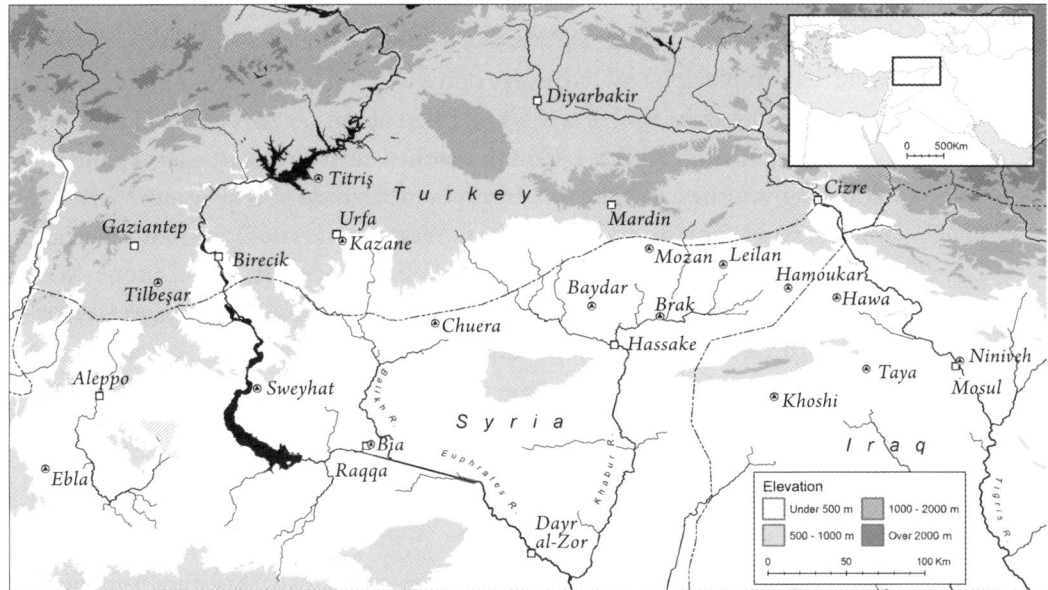

FIGURE 1. *Cities and towns in northern Mesopotamia in the Early Bronze Age (c. 2600–2000 BC).*

Even in areas that receive on average over 300 mm per year, the amount of rainfall fluctuates dramatically each year. For example, the rainfall station at Mosul in northern Iraq shows very high interannual variability in the 20th century AD. Such variability means that some areas might have one failed crop every three years, even if the long-term average is high enough for successful cultivation. We must consider the possibility that ancient societies were faced with such routine climatic variability as well.

In the early 3rd millennium, most settlements were small villages of 1–2 hectares, with a few larger settlements around 15 hectares. Beginning around 2600 BC, several of these settlements grew rapidly to sizes of 60 to 100 hectares. Even at the earliest stages of this expansion, they hosted large institutional households such as palaces and temples on monumental scales. Outside of these elite core areas, the expanding lower towns contained narrow streets and narrower alleys running between densely packed domestic residential neighbourhoods. The basics of political structure are beginning to emerge from scattered tablet finds at Ebla, Mari and Beydar. The major centres were Ebla (Tell Mardikh), Mari (Tell Hariri) and Nagar (Tell Brak), and their rulers engaged each other in a combination of diplomacy and warfare (Sallaberger 2007; Archi and Biga 2003; Eidem et al. 2001). Archaeological and textual evidence attests to mass production of ceramics and specialised production and exchange of metals and textiles (for recent reviews, see Akkermans and Schwartz 2003; Stein 2004; Ur 2010a).

The economic backbone of these cities and their hinterlands was, however, cereal agriculture and sheep and goat pastoralism (Weiss 1986; Wilkinson 1994; Deckers et al. 2010). Paleobotanical studies demonstrate the importance of barley and wheat (Charles and Bogaard 2001; Riehl 2009). Recovered animal bones are mostly sheep and goat, with lesser amounts of cattle and pig (Zeder 2003). Archaeologists give disproportionate attention to the manufacturing aspects of their economies, but at their base, these cities were overwhelmingly agro-pastoral in nature (Danti 2010), and as a result closely connected to their local ecologies.

Climatic events and variability in the later 3rd millennium BC

There is general agreement that the Near Eastern climate changed during the course of the 3rd millennium BC (see recent reviews in Roberts et al. 2011b; Bar-Matthews and Ayalon 2011). There is, however, considerable debate regarding its timing and duration (fig. 2). The earliest hypothesis, by Harvey Weiss and colleagues, proposed that an abrupt event changed the climate to a new arid state at 2200 BC, an event referred to as the 4.2k BP event. The empirical core of the initial hypothesis was micromorphological evidence for volcanic activity, found in layers covering the final occupation of Tell Leilan (Weiss and Courty 1993; Weiss et al. 1993; Courty and Weiss 1997). More recently, the micromorphological evidence has been disregarded in favour of changes in atmospheric circulation (Staubwasser and Weiss 2006). Although the cause of this event has evolved since its initial publication in 1993, the hypothesised human response has not: it is still considered to be a "collapse to less extractive political organisation, directed habitat-tracking to regions where agriculture was sustainable, and the abandonment of reduced-production cultivation for pastoralism" (Weiss 2000a: 88). It is thus proposed that urbanism collapsed at Leilan, almost all other sedentary settlement in northern Mesopotamia was abandoned, and the resulting displaced populations forced the fragmentation of the Akkadian empire. This same 4.2 kya event is now implicated in the demise of the Old Kingdom in Egypt and the end of the Harappan civilisation in the Indus (Weiss 2000b).

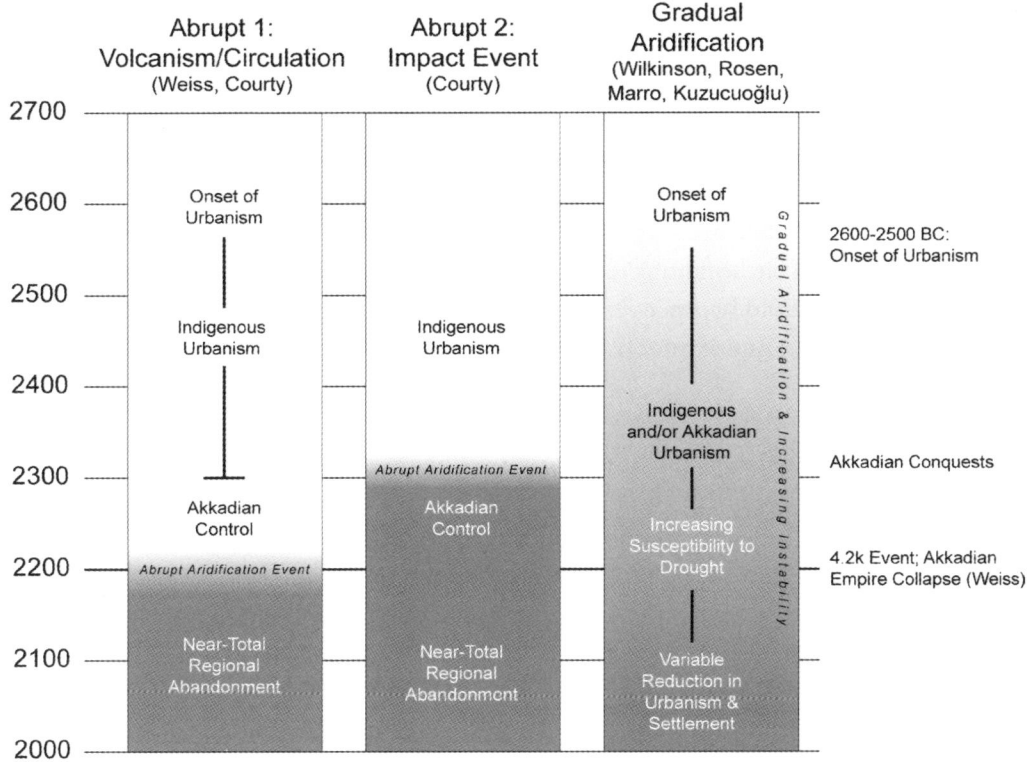

FIGURE 2. *Models of climate and social development in the later Early Bronze Age.*

The paleoclimatic reconstruction of Weiss and colleagues is not unchallenged, however. The same micromorphological tephra signature was found in layers at Tell Brak that could be as much as a century earlier (c. 2300 BC), and has been interpreted as the result of an extra-terrestrial impact event (Courty 1998; Courty 2001). The context of the Brak samples, in levels predating the fortress constructed by the Akkadian king Naram-Sin, places this event in the middle of the span of urban development, not at the time of its demise.

Others see a gradual climatic drying, based on a variety of proxy indicators. A mid-Holocene humid phase was replaced by more arid conditions throughout the 3rd millennium, reaching a driest phase around 2000 BC and remaining that way for the remainder of the Bronze Age (Wilkinson

1997; Hole 1997; Rosen 2007; Marro and Kuzucuoğlu 2007; Bar-Matthews and Ayalon 2011). These models also assume substantial interannual variability in temperature and rainfall, and a long-term human adaptation to such variability. Under this climatic model, 3rd millennium urbanism appeared and developed in the context of increasing aridity.

In conclusion, there is disagreement as to the nature and timing of climatic events in the late 3rd millennium BC. In one model, volcanoes, meteorites and atmospheric circulation shifts are blamed for sudden aridification on the scale of years or decades; in the other, climatic drying was a centuries-long process that may not have even been perceptible within the normal range of interannual variation.

The landscapes of EBA cities

Whether abrupt or gradual, aridification must have had an impact on the staple economy, the foundation upon which EBA urbanism was constructed. Fortunately, in approaching questions of settlement and economy, we have more than just sites with carbonised plants and animal bones. The landscape of northern Mesopotamia preserves an array of off-site features that demonstrate the extent and intensity of agricultural and pastoral activities at this time. Such features are crucial for approaching questions of human-environmental relationships and adaptation to climatic variability.

The first type of features is a set of shallow linear depressions that radiate outward from mounded sites. They are difficult to see on the ground, but under proper conditions, they are clear in remote sensing datasets. These features, called "hollow ways", represent the traces of past trackways. They were first mapped by Willem van Liere in the 1950s (van Liere and Lauffray 1954–55) and later by Tony Wilkinson in the 1980s (Wilkinson 1993). Using declassified intelligence satellite photographs, over 6,000 kilometres of hollow ways have been mapped across northeastern Syria and northern Iraq (Ur 2003, 2009, 2012a; Wilkinson et al. 2010). Hollow ways are mostly associated with sites of this 3rd millennium urban phase, both large cities and small villages; a smaller, morphologically distinct set can be associated with sites of the early Islamic period.

Hollow ways demonstrate routes of communication between settlements, but they also serve as a proxy for extensification of ancient agriculture and pastoralism (Wilkinson 1994: 492–493). The majority of traffic consisted of farmers and their plough animals moving to and from their fields adjacent to the settlement, and herders and their flocks moving from the settlement to the pasture beyond those fields. Within the agricultural zone, their movement was constrained onto these paths by the growing crops; hence the disturbance that causes the paths to sink was also constrained, resulting in the formation of hollow ways. Beyond the edge of the agricultural zone, no such constraints existed, and movement would have dispersed; in this zone, no hollow ways formed. What survives for archaeology are the settlement, now in the form of a mound, and the former paths where constraint caused the deepest incision. With this model of hollow way formation in mind, we can reconstruct agricultural zones by connecting the terminal ends of hollow ways.

When viewed on a regional scale, the areas of densest settlement were nearly completely cultivated, since the reconstructed zones almost abut each other. By estimating settlement population and the productivity of their associated agricultural catchments, we find that most villages and small towns would have been self-sufficient or surplus-producing. However, larger towns and cities would have required some of this surplus to sustain their populations. In the region of Tell Beydar, for example, the deficit at Beydar would have been more than offset by exchanges with its surplus-producing satellite villages (Ur and Wilkinson 2008: 312–314).

The extensification of agriculture, as demonstrated by hollow ways, was accompanied by intensification, indicated by a second class of landscape feature. Across the alluvial plains of northern Mesopotamia, the landscape between sites is covered by a carpet of small and abraded potsherds. The distribution is nearly continuous but of variable density. This phenomenon was initially recognised by Williamson (1974: 90) in Oman, systematically studied by Wilkinson (1982, 1989), and is now documented throughout the Near East and beyond (Bintliff and Snodgrass 1988). In northern Mesopotamia, the densest concentrations, in the range of 60–100 sherds per hundred square metres, are found in proximity to urban sites of the later 3rd millennium (Wilkinson 1994: 491–492). Lighter densities are found around smaller sites of the same time period. Around sites of other periods, scatters are low, between five and twenty sherds per hundred square metres. The scatters themselves are difficult to date because of their small and abraded morphologies, but identifiable diagnostics are predominantly later 3rd millennium in date, and fabrics are consistent with this time as well.

There are many ways that potsherds can find their way into the fields around sites, but only one interpretation can explain their extent and ubiquity. Wilkinson (1982) has proposed that they represent the inorganic component of manuring. Cultivation removes moisture and nutrients from soils, especially when fallow cycles are shortened or skipped. The deliberate addition of organic materials to fields ameliorates this process. In Wilkinson's model, organic refuse in the form of animal dung, cooking debris and food remains was collected and composted in settlements, along with other household waste, including discarded ceramics. Subsequently this material was deposited onto the fields. The organic component has been washed through the soil but the inorganic elements remain and have been kept on the surface by over four millennia of subsequent ploughing.

By the end of the EBA, population nucleation in the great cities of northern Mesopotamia placed them in an ecologically precarious situation. The largest settlements were incapable of sustaining themselves and would have relied on exchange or transfers from settlements in their hinterlands. Because of this reliance on surplus production, this system was particularly attuned to fluctuations in climate. Under modern climatic conditions, this economic system would have existed at or near the limits of viability, since a climatic drying or merely a run of particularly dry years could have resulted in its collapse (Wilkinson 1994, 1997). The record of the archaeological landscape shows that the inhabitants of these cities and villages made technological adaptations to such climatic variation by extensifying and intensifying cultivation. A closer look at individual cities reveals, however, that such adaptations were not uniform and may explain the variation in settlement trajectories at the close of the 3rd millennium.

Adaptive variability in three urban communities

In light of these conflicting paleoclimatic reconstructions, it is worthwhile to consider urban demography and land use at three large, and therefore ecologically fragile, urban settlements of the time: Hamoukar (98 ha), Tell Brak (70 ha) and Tell Leilan (90 ha). Each of these places has a full component of excavation, survey and off-site landscape data.

The largest is Hamoukar, situated close to the Syrian-Iraqi frontier. At the start of the 3rd millennium, settlement was constrained to a 15 hectare mound, but around 2600 BC, an extensive lower area of settlement grew around it to the south. The 1999 surface collection showed ceramics from the later 3rd millennium BC covering the entire lower town, making Hamoukar a 98 hectare city (Ur 2010b). Settlement appears to have survived any abrupt events around 2200 BC, since post-Akkadian diagnostic types were found throughout the lower town (Ur 2012b).

Third millennium excavation areas on the high mound have been spatially limited, but lower town trenches have revealed dense domestic occupation (Ur and Colantoni 2010; Colantoni and Ur 2011). The associated ceramics include sherds with comb-incised decoration that are most common in the post-Akkadian period, in other words the time after the proposed abrupt climatic event of 2200 BC. The frequency of baked brick in all lower town trenches suggests that not only did urban settlement at Hamoukar last beyond 2200 BC but its inhabitants also remained considerably wealthy. The end of Hamoukar was violent: all lower town trenches show sudden abandonment, with complete ceramic inventories left on floors, some with burning. In one area, semi-articulated human remains were found where they had been left by scavengers. Hamoukar's end was abrupt, but its immediate cause was social rather than environmental.

Extensive agriculture in Hamoukar's hinterland can be reconstructed from the presence of hollow ways (fig. 3). Several trackways converge at a point on Hamoukar's western edge, likely an ancient gate into the settlement. Others radiate to the southeast. The proposed agricultural catchment of Hamoukar includes only 5,200 hectares, far less than would be required to feed its estimated population. Almost 1,000 field scatter collections were made in Hamoukar's hinterland, and they describe dense scatters immediately surrounding the site. Thus Hamoukar's cultivators extensified and intensified their agricultural production. Scatters also surround Hamoukar's 3rd millennium satellites, although at lower density. Almost all of these satellites were also settled in the post-2200 BC period.

Tell Brak has been excavated intermittently since the 1920s, and is the primary site for the archaeology and chronology of the later 3rd millennium BC in the region (Oates, Oates and McDonald 2001). After an urban expansion in the 4th millennium (Ur, Karsgaard and Oates 2007), Brak had shrunk to around 40 hectares in the early 3rd millennium. Its initial growth was later than at Leilan or Hamoukar. Around 2500 BC, settlement expanded to the south to create a lower town, totalling approximately 70 hectares (Ur, Karsgaard and Oates 2011). By around 2400 BC, Brak was known as Nagar, and it was the dominant political centre in northern Mesopotamia (Eidem,

FIGURE 3. *Hollow ways and field scatters around Hamoukar.*

Finkel and Bonechi 2001). We know nothing of the nature of settlement on its lower town, but the ancient high mound contained a sequence of monumental scale buildings, most of which were hybrid temple and administrative complexes (Oates, Oates and McDonald 2001). One such structure was the so-called Naram-Sin fortress, constructed of bricks stamped with the name of the Akkadian conqueror.

Brak's landscape also shows strong evidence for both extensive cultivation and intensification through manuring (fig. 4). The radial system of hollow ways around Brak (Ur 2003; Wilkinson et al. 2010) is the largest and most deeply etched in the basin. Some linear features extend more than five kilometres. Many coalesce around particular points on the mound's edge, near gullies that might represent the positions of gates. The fields northeast and southwest of Brak were extensively sampled for field scatters, which occur in high numbers, especially close to Brak itself (Ur, Karsgaard and Oates 2011).

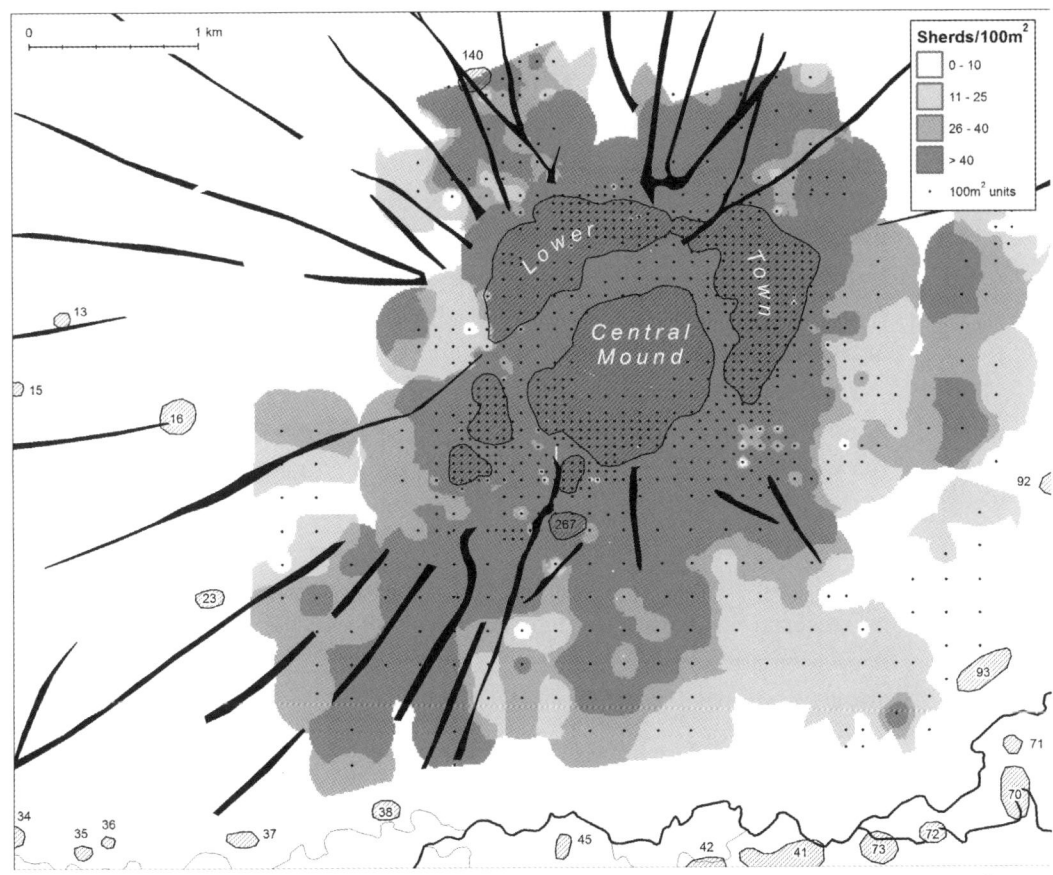

FIGURE 4. *Hollow ways and field scatters around Tell Brak (sites from Wright et al. 2006–2007).*

Like Hamoukar, Brak also survived the proposed abrupt aridification event, although substantial changes appear to have been roughly coincident with 2200 BC. Around this time, many of the large institutional complexes were ritually sealed and went out of use (Oates, Oates and McDonald 2001: 90–91, 389–390). The succeeding occupations were all domestic in origin. In the absence of the distinctive comb-incised decorated sherds, it appears that Brak's lower town ceased to be settled around this time as well (Ur, Karsgaard and Oates 2011; Ur 2012b). Brak's post-Akkadian occupation was thus reduced in scale and socially reorganised.

Yet a third settlement trajectory can be described for Tell Leilan, the point of origin of the 2200 BC abrupt aridification event hypothesis. Leilan's urban beginnings are similar to Hamoukar's. Building on several prehistoric settlements, Leilan covered 15 hectares in the early 3rd millennium BC before rapidly expanding to 90 hectares around 2600 BC. For the next four centuries, Leilan's high mound hosted elite households while the lower town was filled with dense residential occupation. Its excavators propose that it was conquered by the Akkadian state of southern Mesopotamia, which initiated a regional demographic reorganisation and closely managed the agricultural and pastoral economy of Leilan itself (Weiss and Courty 1993).

Landscape evidence for extensive agriculture takes the form of abundant hollow ways associated with Leilan and also its 3rd millennium satellites to the southeast, which have been mapped from CORONA satellite imagery (fig. 5; see also Ur 2003, fig. 10; Ur 2010b, Map 2). However, the Leilan area lacks the extensive field scatters seen elsewhere in the basin. In radial transects around several major late 3rd millennium sites, scatters were light and no material was found further than 500 metres away. These scatters have been interpreted as the result of post-depositional processes, rather than any ancient activities (Ristvet 2005: 25).

Urbanisation at Leilan and its hinterland was based on extensification of agricultural fields, as demonstrated by the hollow ways, but its farmers did not adopt intensification methods in the form of manuring.[1] They may have attempted to raise yields by reducing the fallow interval, for example by cultivating a field annually. This technique provides a short-term yield increase but would deprive the soil of the moisture carry-over that a fallow year provides, increasing the likelihood of crop failure. The combination of nucleated population, declining precipitation and fallow violation would have made Leilan's staple economy particularly fragile.

Leilan and its hinterland was rapidly abandoned around 2200 BC, based on radiocarbon dates and ceramic chronology (Weiss et al. 1993: 999–1002). The ceramic type used to identify the few surviving settlements is a style of comb-incised decoration. Based on the presence of sherds of this type, the

---

[1] It is possible that Leilan's farmers manured their fields by allowing their animals to drop dung directly onto the fields, in which case it would be archaeologically invisible without geochemical analysis. However, raw dung left directly on fields in semi-arid climates is not beneficial as a manure because it oxidises rapidly; this problem can be avoided by settlement-based animal penning and composting (Wilkinson 1982: 324).

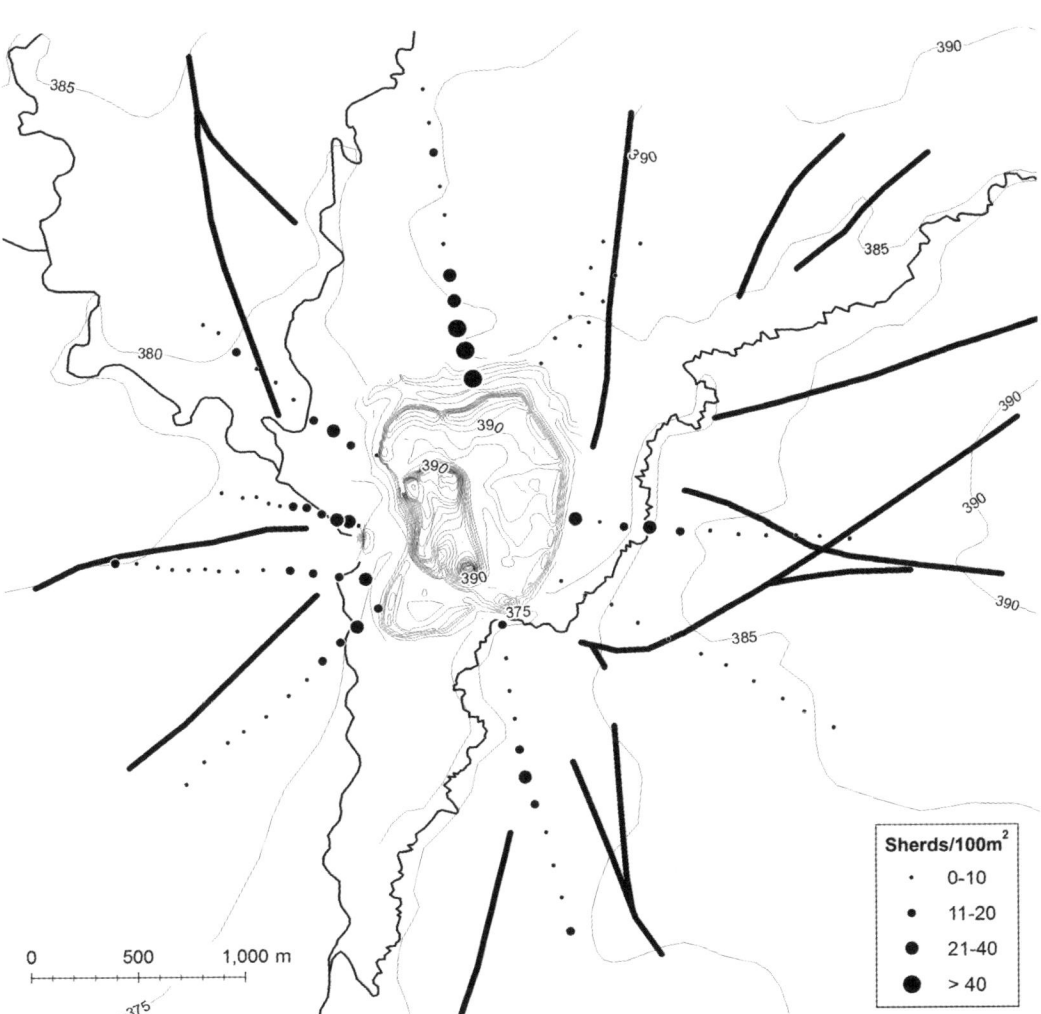

FIGURE 5. *Hollow ways and field scatters around Tell Leilan (field scatters from Ristvet 2005, fig. 1.3).*

nucleated pattern of the Akkadian period dissipated into a few small and dispersed villages; 69 per cent of sites were abandoned, and the total settled area declined by 72 per cent (Ristvet 2005: 120).

The focus here has been on three of the largest EBA cities, but a similar variability in demographic trajectories and adaptive responses can be seen in other places, including smaller towns. Tell es-Sweyhat on the Syrian Euphrates reached its maximum extent of 45 ha precisely at the time of the alleged abrupt event, and settlement continued for an additional two centuries (Danti 2010). Field scatters and hollow ways (Wilkinson 2004: 55–82) demonstrate that Sweyhat's farmers employed the same agricultural strategies that were used at Brak and Hamoukar, but its emphasis on its animal economy is perhaps the reason why it persevered at a time of arid and unstable climate (Danti 2010). As the climate dried, the small towns and villages of the Balikh drainage underwent a simultaneous nucleation and population reduction; the former is interpreted as a social adaptation via new land tenure rules to reduced resource availability, while the latter reflects the decision by some farmers to embrace a non-sedentary pastoral economy (Wossink 2010: 187–188).

In some places, no social changes or adaptations have been detected. At the large city of Urkeš (modern Tell Mozan), crop choices, vegetation patterns and animal husbandry were largely unchanged in the last century of the 3rd millennium (Deckers et al. 2010). Further to the north around Titriş Höyük, a 43 ha site in southeastern Turkey, survey and excavation revealed a similar lack of correlation between climatic events and settlement history (Algaze and Pournelle 2003). Both of these sites are on the wettest northern fringe of the Fertile Crescent today, and it may be that 3rd millennium precipitation never decreased to the point where their farmers were forced to adopt new agricultural technologies or modify their staple economies.

Thus far the discussion has considered successful responses that allowed the perseverance of sedentism or, in the case of Tell Leilan, the abandonment of sedentism in favour of pastoral nomadism. It is probable, however, that societies which perceived the climatic changes may have responded unsuccessfully, or, to a modern Western observer, inappropriately. For example, the Levantine towns of the Early Bronze (EB) III period may have turned to the gods for help (Rosen 2007). Farmers invested more authority in elite managers who were capable of interceding with the gods. One archaeological manifestation appears to be an increase in temple building at the end of the EB III, which ultimately proved to be unsuccessful; the subsequent response of most settlements by the EB IV was abandonment in favour of pastoral nomadism.

Summary and conclusions

To summarise, we have examined three major cities and seen three different settlement trajectories (table 1). In terms of demography and structure, Leilan's trajectory is the most dramatic: it was abruptly and completely abandoned relatively early, around 2200 BC. Hamoukar survived in a fully urbanised state perhaps as late as 2000 BC, and while the nature of settlement in its earliest urban phases is uncertain, its final phase appears to be one of substantial wealth, so a post-drought impoverishment seems unlikely. Brak's settlement was reduced in scale from 70 hectares to approximately 40, and it underwent substantial structural changes. The excavations have not, however, revealed evidence that this transformation was climatically induced. By the early 2nd millennium, settlement at Brak had been reduced further, but on present evidence it appears never to have been abandoned.

The nature of EBA climatic change is not agreed upon, but the continuation of settlement at Hamoukar, Brak, Sweyhat, Titriş, Mozan and others suggests that the severity of the region-wide event at 2200 BC has been overestimated. More likely, the climate of northern Mesopotamia experienced a gradual aridification that was probably only perceptible through the social memory of a past time when the rains were more reliable and crop yields were greater.

One response to reduced agricultural yields is simply to cultivate more land. At all three sites, agriculture was extensified with urban expansion, creating movement-constraining conditions that caused the formation of incised patterns of hollow way routes. In agent-based simulations, farmers who have the option to expand cultivation will do so if the productivity of their existing fields has declined, whether due to arid conditions or some other yield-reducing circumstance (Wilkinson et al. 2007: 65–66). A second possible response is to intensify cereal production to obtain more yields per area unit. The farmers in some, but not all, northern Mesopotamian cities chose to intensify by manuring their fields with settlement-derived debris. The fields around Brak, Hamoukar and Sweyhat are littered with the remaining inorganic component of this ancient practice. Farmers around Leilan, however, appear not to have adopted manuring as an intensification technology.

	Hamoukar	Brak	Leilan
Nature of late EBA Transformation	City-wide violent event	Demographic reduction, institutional reorganisation	Rapid and extensive urban abandonment
Agricultural Extensification (Hollow way proxy)	Yes	Yes	Yes
Agricultural Intensification (Field scatter proxy)	Yes	Yes	No
Abandonment	Ca. 2100–2000 BC	(not abandoned)	2200 BC

TABLE 1. *Settlement trajectories and agricultural adaptations at three Early Bronze Age cities.*

It could be proposed, therefore, that the different settlement trajectories of the sites and their differential adoption of agricultural technologies were related. On the one hand, around Hamoukar, Brak and Sweyhat, farmers extensified and intensified agricultural production, and urban settlement was sustained for a long duration through an increasingly arid and unstable climate, and also in the face of possible abrupt events. On the other hand, without such techniques, the staple economy of Leilan and its neighbours was more susceptible to an abrupt dry phase. It may not be coincidental, therefore, that when Leilan was deurbanised and its region nearly completely abandoned at 2200 BC, settlement persevered at Hamoukar, Brak, Sweyhat, Mozan, Titriş and other cities and towns. Because their farmers employed techniques to increase agricultural productivity, these other urban agglomerations were better suited to weather droughts.

We must ask to what extent were these agricultural adaptations *responses* to climatic changes. As is common for syntheses of environmental and archaeological data, causation is rendered difficult by chronological imprecision. Hollow ways and field scatters can only be generally dated to this time of urban expansion, and not to any particular phase of it. Thus we cannot say whether extensification and manuring were responses to suddenly poorer conditions for dry farming, or if these techniques emerged along with the initial expansion of urban settlement, and hence predated the climatic aridification. It has been argued, for example, that intensification was driven by the need for social surpluses to supply commensal events (Ur and Colantoni 2010), or that population growth and nucleation may have necessitated it (Wilkinson 1994). The adoption of these techniques need not have been due to a single factor, however. One can imagine a scenario in which a household manured its fields to support its social ambitions but also did well in dry years when other households' crops failed. The success of that household would be noticed by its neighbours, and its techniques emulated. If climatic drying was a process that unfolded over several centuries, it is possible that these techniques were increasingly adopted over a long period in some regions, but not at all in others. The most significant question is why Leilan's farmers elected not to employ manuring on their fields, despite being only a day's walk away from the cities at Brak and Hamoukar, whose farmers did manure.

Although they proved to be more durable than Leilan, the cities that continued past 2200 BC were transformed or eventually abandoned. Hamoukar survived for another one or two centuries before its rapid abandonment as violence and looting swept the city. Settlement at Brak contracted and was restructured, with major temple institutions sealed off and replaced by domestic structures. These variable and non-synchronic events might very well be interrelated, and an increasingly arid and unstable climate might be a part of this story. Rather than a simplistic scenario in which climate changes and human communities respond uniformly, we could envision a non-linear sequence of events in which the collapse of one major urban settlement created political, economic and demographic waves that impacted other cities and regions in a variety of ways (Coombes and Barber 2005). Military violence was pervasive throughout northern Mesopotamia in the late EBA and certainly played a role (Archi and Biga 2003; Sallaberger 2007).

Thus, at the present state of knowledge, the trajectories of urban settlement at the end of the 3rd millennium BC are far more variable than many reconstructions would suggest, and the same is true for the social response. The variability in settlement histories in the northern Fertile Crescent is now matched by variation elsewhere in western Syria and southeastern Turkey (e.g. Schwartz 2007; Danti 2010). Whether climate change was abrupt or gradual, the responses of human groups varied dramatically, sometimes within a hundred kilometres of each other. In this case, plausible responses that have been proposed include political and demographic collapse and regional abandonment, alteration of settlement patterns and occupational density, emphasis on a pastoral economy, and the development or expansion of agricultural techniques for ameliorating climatic impacts. It seems quite likely that all of these responses appeared in some form or combination at the end of the 3rd millennium BC in northern Mesopotamia.

Acknowledgements

I thank the Directorate General of Antiquities and Monuments, particularly Michel al-Maqdissi in Damascus and Abd al-Massieh Baghdo in Hassake, for permission to undertake the field research at Brak and Hamoukar presented here. At Hamoukar, the Tell Hamoukar Survey was entrusted to me by the expedition directors McGuire Gibson (Chicago) and Amr al-Azm (Damascus), and I was assisted by Carlo Colantoni and Lamya Khalidi. The Suburban Survey surface collection at Tell Brak was undertaken in collaboration with Philip Karsgaard (Edinburgh) under the auspices of the Tell Brak Excavation project, directed by David and Joan Oates. I also thank Henry Wright, field director of the Tell Brak Sustaining Area Survey, for his advice and encouragement. Funding was provided by the University of Chicago, Harvard University, the British Academy, the McDonald Institute, and the University of Michigan. This paper was strengthened by comments and critiques from Tony Wilkinson, Michael Danti and an anonymous reviewer.

This chapter is dedicated to the memory of Dr Stine Rossel, a gifted and energetic young scholar who stood poised to make a lifetime of scholarly contributions. The memory of our student, colleague and friend is alive at Harvard and Copenhagen.

Bibliography

Akkermans, P. M. M. G. and G. Schwartz. 2003. *The Archaeology of Syria: From Complex Hunter-Gatherers to Early Urban Societies (ca. 16,000–300 BC)*. Cambridge University Press. Cambridge.

Algaze, G. and J. Pournelle. 2003. "Climatic Change, Environmental Change, and Social Change at Early Bronze Age Titriş Höyük." In M. Özdoğan, H. Hauptmann and N. Başgelen (eds.), *From Village to Towns: Studies Presented to Ufuk Esin*, Arkeoloji ve Sanat, 103–128. Istanbul.

Archi, A. and M. G. Biga. 2003. "A Victory over Mari and the Fall of Ebla." *Journal of Cuneiform Studies* 55: 1–44.

Bar-Matthews, M. and A. Ayalon. 2011. "Mid-Holocene Climatic Variations Revealed by High-Resolution Speleothem Records from Soreq Cave, Israel and their Correlation with Cultural Changes." *Holocene* 21: 163–171.

Bintliff, J. L. and A. M. Snodgrass. 1988. "Off-Site Pottery Distributions: A Regional and Interregional Perspective." *Current Anthropology* 29: 506–513.

Charles, M. and A. Bogaard. 2001. "Third-Millennium BC Charred Plant Remains from Tell Brak." In D. Oates, J. Oates and H. McDonald (eds.), *Excavations at Tell Brak, Vol. 2: Nagar in the Third Millennium BC*, McDonald Institute for Archaeological Research and the British School of Archaeology in Iraq, 301–326. Cambridge and London.

Colantoni, C. and J. A. Ur. 2011. "The Architecture and Pottery of a Late 3rd Millennium BC Residential Quarter at Tell Hamoukar, Northeastern Syria." *Iraq* 73: 21–69.

Coombes, P. and K. Barber. 2005. "Environmental Determinism in Holocene Research: Causality or Coincidence?" *Area* 37: 303–311.

Courty, M.-A. 1998. "Soil Record of an Exceptional Event at 4000 BP in the Middle East." In B. J. Peiser, T. Palmer and M. E. Bailey (eds.), *Natural Catastrophes during Bronze Age Civilisations*, BAR International Series 728, Archaeopress, 93–108. Oxford.

Courty, M.-A. 2001. "Evidence at Tell Brak for the Late EDIII/Early Akkadian Air Blast Event (4 kyr BP)." In D. Oates, J. Oates and H. McDonald (eds.), *Excavations at Tell Brak, Vol. 2: Nagar in the Third Millennium BC*,

McDonald Institute for Archaeological Research and the British School of Archaeology in Iraq, 367–372. Cambridge and London.

Courty, M.-A., and H. Weiss. 1997. "The Scenario of Environmental Degradation in the Tell Leilan Region, NE Syria, During the Late Third Millennium Abrupt Climate Change." In H. N. Dalfes, G. Kukla and H. Weiss (eds.), *Third Millennium BC Climate Change and Old World Collapse*, Springer Verlag, 107–147. Berlin.

Danti, M. 2010. "Late Middle Holocene Climate and Northern Mesopotamia: Varying Cultural Responses to the 5.2 and 4.2 ka Aridification Events." In A. B. Mainwaring, R. Giegengack and C. Vita-Finzi (eds.), *Climate Crises in Human History*, American Philosophical Society, 139–172. Philadelphia.

Deckers, K., M. Doll, P. Pfälzner and S. Riehl. 2010. *Development of the Environment, Subsistence and Settlement of the City of Urkeš and its Region*. Studien zur Urbanisierung Nordmesopotamiens Serie A Band 3. Harrassowitz. Wiesbaden.

Eidem, J., I. Finkel and M. Bonechi. 2001. "The Third Millennium Inscriptions." In D. Oates, J. Oates and H. McDonald (eds.), *Excavations at Tell Brak, Vol. 2: Nagar in the Third Millennium BC*, McDonald Institute for Archaeological Research and the British School of Archaeology in Iraq, 99–120. Cambridge and London.

Geyer, B., and Y. Calvet. 2001. "Les steppes arides de la Syrie du Nord au Bronze ancien ou 'la premiere conquête de l'est'." In B. Geyer (ed.), *Conquête de la steppe et appropriation des terres sur les marges arides du Croissant fertile*, Maison de l'Orient Méditerranéen-Jean Pouilloux, 55–68. Lyon.

Hole, F. 1997. "Evidence for Mid-Holocene Environmental Change in the Western Khabur Drainage, Northeastern Syria." In H. N. Dalfes, G. Kukla and H. Weiss (eds.), *Third Millennium BC Climate Change and Old World Collapse*, NATO ASI Series I Vol. 49, Springer Verlag, 39–66. Berlin.

Hole, F. 1998. "Paleoenvironment and Human Society in the Jezireh of Northern Mesopotamia 20,000–6,000 BP." *Paléorient* 23: 17–29.

Kuzucuoğlu, C., and C. Marro (eds). 2007. *Human societies and climate change at the end of the Third Millennium: Did a crisis take place in Upper Mesopotamia? (Sociétés humaines et changement climatique à la fin du Troisième Millénaire: une crise a-t-elle eu lieu en Haute Mésopotamie?)*, Varia Anatolica XIX, IFEA (Istanbul) and de Boccard (Paris).

Marro, C., and C. Kuzucuoğlu. 2007. "Northern Syria and Upper Mesopotamia at the End of the Third Millennium B.C.: Did a Crisis Take Place?"

In C. Kuzucuoğlu and C. Marro (eds.), *Sociétés humaines et changement climatique à la fin du troisième millénaire: une crise a-t-elle eu lieu en haute Mésopotamie?*, Varia Anatolica 19, Institut Français d'études anatoliennes Georges-Dumézil, 583–590. Istanbul.

McIntosh, R. J., J. A. Tainter and S. K. McIntosh. 2000a. "Climate, History, and Human Action." In R. J. McIntosh, J. A. Tainter and S. K. McIntosh (eds.), *The Way the Wind Blows: Climate, History, and Human Action*, Columbia University Press, 1–42. New York.

McIntosh, R. J., J. A. Tainter and S. K. McIntosh (eds.). 2000b. *The Way the Wind Blows: Climate, History, and Human Action*. Columbia University Press. New York.

Oates, D., J. Oates and H. McDonald, 2001. *Excavations at Tell Brak, Vol. 2: Nagar in the Third Millennium BC*. McDonald Institute for Archaeological Research and the British School of Archaeology in Iraq. Cambridge and London.

Riehl, S. 2009. "Archaeobotanical Evidence for the Interrelationship of Agricultural Decision-Making and Climate Change in the Ancient Near East." *Quaternary International* 197: 93–114.

Ristvet, L. 2005. "Settlement, Economy, and Society in the Tell Leilan Region, Syria, 3000 1000 BC." PhD dissertation, University of Cambridge.

Roberts, N., D. Brayshaw, C. Kuzucuoğlu, R. Perez and L. Sadori. 2011a. "The Mid-Holocene Climatic Transition in the Mediterranean: Causes and Consequences." *Holocene* 21: 3–13.

Roberts, N., W. J. Eastwood, C. Kuzucuoğlu, G. Fiorentino and V. Caracuta. 2011b. "Climatic, Vegetation and Cultural Change in the Eastern Mediterranean During the Mid-Holocene Environmental Transition." *Holocene* 21: 147–162.

Rosen, A. Miller. 2007. *Civilizing Climate: Social Responses to Climate Change in the Ancient Near East*. Altamira Press. Lanham, Maryland.

Rosen, A. Miller, and S. A. Rosen. 2001. "Determinist or Not Determinist? Climate, Environment, and Archaeological Explanation in the Levant." In S. R. Wolff (ed.), *Studies in the Archaeology of Israel and Neighboring Lands in Memory of Douglas L. Esse*, Studies in Ancient Oriental Civilization 59, Oriental Institute Publications, Oriental Institute of the University of Chicago, 535–549. Chicago.

Ryan, W. B. F., and W. C. Pitman. 1998. *Noah's Flood: The New Scientific*

Discoveries about the Event that Changed History. Simon & Schuster. New York.

Sallaberger, W. 2007. "From Urban Culture to Nomadism: A History of Upper Mesopotamia in the Late Third Millennium." In C. Kuzucuoğlu and C. Marro (eds.), *Sociétés humaines et changement climatique à la fin du troisième millénaire: une crise a-t-elle eu lieu en haute Mésopotamie?*, Varia Anatolica 19, Institut Français d'études anatoliennes Georges-Dumézil, 417–456. Istanbul.

Schwartz, G. M. 2007. "Taking the Long View on Collapse: A Syrian Perspective." In C. Kuzucuoğlu and C. Marro (eds.), *Sociétés humaines et changement climatique à la fin du troisième millénaire: une crise a-t-elle eu lieu en haute Mésopotamie?*, Varia Anatolica 19, Institut Français d'études anatoliennes Georges-Dumézil, 45–67. Istanbul.

Staubwasser, M., and H. Weiss. 2006. "Holocene Climate and Cultural Evolution in Late Prehistoric-Early Historic West Asia." *Quaternary Research* 66: 372–387.

Stein, G. J. 2004. "Structural parameters and sociocultural factors in the economic organization of north Mesopotamian urbanism in the third millennium BC." In G. M. Feinman and L. M. Nicholas (eds.), *Archaeological Perspectives on Political Economies*, University of Utah Press, 61–78. Salt Lake City.

Ur, J. A. 2003. "CORONA Satellite Photography and Ancient Road Networks: A Northern Mesopotamian Case Study." *Antiquity* 77: 102–115.

Ur, J. A. 2009. "Emergent Landscapes of Movement in Early Bronze Age Northern Mesopotamia." In J. E. Snead, C. Erickson and W. A. Darling (eds.), *Landscapes of Movement: Paths, Trails, and Roads in Anthropological Perspective*, University of Pennsylvania Museum Press, 180–203. Philadelphia.

Ur, J. A. 2010a. "Cycles of Civilization in Northern Mesopotamia, 4400–2000 BC." *Journal of Archaeological Research* 18: 387–431.

Ur, J. A. 2010b. *Urbanism and Cultural Landscapes in Northeastern Syria: The Tell Hamoukar Survey, 1999–2001*. Oriental Institute Publications 137, Oriental Institute of the University of Chicago. Chicago.

Ur, J. A. 2012a. "Landscapes of Movement in the Ancient Near East." In R. Matthews and J. Curtis (eds.), *Proceedings of the 7th International Congress on the Archaeology of the Ancient Near East, 12–16 April 2010, the British Museum and UCL, London, Volume 1*, Harrassowitz, 521–538. Wiesbaden.

Ur, J. A. 2012b. "Spatial Scale and Urban Evolution at Tell Brak and Hamoukar at the End of the 3rd Millennium BC." In N. Laneri, P. Pfälzner and S. Valentini (eds.), *Looking North: The Socio-Economic Dynamics of the Northern Mesopotamian and Anatolian Regions during the Late Third and Early Second Millennium BC*, Studien zur Urbanisierung Nordmesopotamiens, Tübingen University, 25–35. Tübingen.

Ur, J. A., and C. Colantoni. 2010. "The Cycle of Production, Preparation, and Consumption in a Northern Mesopotamian City." In E. Klarich (ed.), *Inside Ancient Kitchens: New Directions in the Study of Daily Meals and Feasts*, University Press of Colorado, 55–82. Boulder.

Ur, J. A., P. Karsgaard and J. Oates. 2007. "Urban Development in the Ancient Near East." *Science* 317: 1188.

Ur, J. A., P. Karsgaard and J. Oates. 2011. "The Spatial Dimensions of Early Mesopotamian Urbanism: The Tell Brak Suburban Survey, 2003–2006." *Iraq* 73: 1–19.

Ur, J. A., and T. J. Wilkinson. 2008. "Settlement and Economic Landscapes of Tell Beydar and its Hinterland." In M. Lebeau and A. Suleiman (eds.), *Beydar Studies I*, Brepols, 305–327. Turnhout.

van Liere, W. J., and J. Lauffray. 1954–55. "Nouvelle Prospection Archeologique dans la Haute Jezieh Syrienne." *Les Annales Archéologiques de Syrie* 4–5: 129–148.

Weiss, H. 1986. "The Origins of Tell Leilan and the Conquest of Space in Third Millennium Mesopotamia." In H. Weiss (ed.), *The Origins of Cities in Dry-Farming Syria and Mesopotamia in the Third Millennium B.C.*, Four Quarters, 71–108. Guildford.

Weiss, H. 1997. "Late Third Millennium Abrupt Climate Change and Social Collapse in West Asia and Egypt." In H. N. Dalfes, G. Kukla and H. Weiss (eds.), *Third Millennium BC Climate Change and Old World Collapse*, NATO ASI Series I Vol. 49, Springer Verlag, 711–723. Berlin.

Weiss, H. 2000a. "Beyond the Younger Dryas: Collapse as an Adaptation to Abrupt Climate Change in Ancient West Asia and the Eastern Mediterranean." In G. Bawden and R. M. Reycraft (eds.), *Environmental Disaster and the Archaeology of Human Response*, Maxwell Museum of Anthropology, 75–98. Albuquerque.

Weiss, H. 2000b. "Causality and Chance: Late Third Millennium Collapse in Southwest Asia." In O. Rouault and M. Wäfler (eds.), *La Djéziré et l'Euphrate Syriens de la protohistoire à la fin du IIe millénaire av. J.-C.: Tendances dans l'interprétation historique des donnés nouvelles*, Subartu 7, Brepols, 207–217. Turnhout.

Weiss, H., and M.-A. Courty. 1993. "The Genesis and Collapse of the Akkadian Empire: The Accidental Refraction of Historical Law." In M. Liverani (ed.), *Akkad: The First World Empire*, History of the Ancient Near East Studies 5, Sargon, 131–155. Padua.

Weiss, H., M.-A. Courty, W. Wetterstrom, F. Guichard, L. Senior, R. Meadow and A. Curnow. 1993. "The Genesis and Collapse of Third Millennium North Mesopotamian Civilization." *Science* 261: 995–1004.

Wilkinson, T. J. 1982. "The Definition of Ancient Manured Zones by Means of Extensive Sherd-Sampling Techniques." *Journal of Field Archaeology* 9: 323–333.

Wilkinson, T. J. 1989. "Extensive Sherd Scatters and Land-Use Intensity: Some Recent Results." *Journal of Field Archaeology* 16: 31–46.

Wilkinson, T. J. 1993. "Linear Hollows in the Jazira, Upper Mesopotamia." *Antiquity* 67: 548–562.

Wilkinson, T. J. 1994. "The Structure and Dynamics of Dry-Farming States in Upper Mesopotamia." *Current Anthropology* 35: 483–520.

Wilkinson, T. J. 1997. "Environmental Fluctuations, Agricultural Production and Collapse: A View from Bronze Age Upper Mesopotamia." In H. N. Dalfes, G. Kukla and H. Weiss (eds.), *Third Millennium BC Climate Change and Old World Collapse*, NATO ASI Series vol. I 49, Springer Verlag. Berlin.

Wilkinson, T. J. 2004. *On the Margin of the Euphrates: Settlement and Land Use at Tell es-Sweyhat and in the Upper Lake Assad Area, Syria*. Oriental Institute Publications 124, Oriental Institute of the University of Chicago. Chicago.

Wilkinson, T. J., J. Christiansen, J. A. Ur, M. Widell and M. Altaweel. 2007. "Urbanization within a Dynamic Environment: Modelling Bronze Age Communities in Upper Mesopotamia." *American Anthropologist* 109: 52–68.

Wilkinson, T. J., C. French, J. A. Ur and M. Semple. 2010. "The Geoarchaeology of Route Systems in Northern Syria." *Geoarchaeology* 25: 745–771.

Wilkinson, T. J., E. Wilkinson, J. A. Ur and M. Altaweel. 2005. "Landscape and Settlement in the Neo-Assyrian Empire." *Bulletin of the American Schools of Oriental Research* 340: 23–56.

Williamson, A. 1974. "Harvard Archaeological Survey in Oman, 1973: III- Sohar and the Sea Trade of Oman in the Tenth Century AD." *Proceedings of the Arabian Seminar,* 4: 78–96.

Wossink, A. 2010. "Climate, History, and Demography: A Case Study from the Balikh Valley, Syria." In H. Alarashi, M.-L. Chambrade, S. Gondet, A. Jouvenel, C. Sauvage and H. Tronchère (eds.), *Regards croisés sur l'étude archéologique des paysages anciens. Nouvelles recherches dans le Bassin méditerranéen, en Asie centrale et au Proche et au Moyen-Orient,* Maison de l'Orient, 181–192. Lyon.

Wright, H. T., E. S. A. Rupley, J. A. Ur, J. Oates and E. Ganem. 2006–2007. "Preliminary Report on the 2002 and 2003 Seasons of the Tell Brak Sustaining Area Survey." *Les Annales Archéologiques Arabes Syriennes* 49–50: 7–21.

Yoffee, N. 1988. "Orienting Collapse." In N. Yoffee and G. L. Cowgill (eds.), *The Collapse of Ancient States and Civilizations,* University of Arizona Press, 1–19. Tucson.

Zeder, M. A. 2003. "Food Provisioning in Urban Societies: A View from Northern Mesopotamia." In M. L. Smith (ed.), *The Social Construction of Ancient Cities,* Smithsonian Institution, 156–183. Washington, D.C. and London.

Cultural Transformation and the 8.2 ka Event in Upper Mesopotamia

Peter M. M. G. Akkermans, Johannes van der Plicht, Olivier P. Nieuwenhuyse, Anna Russell and Akemi Kaneda

Abstract

The Late Neolithic site of Tell Sabi Abyad in the Balikh region, northern Syria, provides the perfect case study for looking at how ancient societies coped with climate change. Extensive excavations at this site have revealed a unique, continuous sequence of seventh and early sixth millennium occupation layers, hitherto unparalleled at any other site in the Near East. The occupation of this site spans the so-called "8.2k cal BP climate event". This abrupt climate anomaly has recently gained much attention from climatologists but until now the archaeological and cultural implications have not been studied in detail. The spectacular data from Tell Sabi Abyad changes existing interpretations of the Late Neolithic and allows the unique opportunity to study the effects of abrupt climate change in the past. This paper will outline the preliminary results of a broad multidisciplinary project investigating all aspects of Neolithic life at the site such as settlement dynamics, material culture, animal exploitation, and use of plants together with a comprehensive programme of high resolution ^{14}C dating.

Introduction

In 2006, the authors embarked on a research programme comprising multidisciplinary research into the material, technological, economic, social and ecological changes in Syria during the seventh and early sixth millennium BC. The period around 6200 cal BC was of particular interest because of the possible relationship between an abrupt climate change at this time, the so-called "8.2 event", and a number of apparently contemporaneous, far reaching material innovations and socioeconomic transformations.[1]

[1] This paper was submitted in 2009 as a short contribution to the conference. Our presenta-

Many scholars are now facing the possibility that climatic and environmental changes can be the root of chain reactions leading to economic recession and resource depletion that could ultimately have been responsible for the decline and collapse of some civilisations. Of course, one does not want to be a climate determinist as many other factors can come into play within a cultural change. Societies are not simply monolithic bodies that roll and flow with environmental tides (Rosen 2007: 4). The problem with linking environmental and cultural change is demonstrating causality. Just because these changes take place at the same time does not mean that one caused the other. In order to establish a link it must be demonstrated that the environmental change was the critical factor (Coombes and Barber 2005; Rosen 2007). This involves carefully dated environmental and cultural changes, the use of models and the consideration of all other factors. This is one of the aims of the current research at Tell Sabi Abyad.

The 8.2 ka event

The 8.2 ka event has been receiving increasing attention in the past decade. This event is thought to have brought about generally cold and dry conditions causing a drought in arid areas (Rohling and Pälike 2005; Alley and Ágústsdóttir 2005; Alley et al. 1997; Weiss and Bradley 2001; Hoek and Bos 2007). The generally accepted explanation for the 8.2 ka event is a slowing down of the ocean thermohaline circulation (THC) as a result of freshwater perturbation, following the drainage of the huge proglacial Laurentide Lakes (Lake Agassiz and Ojibway) in front of the Laurentide Ice Sheet (Barber et al. 1999; Bauer et al. 2004; Klitgaard-Kristensen et al. 1998; Renssen et al. 2002; Teller et al. 2002; Wiersma and Renssen 2006). Oxygen records from Greenland ice cores have provided convincing evidence for this abrupt cold event with an estimated cooling of $6 \pm 2°C$ over Greenland (Alley et al. 1997). This event was not restricted to the Northern Hemisphere with evidence for a sudden and marked reduction in precipitation in Africa, the Near East and Asia at this time (Renssen et al. 2002; Gasse 2000; Bar-Matthews et al. 1999; Migowski et al. 2006; Enzel et al. 2003). There appears to be a link between cold temperatures in the North Atlantic brought about by the

_{tion was a preliminary summary of the project "Abrupt Climate Change and Cultural Transformation in Syria in Late Prehistory (c. 6800–5800 BC)". The views and information presented in this preliminary paper no longer adequately represent the perspective held by the team at the completion of the project.}

8.2 event and drought in Asian and African monsoonal regions (Alley and Ágústsdóttir 2005).

Water availability is one of the key climatic determinants for people living in semi-arid areas such as the Near East (deMenocal 2001) and an increase in aridity can be fatal. Discussions of the effects of climate change on the archaeological record are replete with putative evidence for the "collapse" of cultures (Berglund 2003; Brooks 2006; Catto and Catto 2004; Clare et al. 2008; Diamond 2002; Weninger et al. 2006; Weiss and Bradley 2001) and the Neolithic in the Near East is no exception. Complex cultures and societies are readily able to adapt to variations in weather and climate but persistent droughts are a challenge that require developed coping strategies (deMenocal 2001). The problem is that these "collapses" have been assumed, but so far never demonstrated convincingly. Our work at Tell Sabi Abyad is now starting to address the range and subtlety of such changes.

The Late Neolithic site of Tell Sabi Abyad

Tell Sabi Abyad is located in the upper Balikh valley of northern Syria approximately 30 km from the Syro-Turkish border. The main focus of this research is an area known as Operation III (fig. 1). The tell at Operation III is in fact composed of three broad occupation mounds known as Sequence A (c. 7100–6200 cal BC), B (c. 6200–5900 cal BC), and C (c. 5900–5500 cal BC). There are noticeable differences between Sequence A and B with current dates suggesting that the transition between these two periods was realised within a very short span of time around 6200 cal BC (Akkermans et al. 2006).

The Balikh Valley forms a narrow irrigated corridor through the arid Jazirah desert steppe (Mulders 1969; Copeland 1979; Wilkinson 1998). Tell Sabi Abyad is situated between isohyets of 300 mm (Tell Abyad) and 200 mm (Raqqa), making the area rather marginal for dry farming and crop failures common (Van Zeist 1988; Wilkinson 1998). Even small changes in the amount of annual precipitation can have drastic consequences as seen in the recent three-year drought that has hit Syria causing both crop failure and livestock loss.

Investigating the cultural effects of a climate change obviously requires a detailed insight into the development of society before and after the event.

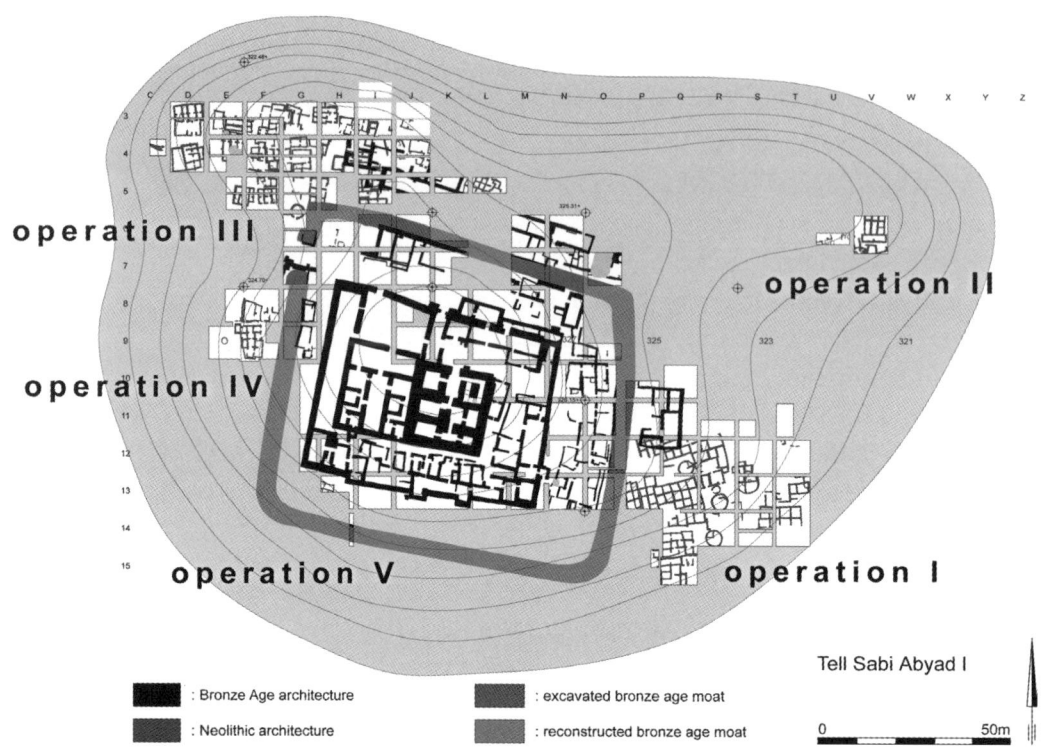

FIGURE 1. *The mound of Tell Sabi Abyad 1 showing the various field operations.*

Tell Sabi Abyad provides the perfect opportunity for archaeologists to do just that. The occupation of Tell Sabi Abyad spans the event and provides a sequence of uninterrupted settlement layers that through extensive excavations have shed light on not only the period of apparent climate change, but also the long trajectory of the seventh millennium cultural development that preceded the event and the centuries that followed it. So far, Tell Sabi Abyad is the only site where the strata of occupation relevant to the study of the 8.2 ka event have been unearthed on a large scale. The wealth of burgeoning new information derived from these excavations allows a diachronic study of Late Neolithic society with a focus on both short- and long-term changes in settlement, material culture, socio-economic organisation and environment.

FIGURE 2. *Large Coarse Ware storage jars.*

Changes in the pottery

One sub-project adopts a long-term perspective on change and innovations in containers made from pottery and other materials. Pottery vessels occurred in great quantities before, during and after the event, but a number of interesting changes occurred at the close of the 7th millennium cal BC. Presently we are investigating the synchronicity of these changes with the 8.2 ka event. How might archaeologists correlate changes in ceramics with climate change? Following a seminal paper by Halstead and O'Shea (1989), the project explores changes in ceramic storage technologies, changes in cuisine and food preparation strategies, and shifts in the symbolic role of ceramics.

A major long-term innovation in the pottery comprises the gradual development of strong and durable voluminous storage containers. This innovation may perhaps be related with changes in the technological realm, such as the development of a strong plant temper that enabled potters to build large storage jars (fig. 2). In terms of changes in cuisine and how food was being prepared, we can point to the introduction of various technologically specialised cooking wares. Symbolic changes, finally, are seen in the rapidly increasing importance of the role of decorated ceramics. Various "styles" or fashions can be identified, some of which can be associated with specific functional categories. Pottery style may have become implicated in social networking over much larger distances than ever before at the end of the 7th millennium.

Settlement Dynamic and High-Precision ^{14}C dating

Both the excavations and stratigraphic analyses indicate that the mound at Operation III is not composed of a single habitation horizon but should instead be viewed as three superimposed occupational mounds referred to as sequences A to C. The earliest occupation, dated from the end of the Pre-Pottery Neolithic B (PPNB) to the end of the Early Pottery Neolithic (c. 7100–6200 cal BC) when it was abandoned, is found at sequence A. Preliminary results suggest the border between Sequences A and B dates from about 6200 cal BC (fig. 3) (Akkermans et al. 2010). It appears that sequence B was established on the eastern edge of the former mound shortly after its desertion. The pottery study concurs: Sequence B dates to the so-called Pre-Halaf and Transitional periods (c. 6200–5900 cal BC); Sequences C and D indicate another phase of settlement took place on the Operation III mound from the Early Halaf to Late Halaf periods (c. 5900–5500 cal BC). Situated in the north-eastern area of the Early Pottery Neolithic mound is the Halaf settlement. In conclusion, it can be seen that the north-eastern mound of Tell Sabi Abyad was occupied over the whole Late Neolithic period. The sequence of the mounds can be summarised as follows.

Stratigraphic analysis of Operation III suggests that the occupation of Sequence A is comprised of at least 12 distinct levels continuing until the Late Neolithic period prior to c. 6200 cal BC (from Initial Pottery Neolithic to Early Pottery Neolithic). Sequence B has at least eight levels after c. 6200 cal BC (Pre-Halaf and Transitional periods) with two distinguished Halaf settlement sites (Early and Middle to Late Halaf) at Sequence C (fig. 4). These occupational sequences of Operation III are defined by the timing of

Date cal. BC	Period	Tell Sabi Abyad I - operations					Tell Sabi Abyad II	Tell Sabi Abyad III
		I	II	III	IV	V		
5700	Middle Halaf			level C-1				
5800	Early Halaf	level 1						
		level 2						
5900		level 3	level 1	level C-2/8				
		level 4	level 2	level B-1				
6000	Transitional	level 5 Burnt Village	level 3	level B-2		phase III		
		level 7	level 4	level B-3				
		level 8		level B-4				
6100		P-15 - 8		level B-5				
	Pre-Halaf	P-15 - 9		level B-6		phase II		
		P15 - 10		level B-7				
6200				level B-8				
				level A-1				
6300				level A-2				
				level A-3	level 1	phase I		
6400	Early	P-15 - 11		level A-4	level 2			
6500	Pottery			level A-5				
	Neolithic			level A-6				
6600				level A-7				
				level A-8				
6700				level A-9			level 1	trench H7
6800				level A-10			level 2	trench H8
	Initial PN			level A-11				
6900				level A-12				
7000							level 3	trench H9
	Late PPNB						level 4	
7100								

FIGURE 3. *Chronology summary.*

the construction and abandonment of architectural remains. We designate these building phases to different levels, with each level being distinct from the stratigraphic difference in architectural phases. Given the dynamic development of the settlement and its complex architectural phases, some levels are divided into sublevels. The results of stratigraphic analyses are being hardened against a comprehensive sample of radiocarbon dating with secure absolute dates provided for each level (see below).

Excavations of Sequence A exposed a large area of 1,662 m². Twelve levels were discovered (A12–A1; natural soil, however, has not yet been reached) revealing an organised ground plan of buildings dispersed over the occupation area. We can presume some compounds consisted of a unit or group with distinct units being located some metres apart. Some buildings were constructed on an uneven surface, such as the slope of the mound. Few buildings have been unearthed at Sequence B, though there are ovens, fire pits and pits indicating activities in an open area, characteristic of a Late Neolithic village.

Excavations on Sequence B from 2005 to 2008 exposed a small area of 760 m², though the settlement area containing buildings has yet to be excavated. Eight levels were recovered (B8–B1). We can see that there is a tendency for the village to shift from west to east; each new level is constructed slightly to the east on the ruins of the previous habitation. Why this has taken place is as yet unclear but studies will address the issue in the near future.

A large set of samples have been selected for isotopic analysis: ^{14}C dating, and ^{13}C/^{15}N investigations of human and animal bones. All isotopic analysis was conducted at the Center for Isotope Research of Groningen University.

For Tell Sabi Abyad, a grand total of almost 300 ^{14}C dates are obtained. Most of these (95%) are AMS dates; only 18 are conventional dates. Originally, more conventional measurements were planned, in particular bones from human burials, but all bones selected for conventional dating yielded too small an amount of datable collagen and had to be rerouted to the AMS laboratory. Of the 18 conventional dates, two samples consisted of large seeds which could be dated to a precision of 15 BP. Typical precision for AMS dates ranges from 30 to 50 BP (1-sigma standard deviation).

About two thirds of the total dataset represents dated charcoal or other botanical matter; about one third is (human and animal) bone.

The many samples from levels A12–A2 and B8–B3 have been analysed using Bayesian analysis (Bronk Ramsey 2009). Such stratified analysis has the advantage that the calibrated ^{14}C dates of events of interest can be established with good precision. This is in contrast to single ^{14}C dates which are not very precise after calibration due to the wiggles in the calibration curve. In our case, the event of interest is obviously the 8.2 ka event. (For a discussion see van der Plicht et al. 2011.)

Changing patterns of animal exploitation

The analysis of the faunal remains is very important in assessing the effects of climate change on a subsistence economy. Moderate fluctuations in cli-

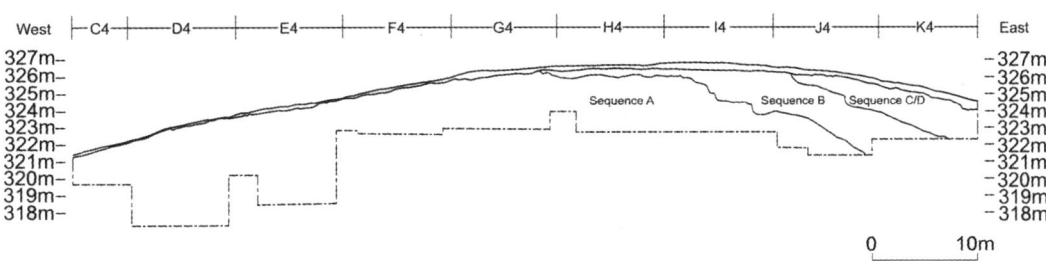

FIGURE 4. *Section through Operation III, taken through trenches C4–K4.*

mate can affect the internal structure of animal communities, i.e. changing the relative frequency of species. Moderate fluctuations can also change the physiological tolerance of a species causing, for example, a change in body size over generations. Abrupt climate induced ecosystem deterioration can on the other hand cause the mass elimination of a species (Tchernov 1982). Each species has an individual physiological response to environmental factors and some animals are better equipped to deal with negative changes in their environment than others. Climate change can also impact on the availability of animal feed due to either crop failure or the lack of good quality pasture (Valtorta 2002). Arguably the most problematic aspect of climate change faced by farmers is uncertainty and unpredictability.

Over 170 kilograms of material was analysed, comprising in total 48,802 bone fragments. Ovicaprids (domestic sheep and goat) dominated the assemblage in all levels and the main temporal changes were seen in the proportions of cattle, pigs and small and large wild game exploited at the site (figs. 5 and 6; see Russell 2010 for the final results of this analysis, undertaken as part of a PhD research project).

The initial results from the faunal analysis show that there were several changes in the economy occurring at level A1, around 6200 cal BC, including an increase in the exploitation of domestic cattle and a shift from the exploitation of domestic pigs to domestic cattle, with pig husbandry apparently no longer practised in levels B8–B1, and a decrease in the exploitation of wild animals. These changes could perhaps be linked to an aridification

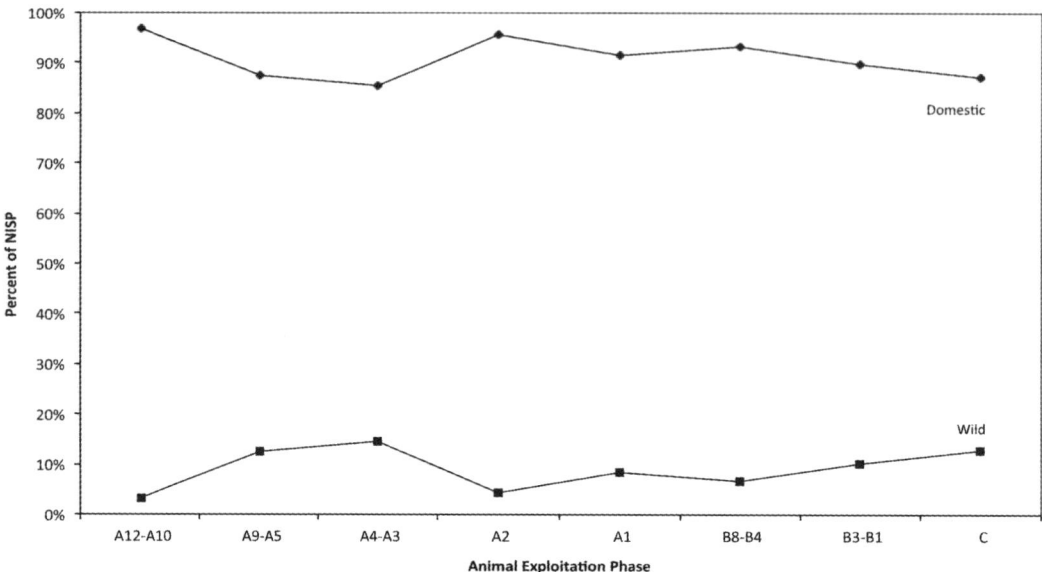

FIGURE 5. *Proportions of domestic and wild animals by animal exploitation phase (based on Number of Identified Specimens).*

of the environment caused by the 8.2 ka event. Despite the clear changes occurring at this time, the analysis of the faunal material showed that the sequence was characterised by continuity with no real rupture, decline or collapse in animal husbandry but instead subtle changes from level to level. The main trend seen was an increasing reliance on domestic animals, particularly pastoral animals such as goat, sheep and cattle. The other changes such as the domestication of cattle and the decline in hunting of both small and large game are also gradual (Astruc and Russell 2013; Russell 2010, in press).

Whether these changes in species proportions and animal husbandry strategies can be put down to deterioration in the local climate caused by an increase in aridity as a result of 8.2 ka event can be debated. Although the results of the faunal analysis offer some support for the hypothesis of climate change, this is not in itself direct evidence.

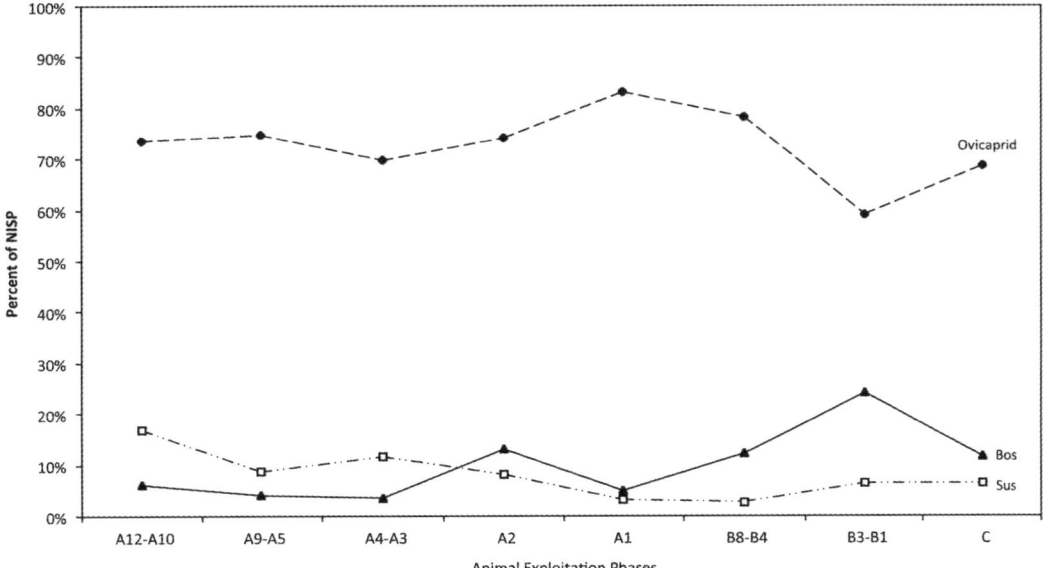

FIGURE 6. *Proportions of ovicaprid, pig and cattle by animal exploitation phase (based on Number of Identified Specimens).*

Plant micro-fossils and climate change

A comprehensive study of the plant macro-fossils from Operation III is currently underway. A large number of plant samples from well-defined deposits have been analysed and they will be used to reconstruct the exploitation of plant resources, as well as the reconstruction of the prehistoric agricultural practices and storage behaviour. An important aspect of this study will be the use of these macro-fossils as proxy evidence for any possible climate change around 6200 cal BC.

Conclusions

Only when a number of separate lines of evidence agree on climate change being a forcing factor are we safe in positing a true climate effect on past societies (Kuniholm 1990). It was for this reason that the issue of climate change within the sequence at Tell Sabi Abyad was tackled with this multidisciplinary approach.

The long and continuous sequence of seventh and early sixth millennium occupation layers at Tell Sabi Abyad provides proof for substantial cultural change around 6200 cal BC. There is increasing evidence for a fundamental, rapid transformation of society at this time, in the form of rearrangements and innovations in the pattern of settlement, architecture, material culture, economy, burial practices and social organisation. The many adaptations indicate that a major shift took place from a primarily sedentary, agrarian lifestyle to a more differentiated, pastoral form of existence around a number of central sites. The radical cultural transformation at about 6200 BC may very well be related to the prominent, abrupt climate change in the northern hemisphere around this time. However, the current archaeological evidence at hand in the case of Tell Sabi Abyad does not suggest a collapse or demise of the local communities as a response to the climate change. Rather, we hypothesise that people deliberately set into motion a range of changes and innovations in order to mitigate the environmental effects of the climate shift. Once the change was made, it may have involved many innovations that were probably not anticipated in the beginning but that appear to have been crucial in the subsequent shaping of complex society at the end of the seventh millennium.

Bibliography

Akkermans, P. M. M. G., R. Cappers, C. Cavallo, O. Niewenhuyse, B. Nilhamn and I. N. Otte. 2006. "Investigating the early pottery Neolithic of Northern Syria: New evidence from Tell Sabi Abyad." *American Journal of Archaeology* 110: 123–156.

Akkermans, P. M. M. G., J. van der Plicht, O. Nieuwenhuyse, A. Russell, A. Kaneda and H. Buitenhuis. 2010. "Weathering Climate Change: How Neolithic people adapted 8200 years ago." *Antiquity 84 (Project Gallery)*.

Alley, R. B., and A. M. Ágústsdóttir. 2005. "The 8k event: Cause and consequences of a major Holocene abrupt climate change." *Quaternary Science Reviews* 24: 1123–1149.

Alley, R. B., P. A. Mayewski, T. Sowers, M. Stuiver, K. C. Taylor and P. U. Clark. 1997. "Holocene climate instability: A prominent, widespread event 8200 yr ago." *Geology* 25(6): 483–486.

Astruc, L. and A. Russell. 2013. "Hunting and lithic specialization in the Balikh Valley: New data from Tell Sabi Abyad (VIIth millennium BC)." In O. Nieuwenhuyse, R. Bernbeck, P. M. M. G. Akkermans and J. Rogasch (eds.), *Interpreting the Late Neolithic of Upper Mesopotamia*. (PALMA Series 9). Brepols. Turnhout 331–343.

Barber, D. C., A. Dyke, C. Hillaire-Marcel, A. E. Jennings, J. T. Andrews, M. W. Kerwin, G. Bilodeau, R. McNeely, J. Southon, M. D. Morehead and J.-M. Gagnon. 1999. "Forcing of the cold event of 8,200 years ago by catastrophic drainage of Laurentide lakes." *Nature* 400: 344–348.

Bar-Matthews, M., A. Ayalon, A. Kaufman and G. J. Wasserburg. 1999. "The Eastern Mediterranean paleoclimate as a reflection of regional events: Soreq cave, Israel." *Earth and Planetary Science Letters* 166: 85–95.

Bauer, E., A. Ganopolski and M. Montoya. 2004. "Simulation of the cold climate event 8200 years ago by meltwater outburst from Lake Agassiz." *Paleoceanography* 19: 1–13.

Berglund, B. E. 2003. "Human impact and climate change – synchronous events and a causal link?" *Quaternary International* 105: 7–12.

Brooks, N. 2006. "Cultural responses to aridity in the Middle Holocene and increased social complexity." *Quaternary International* 151: 29–49.

Bronk Ramsey, C. 2009. "Bayesian analysis of Radiocarbon dates." *Radiocarbon* 51: 337–360.

Catto, N., and G. Catto. 2004. "Climate change, communities, and civilizations: Driving force, supporting play, or background noise?" *Quaternary International* 123–125: 7–10.

Clare, L., E. J. Röhling, B. Weninger and J. Hilpert. 2008. "Warefare in Late Neolithic/Early Chalcolithic Pisidia, Southwestern Turkey. Climate induced social unrest in the Late 7th Millennium cal BC." *Documenta Praehistorica* 35: 65–92.

Coombes, P., and K. Barber. 2005. "Environmental determinism in Holocene research: Causality or coincidence?" *Area* 37(3): 303–311.

Copeland, L. 1979. "Observations on the prehistory of the Balikh Valley, Syria, during the 7th to 4th Millennium B.C." *Paléorient* 5: 251–275.

deMenocal, P. 2001. "Cultural responses to climate change during the Late Holocene." *Science* 292: 667–673.

Diamond, J. 2002. "Evolution, consequences and future of plant and animal domestication." *Nature* 418: 700–707.

Enzel, Y., R. Bookman, D. Sharon, H. Gvirtzman, U. Dayan, B. Ziv and M. Stein. 2006. "Late Holocene climates of the Near East deduced from Dead Sea level variations and modern regional winter rainfall." *Quaternary International* 60: 263–273.

Gasse, F. 2000. "Hydrological changes in African tropics since the Last Glacial Maximum." *Quaternary Science Reviews* 19: 189–211.

Halstead, P., and J. O'Shea (eds.). 1989. *Bad year economics: Cultural responses to risk and uncertainty*. Cambridge University Press. Cambridge.

Hoek, W. Z. and J. A. A. Bos. 2007. "Early Holocene climate oscillations – causes and consequences." *Quaternary Science Review* 26: 1901–1906.

Klitgaard-Kristensen, D., H. P. Sejrup, H. Haflidason, S. Johnsen and M. Spurk. 1998. "A regional 8200 cal. yr BP cooling event in Northwest Europe, induced by final stages of the Laurentide Ice-Sheet deglaciation?" *Journal of Quaternary Science* 13(2): 165–169.

Kuniholm, P. I. 1990. "Archaeological evidence and non-evidence for climatic change." *Philosophical Transactions of the Royal Society of London* 330: 645–655.

Migowski, C., M. Stein, S. Prasad, J. F. W. Negendank and A. Agnon. 2006. "Holocene climate variability and cultural evolution in the Near East from the Dead Sea sedimentary record." *Quaternary Research* 66: 421–431.

Mulders, M. A. 1969. *The arid soils of the Balikh Basin (Syria)*. Unpublished PhD thesis, Utrecht University.

Renssen, H., H. Goosse and T. Fichefet. 2002. "Modeling the effect of freshwater pulse on the early Holocene climate: The influence of high-frequency climate variability." *Paleoceanography* 17(2): 1–15.

Rohling, E. J., and H. Pälike. 2005. "Centennial-scale climate cooling with a sudden cold event around 8,200 years ago." *Nature* 434: 975–979.

Rosen, A. Miller. 2007. *Civilizing Climate: Social Responses to Climate Change in the Ancient Near East*. Alta Mira Press. Lanham, Maryland.

Russell, A. 2010. *Retracing the Steppe: A Zooarchaeological analysis of changing subsistence patterns in the Late Neolithic at Tell Sabi Abyad, Northern Syria, c. 6900 to 5900 BC*. Leiden University. Leiden.

Russell, A. (in press) "Using zooarchaeology to assess climate change: Changing patterns of animal exploitation and the 8.2k BP climate event: Preliminary findings from the faunal remains of Tell Sabi Abyad, Syria c. 6800–5800 BC." *Proceedings of the 9th meeting of ASWA*.

Tchernov, E. 1982. "Faunal response to environmental changes in the Eastern Mediterranean during the last 20,000 years." In J. L. Bintliff and W. van Zeist (eds.), *Palaeoclimates, palaeoenvironments and human communities in the Eastern Mediterranean region in later prehistory*, BAR International Series 133: 105–129. Oxford.

Teller, J. T., D. W. Leverington and J. D. Mann. 2002. "Freshwater outbursts to oceans from glacial Lake Agassiz and their role in climate change during the last deglaciation." *Quaternary Science Reviews* 22: 879–887.

van Zeist, W. 1988. "Some notes on the plant husbandry of Tell Hammam et-Turkman." In M. N. Van Loon (ed.), *Hamman et-Turkman I*, NHAI: 705–715. Istanbul.

Van der Plicht, J., P. M. M. G. Akkermans, O. P. Nieuwenhuyse, A. Kaneda and A. Russell. 2011. "Tell Sabi Abyad, Syria: Radiocarbon chronology, cultural change, and the 8.2 ka event". *Radiocarbon* 53/2: 229-243.

Weiss, H. and R. S. Bradley. 2001. "What drives societal collapse?" *Science* 291: 609–610.

Valtorta, S. E. 2002. "Animal production in a changing climate: Impacts and mitigation". *Proceedings of the 15th International Congress on Biometeorology*: 98-101.

Weninger, B., E. Alram-Stern, E. Bauer, L. Clare, D. Danzeglocke, O. Joris, C. Kubatzki, G. Rollefson, H. Todorova and T. van Andel. 2006. "Climate forcing due to the 8200 cal yr BP event observed at Early Neolithic sites in the Eastern Mediterranean." *Quaternary Research* 66: 401–420.

Wiersma, A. P., and H. Renssen. 2006. "Model-Data comparison for the 8.2 ka BP Event: Confirmation of a forcing mechanism by catastrophic drainage of Laurentide lakes." *Quaternary Science Reviews* 25: 63–88.

Wilkinson, T. J. 1998. "Water and human settlement in the Balikh Valley, Syria: Investigations from 1992–1995." *Journal of Field Archaeology* 25(1): 63–87.

Climate and Social Change during the Transition between the Late Neolithic and Early Chalcolithic in Central Anatolia

Peter F. Biehl

Abstract

This article scrutinises the process of cultural, social, economic and symbolic transition between the Neolithic and the Chalcolithic in Central Anatolia as revealed at the Çatalhöyük East and Çatalhöyük West Mounds. It situates the transition in the palaeo-environmental changes in the Konya plain and presents a preliminary interpretation of the social changes and continuities between the East and West Mound around 6,000 cal BC. The article also re-evaluates possible reasons for the move from the Neolithic megasite of the Çatalhöyük East Mound to the much smaller and short-lived settlement of the West Mound which became one of several Early Chalcolithic settlements on the Konya Plain.

Introduction

How humans respond to climate change plays a crucial part in the formation of society. When climate changes, societies react by either adapting or transforming. This chapter examines whether and how humans responded to possible effects in Central Anatolia of the global climate change that occurred during the 8200 cal BP "climate event" (Weninger et al. 2005). The key hypothesis is that the change in climate and environment was at least one cause for people to move westwards into Western Anatolia and southeast Europe, and eventually across Europe. Çatalhöyük offers a microcosm that may help us unlock some of the key questions surrounding this time period (Biehl and Rosenstock 2009). At Çatalhöyük, settlement shifted from the East to the West Mound at around this time. The two tell sites

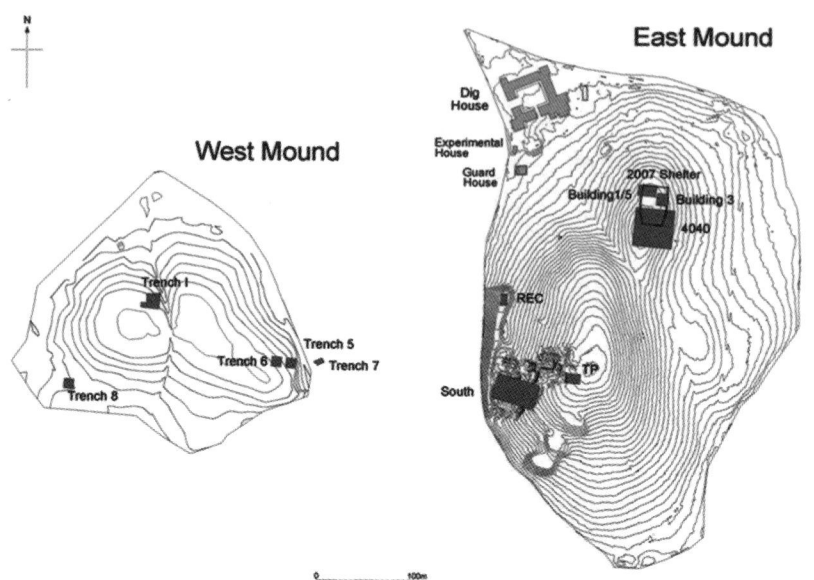

FIGURE 1.
Site plan with the excavation on the East and West Mounds.

sitting side by side offer an exceptional chance to analyse human responses to this event on a micro-scale and may give us the possibility of answering the question of why and how the shift from the East Mound to the West Mound took place at this time. Once we understand the regional process, we can widen our lens and try to determine the broader effects the shift had on the Near East and Europe.

We will focus on the West Mound, which lies separated by the ancient riverbed of the Çarşamba river and ca. 300 m west of the larger and better known Neolithic East Mound at Çatalhöyük in the Konya high plateau in Central Anatolia (Hodder 2005, 2006). The excavations and research have clearly shown that the diversity of resources exploited was from a very different landscape to what we see today. The story of Çatalhöyük began beside a river, today only shown by a line of trees along the ancient river course. It was here that the Neolithic megasite was located on the alluvial fan of the Çarşamba river in a rich wetland environment (fig. 1).

The West Mound at Çatalhöyük

The West Mound at Çatalhöyük covers ca. 8 ha and was first excavated by James Mellaart in 1961 (Mellaart 1965). He dug two small trenches (I and II) and distinguished – based on the pottery analysis – two phases of the Early Chalcolithic (EC I and EC II). In 1998–2003, the excavation was resumed by Jonathan Last and Catriona Gibson (Gibson and Last 2003). Their excavation focused on the areas close to the two trenches previously excavated by Mellaart in 1961.

Until recently it was believed that occupation across the river on the West Mound took place up to 400 years after the abandonment of the Neolithic site. In recent years, however, excavations on the Late Neolithic periods on the East Mound by the team from the University of Poznań (Czerniak and Marciniak 2009) and our work on the Early Chalcolithic West Mound (Biehl et al. 2010) have been closing this gap, and with our West Mound Research Project we are hoping to prove a gradual transition from one period to the next (Biehl and Rosenstock 2009). Whilst this still does not provide an answer as to why the site was finally abandoned in the Middle Chalcolithic period, it fits the results of the Konya Plain Survey Project conducted by Douglas Baird, which indicate a significant increase of sites in the Early Chalcolithic and a general decline of Middle Chalcolithic settlements on the Konya Plain (Baird 2002). Although only a few architectural remains were encountered, large quantities of painted pottery allowed Mellaart to divide the occupation into two phases, Early Chalcolithic I and II (Mellaart 1965; Last 1996; Franz 2010). Early Chalcolithic I is characterised by red-on-white painted pottery with geometric designs based on chevron/zigzag, banded and lattice motifs (fig. 2). Early Chalcolithic II pottery, which came only from a group of large pits in a small area of the site, has darker paint colours, finer lines and a greater use of fine lattice motifs.

While the appearance of painted pottery shows a clear difference from the earlier East Mound, questions remained about how distinct the West Mound was, both culturally and chronologically. Indeed, what does the term "Chalcolithic" mean and is it really valid or helpful to assign the site to a different period from the East Mound? Mellaart himself said that the period "in economic terms ... is as Neolithic as the preceding one" (Mellart 1975: 111). Why, then, use the label? As Ulf Schoop has pointed out in his recent book on the Anatolian Chalcolithic (Schoop 2005: 15), it derives

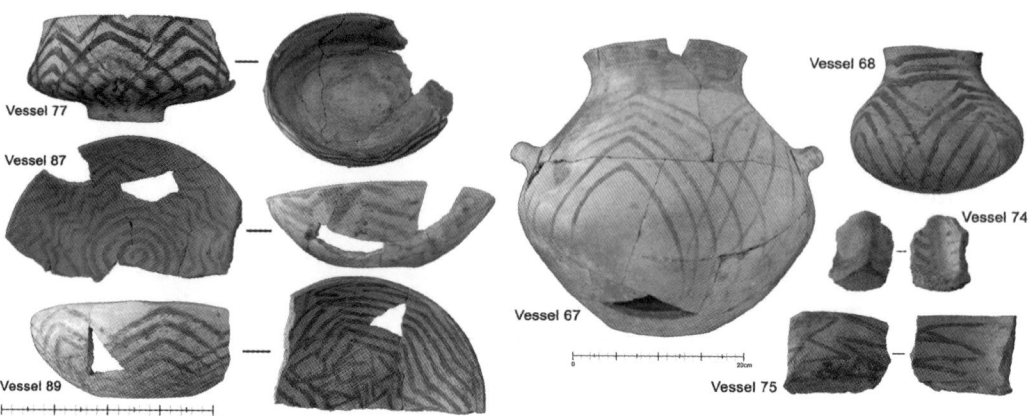

FIGURE 2. *Pottery from the West Mound (Franz 2010).*

from Mesopotamia where the separation of Neolithic and Chalcolithic was equated in a rather arbitrary way with the introduction of painted pottery in the Hassuna culture (Biehl and Rosenstock 2009: 472). But is the "Early Chalcolithic" a separate culture of no relevance to understanding the earlier period? Or is it simply a continuation of the Neolithic – with fancier pots? To understand the significance of the West Mound finds we need to look at the other changes taking place around the same time, and consider the question of architecture.

West Mound excavations

From the recent excavations we can now suggest that the built environment was different as well, with West Mound architecture showing a distinctive spatial organisation and a lesser degree of symbolic elaboration (Biehl and Rosenstock 2009; Biehl et al. 2010) (fig. 3). Conclusions based on one fully excavated (Building 98) and five partially excavated buildings from Trench 5, as well as comparisons with Building 25 (Gibson and Last 2003), 78 and 94 (last Erdoğu 2010) in other parts of the West Mound, must be treated with caution. But in contrast to the self-contained East Mound houses with one or two rooms, the structures here comprise a series of small, cell-like spaces,

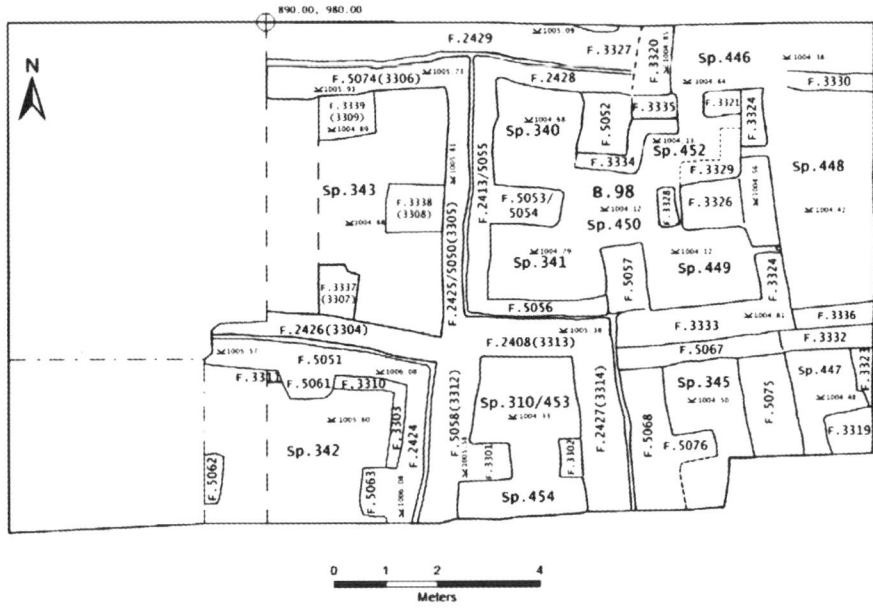

FIGURE 3. *Photo of Building 98 and architecture in Trench 5 of the West Mound.*

probably basements used for storage and working areas, below a larger central "living room" on the first/ground floor. While the central space has plastered walls and surfaces, so far the buildings seemingly lack the decorative architectural elements (bucrania, mouldings and figurative wall-paintings) and intra-mural burials inside a platform – except two infant burials found in the fill of a building – which are characteristic of the East Mound.

Also, the wall construction is different: on the West Mound the houses have either single or double perimeter walls divided by gaps of 5–7 cm, which underlines that there does seem to be more variation in wall construction on the West Mound than on the East Mound. Walls are made of at least two varieties of grey and brown compact mudbricks with or without the use of mortar with a brick format of ca. 80 × 40 cm and with a north-south orientation.

The similarities with roughly contemporary Can Hasan I layer 2B are striking, not only in the overall layout, orientation and dimensions of the buildings (fig. 4). The general brick size in this site was also ca. 0.8 m × 0.4 m, and the plan of the buildings in Trench 5 closely resembles so-called structure 5 in Can Hasan I 2B (French 1998). The internal buttresses are a common characteristic of architecture of the 6th millennium BC, not only in Central Anatolia (Biehl and Rosenstock 2009: 477).

If we accept that climate influences architecture and building materials, we can also see this construction further in the west, where we find mudbrick architecture on stone foundations e.g. in Hacılar I (Mellaart 1970: 84, fig. 29) and Kuruçay 7 (Duru 1994, levha 24) in the Lake District, and related architectural patterns with internal buttresses can be seen in Aktopraklık (Karul 2007: fig. 9) and Ilıpınar (Roodenberg 2001: fig. 13) in the Marmara region, and even in Greece at Otzaki (Milojčić 1983: pl. 3) and Tsangli (Wace and Thompson 1912: fig. 65), the eponymous site for the Tsangli house type (Weinberg 1965: 33).

Other major changes include a faunal assemblage more heavily dominated by sheep/goat, but including cattle which now appear to be fully domesticated. This is also in contrast to the East Mound, prompting questions as to whether the East Mound is, in fact, not yet as "really Neolithic" as the West Mound is (Orton 2010). Pollen evidence from the Acıgöl (Nevşehir Province) profile shows increasing cereal pollen after 6000 BC, coinciding with signs for lake eutrophication and forest decrease, thus reflecting intensified agriculture and animal husbandry (Woldring 2002). Moreover, recent

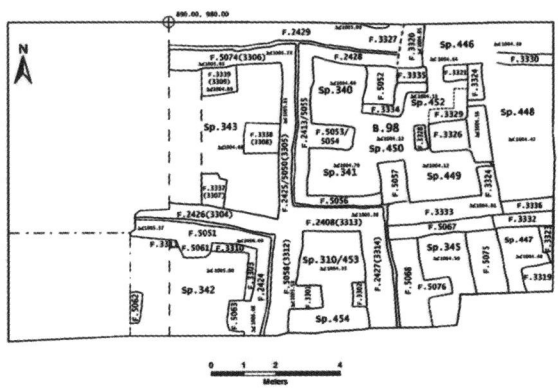

FIGURE 4. *Comparison between the architectural layout in Trench 5 of the Çatalhöyük West Mound and Can Hasan I layer 2B (after French 1998).*

findings of milk protein in assemblages from the Marmara region dating to the 2nd half of the 7th and the first half of the 6th millennium (Evershed et al. 2008) hint, together with the explosion of the pottery production in the Late Neolithic–Early Chalcolithic transition, to an increased consumption of non-solid food, which might have been milk and milk derivatives such as yoghurt and cheese. A good correlation with domesticated cattle bone remains poses the question whether the introduction of domesticated cattle, with its larger output of milk compared to sheep and goat, prompted this ceramic explosion. Further research has to clarify how some of these changes, and the intensification of the Neolithic lifestyle, were prompted by the 8.2. cal BP event, as can be seen also at other sites such as Sabi Abyad (Akkermans 1996; Akkermans et al., this volume).

Also, increased emphasis on central storage can be seen in roughly contemporary Near Eastern sites such as Tell Sabi Abyad I (Akkermans 1996),

Tell Oueili (Huot 1989) and Yarım Tepe (Merpert and Munchaev 1993) and is discussed as a trend in the 6th millennium (Bartl 2004: 526–528). The West Mound data suggests waterproof portable storage and storage sharing as opposed to the immobile bins and baskets of the East Mound and would also explain the emergence of large storage vessels, which were found outside a building in Trench 7 on a plastered surface together with clay and stone balls, indicating an outdoor activity zone or work place (Biehl and Rosenstock 2008). But only if we continue to excavate these buildings completely and also understand the ground plans of their neighbouring buildings, can new light be shed on the so-called secondary products revolution (Sherratt 1981). Although there are elements of continuity in several aspects of material culture and economy, e.g. some pottery forms, worked bone, obsidian and botanical evidence, these only serve to emphasise the scale of the changes in other areas (Ostaptchouk 2010; Bogaard and Charles 2010).

Beyond the site, further evidence of differences between East and West comes from the Konya Plain Survey project, which has discovered at least 14 contemporary but smaller sites within 20 km of Çatalhöyük West, contrasting with the preceding period when the East Mound appears to have existed virtually in isolation (Baird 2006; Düring 2011: 120). It seems that Çatalhöyük remained a dominant site but at the time of the West Mound occupation it lay within a dispersed settlement network, with relationships between neighbouring sites to some extent reflected in their ceramic assemblages. This provides a clue to the appearance of painted pottery, which we suggest had an important role in articulating the relationships between these sites as well as with more far-flung communities beyond the Konya Plain.

Recently, Schoop (2005: 129–131) has once more argued for a hiatus between the occupations of the two mounds, based on attempts to synchronise the Çatalhöyük sequence with that from Can Hasan to the south and especially with the long sequence from Mersin-Yumuktepe on the Cilician coast. Schoop argues that the majority of pottery of the West Mound type is not found before Mersin level XXIV, while comparisons with the East Mound sequence end before Mersin XXVI. However, the fact that level XXIV already has pottery similar to Early Chalcolithic II should not be underestimated, and the radiocarbon determinations from level XXV correspond better with those from parts of buildings in Trench I in the centre of the West Mound. Furthermore, the notion of a gap, initially put forward by

French (1967; 2005: 15) also depends on the stratigraphy of Hacılar (Mellaart 1970). As several attempts to re-evaluate Mellaart's publication have rendered the sequence of Hacılar VI to I doubtful (Thissen 2000; Reingruber 2008: 420; Rosenstock 2010), it is indeed more likely there is no gap at all (Biehl and Rosenstock 2009: 476).

The new investigations on the West Mound are aimed at clarifying this sequence, but the three ^{14}C-samples taken from a deep sounding in T7 of the Early Chalcolithic I structures cannot draw a conclusive picture yet. Nevertheless one date, 5980 to 5810 cal BC (68% probability), is at least contemporary or even slightly older than the Early Chalcolithic I dates obtained so far. In addition, this EC I building still sits on more than two metres of cultural layers above natural deposits. It should also be highlighted that the pottery found within these two metres below the Early Chalcolithic I building was no longer painted and is comparable in form and fabric with the pottery from the latest levels of the East Mound. However, it is still too early to contextualise the unpainted (and incised) pot sherds with a building and, consequently, to exclude the possibility that they were washed down from the centre of the pre-Early Chalcolithic I surface of the West Mound.

However, the changes in pottery production – the quantity of painted sherds decreases considerably towards the lower levels of the stratigraphy, while the number of incised so-called "cut- and prick-ornamented", "Exotic ware" or "Gelveri-like" sherds seem to increase (fig. 5) – can be connected to an increasing quantity of obsidian towards the lower layers. All this seems to speak for a continuous tradition connecting the two mounds and against a hiatus separating them.

The deep sounding we undertook in 2007 and 2008 provides a stratigraphy from the beginning of the settlement on the West Mound to its abandonment and indicates that at the beginning of its sequence changes in the environment occurred (fig. 6). The river, which ran along the edge of the East Mound, and which was the lifeline of its settlement for almost 1,000 years, transformed around 8200 cal BP into oxbow lakes, necessitating a completely new subsistence strategy and settlement pattern at Çatalhöyük and probably across the Konya Plain. We are just at the beginning of this part of the project but can already predict that it will provide the possibility to examine how rapidly this environmental change happened and to demonstrate how the process of adaptation or transformation manifested itself.

Figure 88:: Selection of re-worked sherds from Catalhöyük West - a: 14213; b: 13821; c: 13822; d: 15100 (Scale ~ 1:1); source Ingmar Franz 2007

Figure 89: Selection of cut- and prick-ornamented sherds from Catalhöyük West (Scale ~ 1:1); source Ingmar Franz 2007

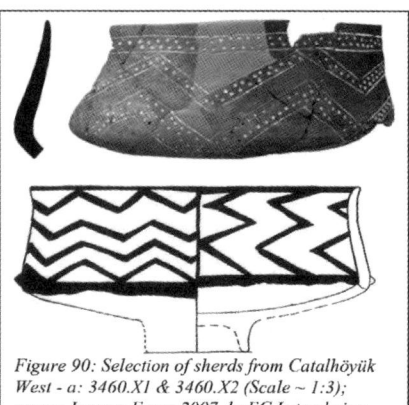

Figure 90: Selection of sherds from Catalhöyük West - a: 3460.X1 & 3460.X2 (Scale ~ 1:3); source Ingmar Franz 2007, b: EC I stand-ring bowl (Scale = 1:5); source James Mellaart 1965

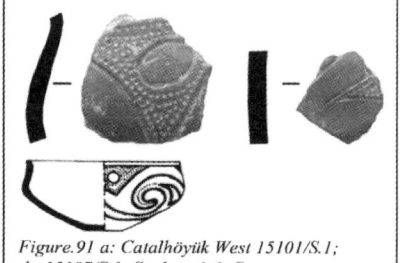

Figure.91 a: Catalhöyük West 15101/S.1; b: 15107/S.1; Scale = 1:1; Source: Ingmar Franz 2007 c: Gelveri-Güzelyurt; Scale = 1:3; Source: Schoop 2005: Taf. 54

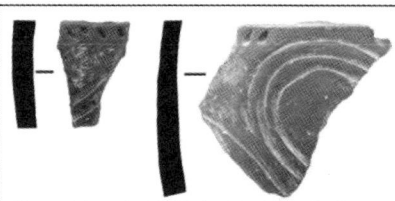

Figure 92:a: 15129/S.2; b: 15115/S.1; Scale = 1:1; Source: Ingmar Franz 2007

Figure 93: Vessel cluster in Trench 7; Source: Naomi Christie 2007

FIGURE 5. Stroke Ornamented so-called "exotic ware" or "Gelveri like ware" from the West Mound (Franz 2007: 130).

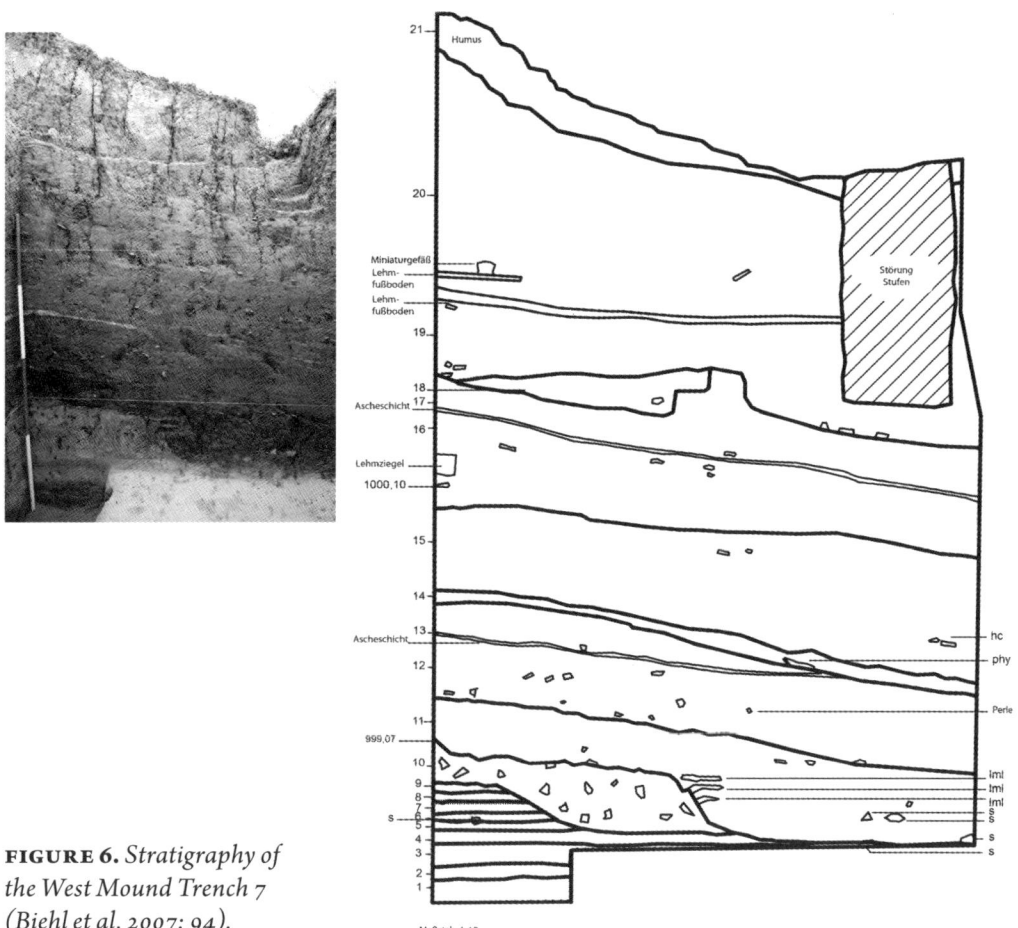

FIGURE 6. *Stratigraphy of the West Mound Trench 7 (Biehl et al. 2007: 94).*

The shift of settlement from the East to the West Mound can thus be viewed as a necessary new start at the beginning of an economically very different period in the history of the site. Similar shifting processes have been observed in roughly contemporary sites ranging from Hacılar VII to VI (Mellaart 1970: 94–95, fig. 41) to Hassuna (Lloyd 1963: 71) and Sabi Abyad (Akkermans 1996) in Upper Mesopotamia and Karanovo (Hiller 1997: 74) in southeastern Europe. Alternatively, it might just be another shift of settlement focus similar to those during the Early Neolithic and Late Neolithic phases of the East Mound (Last 1996: 132), a scenario we can approach only through investigation of the sediments.

Long-term vs. abrupt climate change

Climate change is a hot topic today – literally. World leaders have been gathering to discuss what can be done to protect our planet and preserve our lifestyles. Some might see this as a sign of evolution; we are fretting about what might happen, what might affect our children, grandchildren, the future of our planet. In past societies, even dating back to prehistory, the fretting only started once the changes actually happened. Many among us today understand and fear what abrupt climate change can do. Others are more skeptical. Whatever one believes, one thing is clear: despite all the technology we possess, we are as vulnerable to climate change today as we were in the past. If our climate changes, we change. In denying the reality of climate change, many among us miss the impact such change had on the past and continues to have in present and future human affairs. They not only miss the lessons of world history, but also the opportunity to examine the social dimensions of climate change critically.

Modern industrial nations tend to put all their faith into industry and science. Most of the current discussions about climate change focus on how to stop it. This is natural. The search for alternate energy sources and ways to reduce carbon footprints is necessary, laudable and potentially more lucrative than anything archaeologists or anthropologists hope to do. Still, all the hype about stopping climate change blinds us to looking at its social repercussions. In other words, how does human behaviour change when the surrounding world changes?

We will use Çatalhöyük as an example to illustrate how important microregional research is in order to better understand large-scale climatic change in the past and the impact it had on the people of the time. The key hypothesis is that the change in climate and environment caused people to move westwards into Western Anatolia and southeast Europe – and eventually across Europe. Çatalhöyük offers a microcosm that can help us unlock some of the key questions surrounding this time period.

Studies of climate change are still dominated by climatologists and environmental scientists, although the last few years have witnessed some overtures for collaboration among social and environmental scientists. It is a given fact that the impact of any climatic event depends on the local ecological setting and the organisational complexity, scale, ideology, technology and social values of the local population. It is only through long-term archaeological and historical analysis and, most importantly, detailed examination of the social dynamics on local and regional scales within an interregional framework, that we can begin to detect the differential impact of the climatic event.

Depending on one's leanings – academic or otherwise – the environment has been viewed either as a determining factor of social change or as not important to social change at all. If one believes in the latter, then one assumes human societies exist independently of their environments and view all social change as the result of shifting relationships between individuals and groups. But do social actors exist in an abstract world without physical environments or cultural landscapes? I believe humans should be placed firmly within their ecosystem and landscape. The environment is a very real factor in the decision-making process of individuals. Just think about how much today's weather affected you. The human-environmental relationship is non-linear. Different societies, and indeed different individuals in a society, can respond to environmental and climate changes in variable ways.

Climate change also occurs at varying speeds – over millennia, over centuries, or sometimes within decades. It is natural that people would remember the change that happened the most abruptly, either in their lifetime or within one or two generations. Conversely, they tend to dismiss – or never fully comprehend – change that occurs slowly over hundreds or thousands of years. It is only after the period of change that we can look back with hindsight and see how different lives were at the start of the change.

Memories will naturally be stronger of events that caused the greatest hardship or loss of life and less of gradual decadal or centennial changes in the environment. That is part of the challenge and the puzzle of the 8.2k event, in which the change lasted approximately 200 years which fundamentally changed the social organisation, architecture, pottery and economy of the people in the long term.

In the past 15 years, ice core records in Greenland have revealed that 8,200 years ago, temperatures in the North Atlantic region decreased abruptly, and then subsequently recovered, during an interval of about 200 years. A large number of contemporaneous climate events are documented in terrestrial records in North America, the Caribbean, Europe, Africa and Western Asia. In central Greenland the surface air temperature dropped by 3–7 °C. A reduction in air temperature of this magnitude is likely to be linked with drier conditions and stronger winds over the North Atlantic and drier monsoon regions. According to Barber and colleagues (Barber et al. 1999), the observed cooling was caused by a catastrophic collapse of the last remaining ice dome of the Laurentide Ice Sheet covering Hudson Bay. During deglaciation, a remnant ice mass blocked the northward drainage of the large glacial lakes Agassiz and Ojibway, which previously discharged southeastwards over sillways to the St. Lawrence river. Around 8,200 years ago, the ice dam collapsed, allowing the lakes to drain swiftly northwards into the Labrador Sea. The release of the freshwater, previously stored in the proglacial lakes through the Hudson Strait, may have substantially weakened the deepwater formation in the North Atlantic. Two research groups have performed dedicated numerical climate simulations of the 8.2k event, both using coupled atmosphere–ocean–biosphere models (Renssen et al. 2001; and Bauer et al. 2004). The studies confirm that the amount of out flowing freshwater stored in the proglacial Lake Agassiz corresponds to a global mean sea level rise of 0.5 m.

In contrast to the 4200 cal BP climate change event, which has aroused considerable interest in archaeology (Weiss 2000), only a few archaeological studies have been undertaken to identify the potential effects of the much larger 8.2k event.

We would like to briefly summarise the results from one study by a research group led by Bernhard Weninger from the University of Cologne, Germany, and which are relevant for our research (Weninger et al. 2005, Weninger et al. 2006). It is important to note that climate change can only be inferred from proxies. Among all potential proxies, those that will be observed are those that are pertinent and relevant within the purview of the culturally conditioned canons of perception and comprehension among other natural phenomena that an individual or a group consider of interest and value. In addition to perceptions of changes in amount of rainfall, seasonality, frost, wind, heat and humidity, changes in climate are likely to lead to noticeable changes in surface water distribution, vegetation, animals, and geomorphological and sedimentary processes such as soil erosion, deflation, or invasion by sand dunes.

Weninger's study is based on tree ring calibrated ^{14}C-ages and all the proxies are synchronised – as closely as achievable by visual comparisons – with the tree ring widths measured for the Central European Oak Chronology. By this measure, an age estimate is obtained for the 8.2k event scaled to the European tree ring chronology, thus producing the highest achievable absolute dating precision for the 8.2k event. The study is based on a substantial radiocarbon database for the Pre-Pottery and Pottery Neolithic periods in southeast Europe and the Near East and focuses on large settlements (mostly tell-mounds) encountered in Turkey (N=649), Greece (N=253) and Cyprus (N=130), for which extensive series of stratified ^{14}C-dates are available (see also Roberts et al. 2001).

The compilation of selected major ^{14}C-dated archaeological sites in Cyprus, Greece, Bulgaria and Anatolia that are good candidates for showing the 8.2k event shows that many major archaeological sites in the eastern Mediterranean are either first occupied at ca. 8200 cal BP (in Northwest Anatolia: Hoca Çeşme IV; in Greece: Nea Nikomedeia, Achilleion, Sesklo; in Bulgaria: Ovcarovo), or else deserted (in Cyprus: Khirokitia and likely Kalavassos-Tenta). Weninger also highlights that in these regions there are no sites with clear stratigraphic evidence for a continuous settlement extending through the 8.2k event. It is also remarkable – following the (more or less) simultaneous desertion of Khirokitia and Kalavassos-Tenta around 8200 cal BP – that the island of Cyprus was apparently deserted and remained uninhabited for more than 1,500 years (fig. 7).

To explain these observations he proposes that the 8.2k event triggered the spread of early farmers out of Anatolia, into Greek Macedonia as well as into the fertile floodplains of Thessaly, and simultaneously into Bulgaria. We can follow this event even further west, as recent research has shown at the famous site of Lepenski Vir at the Danube Iron Gates (Bosnall et al. 2002). Here, a conspicuous gap in the radiocarbon record suggests that many riverbank sites were abandoned at the Meso-Neolithic between c. 8250 and 7900 cal BP. This period of site abandonment is linked to increased flooding along the Danube, which can be correlated with the described distinct global climatic oscillation and its effects on river systems. This has major implications for the understanding of the European archaeological record, since some of the largest concentrations of ancient human populations and settlements occurred in river valleys. The Iron Gates show how climate-related flooding impacted on human settlement and use of riverine environments, and may have been an important stimulus of culture change.

We can also follow this event further east in Tell Sabi Abyad in the Balikh Valley in northern Syria (Akkermans et al. 1996). This tell site has also provided evidence of a significant reorganisation of its structure in the transition from the Early Pottery Neolithic (Balikh IIA/B) to the Pre-Halaf period

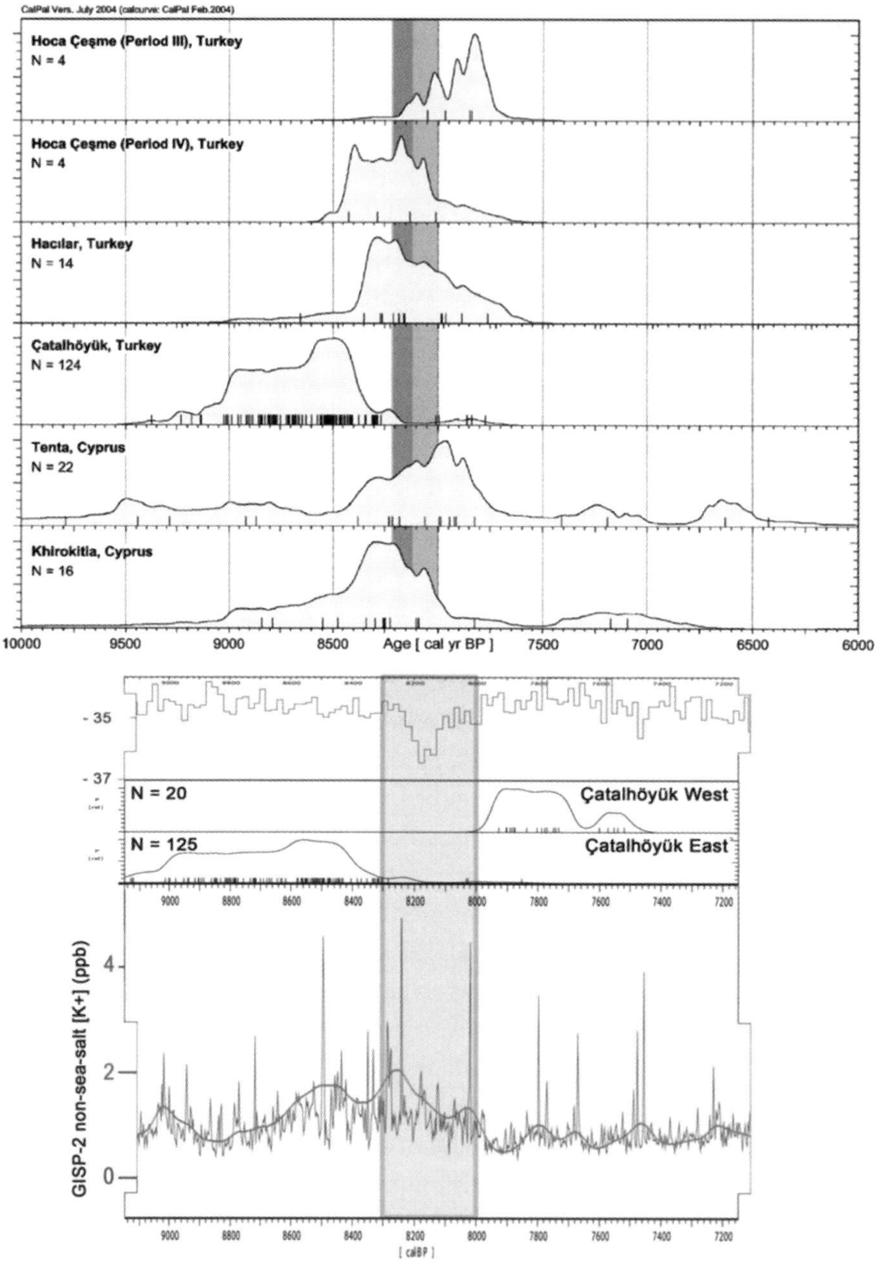

FIGURE 7.
Climate diagrams after Weninger 2008.

FIGURE 8. *Illustration of the Çatalhöyük West Mound (foreground) and the abandoned East Mound (background); illustration by Mesa Schumacher 2011; copyright Peter F. Biehl.*

(Balikh IIC), i.e. at around 6300/6200 cal BC. This was connected with a substantial decline in the size of the settlement, during which, as the excavator Peter Akkermans writes: "the formerly densely occupied area on the western side of Tell Sabi Abyad was abandoned". In the wake of this reorganisation, settlement appears to have continued at two separate occupations on the eastern side of the mound. The subsequent "Transitional" period (Balikh IIIA; c. 6050–5900 cal BC) is associated with a number of major developments marking the run-up to the emergence of the Halaf culture in the early 6th millennium cal BC.

The coincidence of the 8.2k event with this period of intensive Neolithic dispersal seems to be apparent. In the semi-arid interior of the Anatolian peninsula and in northern Syria, communities may have been struck by particularly cold/severe winters and springs, combined with either drought conditions or the effects of more extensive rainfalls and snowfalls. These would have placed an increased strain on resources, leading ultimately to food shortages and famine situations. Could this then also be a likely scenario for the late 7th and early 6th millennium cal BC transition between the Neolithic and Chalcolithic in Central Anatolia? Had the climate in central and eastern parts of the peninsula become too unpredictable? Did parts of the population from these regions seek alternative territories in adjacent, less afflicted areas? And did this, indeed, lead to a reorganisation of their own communities? Finally, and perhaps most controversially, are we witnessing a climate induced intensification of Neolithisation processes from the semi-arid "formative zone" of Neolithic genesis westwards into both the Aegean region and into centres of southeastern European "Early Neolithic" development, e.g. in Thessaly, in the Strumon valley, and beyond?

Conclusions

To sum up, the evidence from the West Mound suggests continuity of occupation alongside rapid social and economic developments taking place after 6200 cal BC (Marciniak and Czerniak 2007). The architecture in the upper occupation layers of Levels I and 0 on the East Mound suggests it is still different from the one on the West Mound. In addition, secure evidence for Early Chalcolithic I painted pottery or the full range of West Mound forms is still missing. But given recent arguments (Boyer et al. 2006) that between 2 and 3 m of archaeological stratigraphy have been eroded from the top of the East Mound, we might not be able to understand and reconstruct the very latest levels of the East Mound settlement.

It would seem to be no coincidence that a decline in the symbolic elaboration of houses occurs alongside a more functional division of household space, an increase in the quantity and complexity of portable artifacts (especially pottery) and the development of a more integrated settlement system on the Konya Plain. The symbolic focus of social integration was shifting. These changes need to be understood as part of a wider Anatolian phenomenon and it is likely that influences from beyond the Konya Plain also played a role. One probable source is the Lakes region to the west, where painted pottery appears at Hacılar earlier than it does at Çatalhöyük.

Çatalhöyük provides the possibility to examine how environmental as well as a climate change effected economic innovation and social transformation starting around 6,200 cal BC in Central Anatolia.

Acknowledgements

I would like to thank Eva Rosenstock as well as Ingmar Franz, David Orton, Jana Rogasch and Sonia Ostaptchouk for their comments on earlier drafts and parts of this paper, and the whole Çatalhöyük Research Project for its support. I would also like to thank Susanne Kerner and her team for the thorough editing of the paper.

Bibliography

Akkermans, P. M. M. G. (ed.). 1996. *Tell Sabi Abyad: the late neolithic settlement: Report on the excavations of the University of Amsterdam (1988) and the National Museum of Antiquities Leiden (1991–1993) in Syria*. Uitgaven van het Nederlands Historisch-Archaeologisch Instituut te Istanbul 76.

Baird, D. 2002. "Early Holocene settlement in Central Anatolia: Problems and prospects as seen from the Konya Plain." In F. Gérard and L. Thissen (eds.), *The Neolithic of Anatolia: Internal developments and external relations during the 9th–6th millennia cal BC*, Ege Yayınları, 139–159. Istanbul.

Baird, D. 2006. "The history of settlement and social landscapes in the Early Holocene in the Çatalhöyük area." In I. Hodder (ed.), *Çatalhöyük Perspectives: Themes from the 1995–1999 Seasons*, McDonald Institute, 55–74. Cambridge.

Barber, D. C., A. Dyke, C. Hillaire-Marcel, A. E. Jennings, J. T. Andrews, M. W. Kerwin, G. Bilodeau, R. McNeely, J. Southon, M. D. Morehead, and J.-M. Gagnon, 1999. "Forcing of the Cold Event of 8,200 Years Ago by Catastrophic Drainage of Laurentide Lakes." *Nature* 400: 344–348.

Bartl, K. 2004. *Vorratshaltung: die spätepipaläolithische und frühneolithische Entwicklung im westlichen Vorderasien; Voraussetzungen, typologische Varianz und sozio-ökonomische Implikationen im Zeitraum zwischen 12,000 und 7,600 BP*. Ex Oriente. Berlin.

Bauer, E., A. Ganopolski and M. Montoya. 2004. "Simulation of the cold climate event 8200 years ago by meltwater outburst from Lake Agassiz." *Paleoceanography* 19: 1–13.

Biehl, P. F., and E. Rosenstock. 2008. "West Mound Excavations." Çatalhöyük 2008 Archive Report. http://www.catalhoyuk.com/downloads/Archive_Report_2008.pdf, 90–103.

Biehl, P. F., and E. Rosenstock. 2009. "Von Çatalhöyük Ost nach Çatalhöyük West: Kulturelle Umbrüche an der Schwelle vom 7. zum 6. Jahrtausend cal BC in Zentralanatolien." In R. Einicke et al. (eds.), *Zurück zum Gegenstand. Festschrift für Andreas E. Furtwängler*. Schriften des Zentrums für Archäologie und Kulturgeschichte des Schwarzmeerraumes 16. Beier and Beran, 471–481. Weissbach.

Biehl, P. F., J. Rogasch and E. Rosenstock. 2010. "West Mound Excavations, Trench 5." Çatalhöyük 2010 Archive Report. http://www.catalhoyuk.com/downloads/Archive_Report_2010.pdf, 44–49.

Bogaard, A., and M. Charles. 2010. "Archaeobotany Preliminary Report on West Mound Trenches 5, 6 and 7." Çatalhöyük 2010 Archive Report. http://www.catalhoyuk.com/downloads/Archive_Report_2010.pdf, 66–69.

Bonsall, C., M. G. Macklin, R. W. Payton and A. Boroneanţ. 2002. "Climate, floods and river gods: Environmental change and the Meso–Neolithic transition in south-east Europe." *Before Farming: The archaeology of Old World hunter-gatherers* 34(2): 1–15.

Boyer, P., R. Roberts and D. Baird. 2006. "Holocene Environment and Settlement on the Çarşamba Alluvial Fan, South-Central Turkey: Integrating Geoarchaeology and Archaeological Field Survey." In *Geoarchaeology* 21(7): 675–698.

Czerniak, L., and A. Marciniak. 2009. Team Poznań 2009 Study Season Report. Çatalhöyük 2010 Archive Report. http://www.catalhoyuk.com/downloads/Archive_Report_2009.pdf.

Düring, B. S. 2011. *The prehistory of Asia Minor: From Complex Hunter-Gatherers to early urban societies*. Cambridge University Press. Cambridge.

Duru, R. 1994. *Kuruçay Höyük I. 1978–1988 kazılarının sonuçları. Neolitik ve erken kalkolitik çag yerleşmeleri*, Türk Tarih Kurumu yayınları 5/44, Ankara 1994.

Erdoğu, B. 2010. "West Mound Trench 8." Çatalhöyük 2010 Archive Report. http://www.catalhoyuk.com/downloads/Archive_Report_2010.pdf, 50–51.

Evershed, R. P. et al. 2008. "Earliest date for milk use in the Near East and southeastern Europe linked to cattle herding." *Nature* 455: 528–531.

Franz, I. 2010. "Pottery Report Trench 5–7." Çatalhöyük 2010 Archive Report. http://www.catalhoyuk.com/pdfs/archive_report_2010.pdf, 77–89.

French, D. 1967. "Excavations at Can Hasan, 1966. Sixth report." *Anatolian Studies* 17: 165–178.

French, D. 1998. *Canhasan sites 1. Canhasan I: Stratigraphy and structures*, British Institute of Archaeology at Ankara Monograph No. 23. London, British Institute at Ankara.

French, D. 2005. *Canhasan sites 2. Canhasan I: The pottery*, British Institute of Archaeology at Ankara Monograph No. 32 London, British Institute at Ankara.

Gibson, C., and J. Last. 2003a. "An early Chalcolithic building on the West Mound at Çatalhöyük." *Anatolian Archaeology* 9: 12–14.

Gibson, C., and J. Last. 2003b. "West Mound Excavations." Çatalhöyük 2003 Archive Report. http://www.catalhoyuk.com/pdfs/archive_report_2003.pdf, 59–63.

Hiller, S. 1997. "Architektur." In S. Hiller and V. Nikolov (eds.), *Karanovo. Die Ausgrabungen im Südsektor 1984–1992*. Berger, Horn/Vienna, 55–92.

Hodder, I. (ed.) 2005. *Çatalhöyük perspectives: reports from the 1995–99 seasons*, McDonald Institute Monographs, Çatalhöyük Research Project Volume 6, British Institute at Ankara BIAA Monograph 40. Cambridge.

Hodder, I. 2006. *Çatalhöyük, the Leopard's Tale: Revealing the mysteries of Turkey's ancient 'town'*. Thames and Hudson. London.

Huot, J. L. 1989. "Ubaidian *Village* of Lower Mesopotamia: Permanence and evolution from Ubaid 0 to Ubaid 4 as seen from Tell el'*Oueili*." In E. Henrickson and I. Thuesen (eds.), *Upon this foundation: The 'Ubaid reconsidered*, 19–42. Museum Tusculanum Press. Copenhagen.

Karul, N. 2007. "Aktopraklık. Kuzeybatı Anadolu'da Gelişkin Bir Köy." In M. Özdoğan and N. Başgelen (eds.), *Türkiye'de neolitik dönem: Anadolu'da Uygarlığın Doğuşu ve Avrupa'ya Yayılımı: yeni kazılar, yeni bulgular*. Istanbul.

Last, J. 1996. "Surface Pottery at Çatalhöyük." In I. Hodder (ed.), *On the surface: Çatalhöyük 1993– 1996*, McDonald Institute Monographs, Çatalhöyük Project 1, British Institute at Ankara Monograph 22, 115–172. Cambridge.

Lloyd, S. 1963. *Mounds of the Near East*. Edinburgh University Press. Edinburgh.

Marciniak, A., and L. Czerniak. 2007. "Social transformations in the Late Neolithic and the Early Chalcolithic periods in central Anatolia." *Anatolian Studies* 57: 115–130.

Mellaart, J. 1965. "Çatal Hüyük West." *Anatolian Studies* 15: 135–156.

Mellaart, J. 1970. *Excavations at Hacılar*. Edinburgh University Press. Edinburgh.

Mellaart, J. 1975. *The Neolithic of the Near East*. Thames and Hudson. London.

Merpert, N. Y., and R. M. Munchaev. 1993. "Yarim Tepe: The Halaf Levels." In N. Yoffee and J. J. Clark (eds.), *Early stages in the evolution of*

Mesopotamian civilization: Soviet excavations in Northern Iraq, 128–205. London and Tucson, The University of Arizona Press.

Milojcic, V. 1983. *Die Deutschen Ausgrabungen auf der Otzaki-Magula in Thessalien. Bd 3/2: III: Das späte Neolithikum und das Chalkolithikum. Stratigraphie und Bauten. II: Das mittlere Neolithikum. Die mittelneolithische Siedlung.* Beiträge zur ur- und frühgeschichtlichen Archäologie des Mittelmeer-Kulturraumes 20 Bonn, Habelt.

Orton, D. 2010. "Animal Bones Report Trench 5." Çatalhöyük 2010 Archive Report. http://www.catalhoyuk.com/downloads/Archive_Report_2010.pdf, 55–58.

Ostaptchouk, S. 2010. "Chipped Stone Report Trench 5." Çatalhöyük 2010 Archive Report. http://www.catalhoyuk.com/downloads/Archive_Report_2010.pdf, 103–110.

Özbaşaran, M., and H. Buitenhuis 2002. "Proposal for a regional terminology for Central Anatolia." In F. Gérard and L. Thissen (eds.), *The Neolithic of Central Anatolia: Internal developments and external relations during the 9th–6th millennia cal BC. Proceedings of the CANeW Table Ronde, Istanbul 23–24 November 2001, Istanbul.* Ege Publishing Co., Istanbul, 67–78.

Mihriban, Ö., and H. Buitenhuis. 2002. "Proposal for a regional terminology for central Anatolia." In F. Gérard and L. Thissen (eds.), *The Neolithic of Central Anatolia. Internal developments and external relations during the 9th–6th millennia cal BC,* Ege Yayınları, 67–77. Istanbul.

Reingruber, A. 2008. *Deutschen Ausgrabungen auf der Argissa-Magula in Thessalien. Bd 2: Das frühe und das beginnende mittlere Neolithikum im Lichte transägäischer Beziehungen.* Beiträge zur ur- und frühgeschichtlichen Archäologie des Mittelmeer-Kulturraumes 35, Habelt, Bonn.

Renssen H., H. Goosse and Fichefet, T. 2002. "Modeling the Effect of Freshwater Pulse on the Early Holocene Climate: The Influence of High-Frequency Climate Variability." *Paleoceanography* 17(2): 1–15.

Roberts N., J. M. Reed, M. J. Leng, C. Kuzucuoğlu, M. Fontugne, J. Bertaux et al. 2001. "The tempo of Holocene climate change in the eastern Mediterranean region: New high-resolution crater-lake sediments data from central Turkey." *The Holocene* 11: 721–736.

Roodenberg, J. J., and C. L. Thissen (eds.), 2001. *The Ilıpınar Excavations II.* PIHANS vol. XCIII, Leiden.

Rosenstock, E. 2010. "Die 'Festung' von Hacılar I. Ein Dekonstruktionsversuch." In J. Šuteková, P. Pavúk, P. Kalábková and B. Kovár (eds.), *PANTA*

RHEI. Studies in the chronology and cultural development of south-eastern and central Europe in earlier prehistory presented to Juraj Pavúk on the occasion of his 75th birthday, Studia Archaeologica et Mediaevalia XI, Comenius University. Bratislava. http://archeologia.fphil.uniba.sk/attachments/FS%20JurajPavuk%202010_Introduction_web.pdf

Schoop, U.-D. 2005. *Das anatolische Chalkolithikum:* eine chronologische Untersuchung zur vorbronzezeitlichen Kultursequenz im nordlichen Zentralanatolien und den angrenzenden Gebieten (Urgeschichtliche Studien 1).

Sherratt, A. 1981. "Plough and pastoralism: Aspects of the secondary products revolution." In I. Hodder, G. Isaac and N. Hammond (eds.), *Pattern of the past: Studies in honour of David Clarke*, Cambridge University Press, 261–306. Cambridge.

Thissen, L. 2000. *Early village communities in Anatolia and the Balkans, 6500–5500 cal BC*. Unpublished PhD Dissertation. Leiden.

Wace, A. J. B., and M. S. Thompson. 1912. *Prehistoric Thessaly: Being some account of recent excavations and explorations in north-eastern Greece from Lake Kopais to the borders of Macedonia*. Cambridge University Press. Cambridge.

Weiss, H. 2000. "Beyond the Younger Dryas: Collapse as Adaptation to Abrupt Climate Change in Ancient West Asia and the Eastern Mediterranean." In Garth Bawden and Richard M. Reycraft (eds.), *Environmental Disaster and the Archaeology of Human Response*, University of New Mexico, 75–98. Albuquerque.

Weninger, B. et al. 2005. "Die Neolithisierung von Südosteuropa als Folge des abrupten Klimawandels um 8200 cal BP." In D. Gronenborn (ed.), *Klimaveränderungen und Klimawandel in neolithischen Gesellschaften Mitteleuropas, 6700–2200 v. Chr.*, Römisch-Germanischen Zentralmuseums, 75–118. Mainz.

Weninger, B., E. Alram-Stern, E. Bauer, L. Clare, U. Danzeglocke, O. Jöris, C. Kubatzki, G. Rollefson, H. Todorova and van T. Andel, 2006. "Climate Forcing Due to the 8200 cal yr B.P. Event Observed at Early Neolithic Sites in the Eastern Mediterranean." *Quaternary Research* 66: 401–420.

Woldring, H. 2002. "Climate change and the onset of sedentism in Cappadocia." In F. Gérard and L. Thissen (eds.), *The Neolithic of Central Anatolia. Internal developments and external relations during the 9th–6th millennia cal BC*, Ege Yayınları, 59–66. Istanbul.

A Narrow Place Can Contain a Thousand Friends: Irrigation as a Response to Climate in the Zerqa Triangle, Jordan

Maurits Ertsen and Eva Kapteijn

Abstract

In the Zerqa triangle in the Jordan Valley, irrigation would have been an important instrument to deal with the arid climate and its associated uncertainties concerning rainfall for societies in different periods. Before irrigation modernisation efforts were started in the 1960s, the people of the Zerqa area used the known ethnohistorical irrigation system, which dates back to the Mamluk period. This system consisted of a number of sub-systems tapping water from the Zerqa river and transporting water to the fields through open canals under gravity. The settlement pattern of the Iron Age points to an irrigation system of similar type being in use during this period. The location of Early Bronze Age settlements along natural watercourses suggests that a form of flood irrigation was practised, without a dedicated canal system. Each of these settings will have had its specific uncertainties in terms of water availability to deal with, which will be discussed. In other words, each setting provided specific materially structuring conditions for societies to develop responses in terms of agriculture, institutions and power relations. This contribution discusses these uncertainties and responses for the different periods. In the discussion, insights from both archaeology and irrigation engineering will be integrated.

Introduction

As much as elsewhere, human survival in the Zerqa triangle in the Jordan Valley depended on human ability to adapt to the natural environment. Adaptation can be seen in a reactive sense, but also in the proactive sense of shaping the environment. A major instrument in reshaping environmental conditions is irrigation, as this enables crop cultivation to continue during periods of drought and climatic fluctuations. Developing systems to bring water from sources to other locations, with different rhythms than the natural hydrological cycle would offer, can profoundly change the natural environment. A clear example is provided on the Peruvian north coast where despite harsh and arid environmental conditions pre-Colombian civilisations prospered. Rivers flowing from the Andes mountains to the west provide the fertile coastal plains with irrigation water. The valleys are oases and, with the exception of some small highly productive areas, exploitation of their agricultural potential depends on irrigation (Kosok 1965; Netherly 1984). Fields under regular cultivation with permanent water availability were located within upstream areas of the canal system, or in areas with high groundwater tables – which were at least partially the result of human-induced changes in hydrology as the redistribution of water from the rivers into the coastal plains fed groundwater on these plains. The new hydrological conditions were not necessarily always beneficial. Too much water could cause water logging or increased salinity of the soils.

Pre-Colombian Peruvians were not the only civilisation in world history who used irrigation to feed their growing population. Intensified production provided a relatively secure food source for a larger population and enabled the peasant population to produce a surplus to support the non-peasant population (Scarborough 2003). The role of irrigation in this area of research has been well studied. Nevertheless, despite much valuable research, it is often not very clear how the irrigation systems underlying the ancient civilisations may have functioned. Understanding the hydrological aspects of ancient irrigation is already challenging, but including the human dimension complicates the picture even further. Irrigation systems are supposed to supply crops with water. This requires physical distribution facilities to transport water and socio-political arrangements to coordinate the actors dealing with the water flows. User strategies have an impact on systems and systems constrain user actions. Consequently, irrigation systems have structuring properties (Ertsen and Van Nooijen 2009).

It is this interplay of hydrology and human activity in irrigation that we will explore here. We focus on the question of how human societies created new physical and social environments through irrigation in the Zerqa area. We explore how irrigation may have been applied as an instrument to decrease direct dependency on climatic fluctuations on shorter timescales (seasons and years) in different societal arrangements and in different periods. Our discussion shows that linking irrigation directly to climate change is not straightforward for several reasons. Firstly, although dry conditions do provide a potentially interesting challenge to overcome, human societies respond to many internal and external pressures, not only climate fluctuations (Orlove 2005). Dry conditions could also provoke strategies of nomadism, for example. Secondly, irrigation decisions on timescales as short as days and weeks are of considerable influence in determining the success of irrigation on longer timescales of years and decades. This makes it more difficult to regard climate as a single factor, because societal arrangements linked to irrigation are as important as climatic conditions. Thirdly, irrigation development in the Zerqa area suggests that decisions made by earlier societies concerning irrigation have set new boundary conditions and opportunities for their successors. This may link climates hundreds years ago to current societies, through the material and cultural remains of the past.

We begin with an overview of irrigation in different periods in the Zerqa triangle, and go on to explain how rain-fed agriculture in this region would not be very successful and how irrigation would be a possible answer. We discuss typical cropping patterns found for the different periods in relation to the capacity to grow food for the population, followed by a discussion in which we link our findings to the overview of irrigation in the Zerqa area. In a final paragraph we conclude that irrigation is indeed a response to climatic conditions, but that climate as such is not the single explanatory factor for the different forms irrigation appears to have had in the Zerqa triangle.

An overview of irrigation in the Zerqa triangle

Temperatures in the Jordan Valley are very high because of the low altitude of 300 m below sea level (BSL). The average winter temperature is 15°C, while the summer average is 32°C and maximum day temperatures of 40–45°C are quite common. This gives very high potential evapotranspiration rates. Rain falls only between November and April, with an average of about 290 mm a year, and there is a high annual variability. Sometimes the rains do not start until as late as January (Nedeco 1969: table B-4) (fig. 1). Theoretically, the average rainfall is just within the limit that is considered the minimum

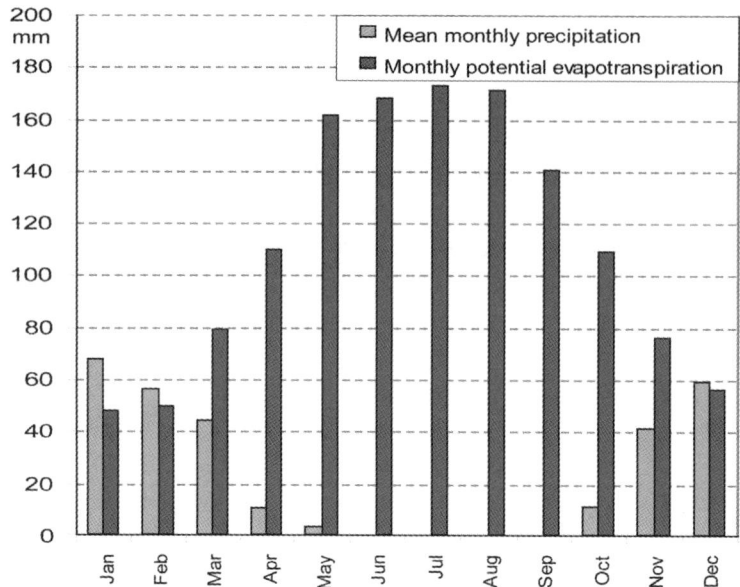

FIGURE 1. *Potential monthly evapotranspiration (Penman-Monteith) and mean monthly precipitation at Deir 'Allā in millimetres.*

for dry farming, but as droughts occur every few years it is impossible to sustain a stable society of size.

Today, irrigation in the Jordan Valley is mostly in the form of drip irrigation. However, before the 1960s a system was used that may well have a long history. From 19th- and early 20th-century itineraries, we know that the few people who lived in the Jordan Valley at this time used canals that tapped water from the river Zerqa (Kooij 2007). Aerial photographs taken in the 1940s clearly show a number of small canals that brought water to fields located at a considerable distance from the river. On unpublished blueprints from the 1950s, the old canal irrigation system is depicted in great detail (fig. 2). From these maps it becomes clear that three main channels tapped water from the Zerqa and that through a series of secondary and tertiary canals a considerable area could be irrigated. Considering the layout of the valley, with a flat plain sloping gently to the west, this seems to be the most logical and relatively easiest method of irrigation.

We not only know the function of the irrigation system in quite some detail, there is also information on the social organisation of the system. In the pre-modern clan-based society, the people in the region were part of the

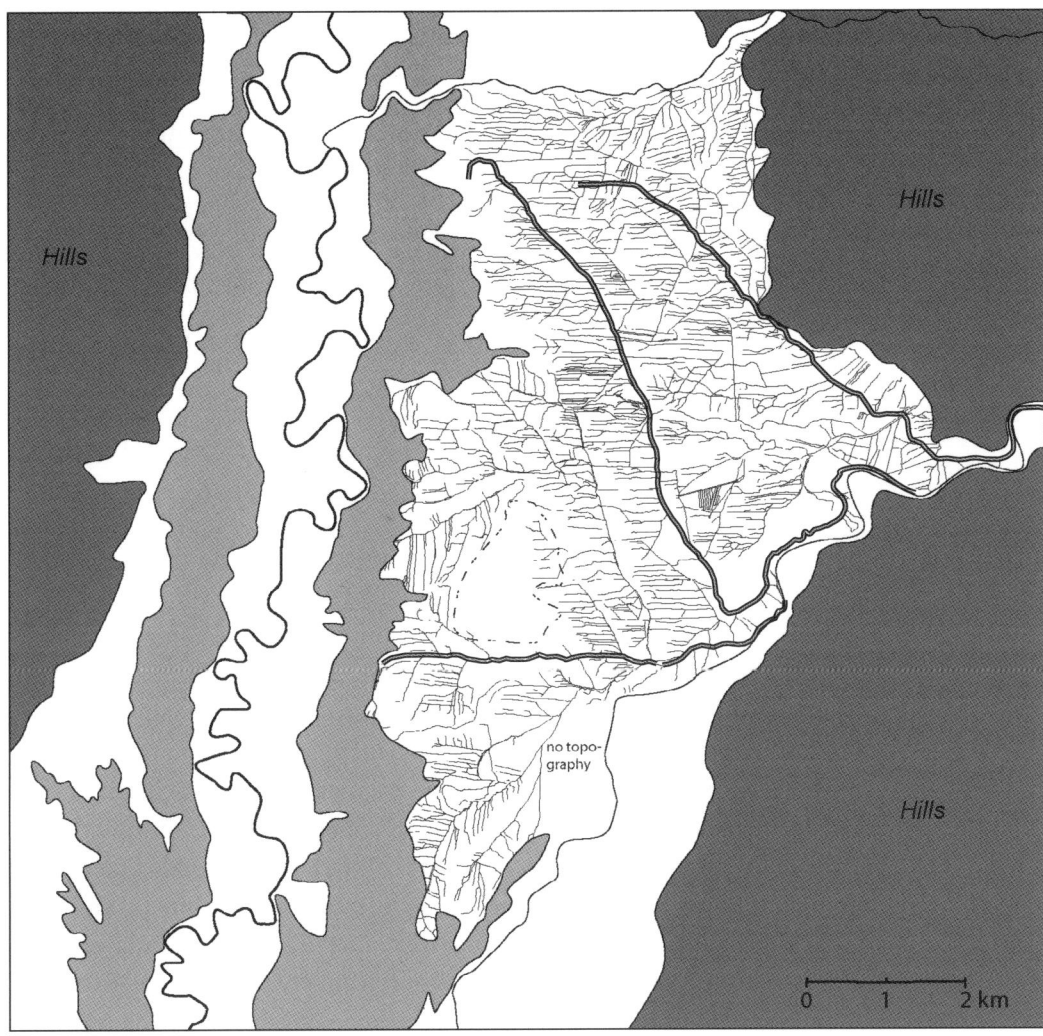

FIGURE 2. *Pre-modern irrigation system.*

Musha' system. Irrigation and land were in collective ownership and a rotation of the fields every few years was practised among the clan members. If a farmer had fields located far away into the valley, he would get more advantageous fields near a primary canal in the next period (Tarawneh 1989: 31). The clan's sheikh acted as a *primus inter pares* within the clan and each day a board of sheikhs from all the clans in the region would gather to decide and check how much water was to be let into which canals. This seems very egalitarian, but only if you were a clan member. There were also some distinctions between the different clans with regards to their access to irrigation. On the one side, there were the powerful Hurr clans who regarded themselves as free and stemming from the Bedouin; on the other, stood the second rate Ghawarneh clans who were regarded as foreign. Considering the size of the territory of one of the Hurr clans, the Mamduh, it is clear that they were the most powerful (fig. 3). The land of the other Hurr clan, the Shararah, is not different in size from the Ghawarneh clans. If, however, attention is paid to the irrigation channels, it becomes clear that the higher position which the Shararah claimed over the Ghawarnah clans is reflected in the irrigation system. Given the location of their territory, the Shararah were in the dominant position as all the other territories were located downstream and were therefore dependent on the Shararah.

Constructing these canals would have been labour intensive and would have required a communal effort as many kilometres of canal were needed to irrigate the entire region. The Jordan Valley was, however, only sparsely populated during the early 20th century and the aerial photographs show that only a small part of the entire system was in use at that time. During the 17th and 18th century, the Valley had been almost devoid of any sedentary population and had been the territory of the Bedouin. It is unlikely, therefore, that the 19th-century farmers constructed a large irrigation system which they used only partially afterwards. Inhabitants of the Zerqa Triangle related in the 1870s that neither they nor their fathers had constructed these channels and that they only cleared existing ones to use (Merrill 1881: 382). This strongly suggests that the system was older. The only likely large-scale farming society from which an irrigation system could have been inherited is that of the Mamluk period, from 1260 to 1500 AD, before sedentary occupation ceased and pastoral nomadism took over.

In the Mamluk period, the Jordan Valley was widely used for sugar cane cultivation. The sugar trade was a very profitable business and sugar was traded with many countries in the Middle East and Europe. In Damascus, there was a separate administrative centre that managed the affairs of the Jordan Valley plantations of the Sultan, who had gained an almost complete monopoly on the sugar trade. There are indications that some of his sugar plantations were located in the Zerqa Triangle (LaGro 2002: 18). Sugar

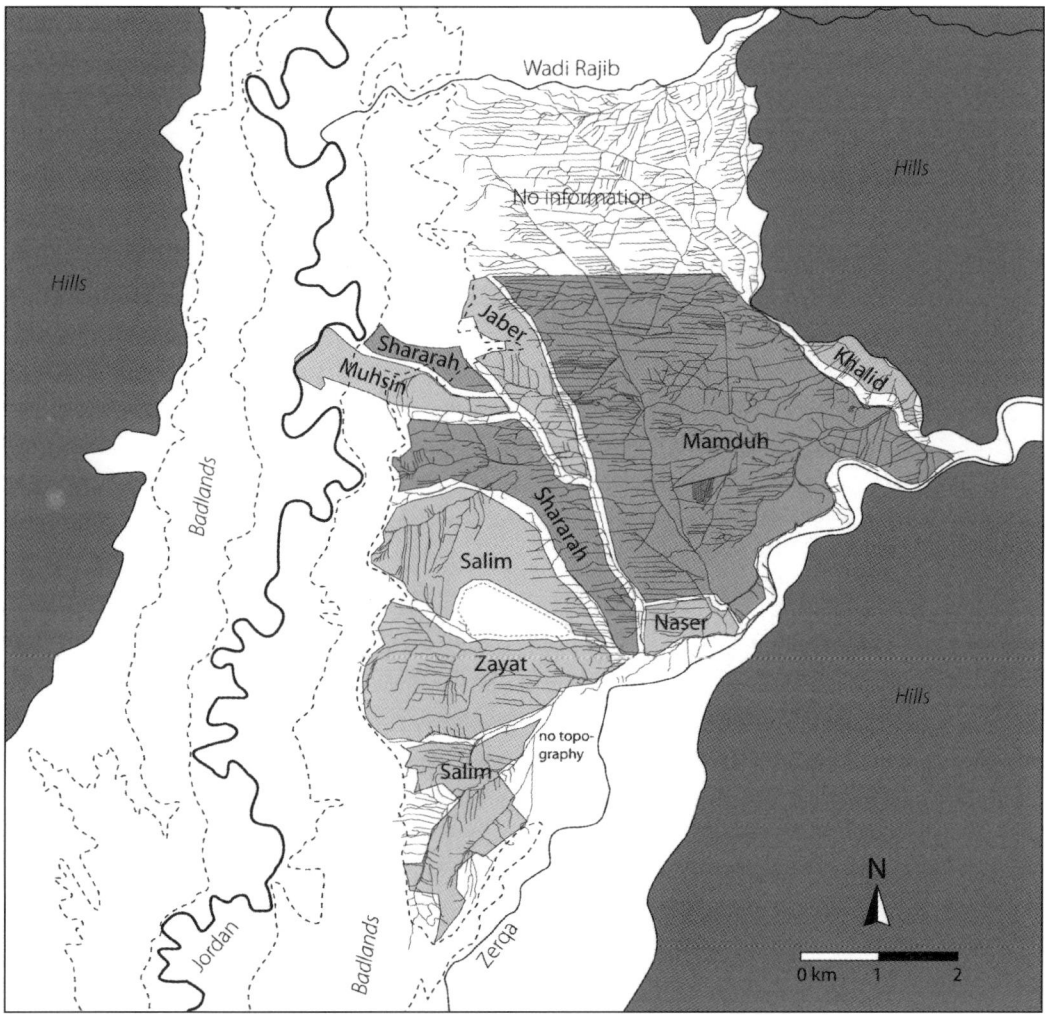

FIGURE 3. *Pre-modern irrigation system and clan territories.*

cane is a tropical crop that grows during the summer and needs much water. Irrigation is thus necessary. Sugar was produced by crushing the cane and boiling the juice down until raw sugar remained; the crushing was done in watermills and the location of these mills is such that human modifications to the hydrology would be needed to operate them. It is clear that all

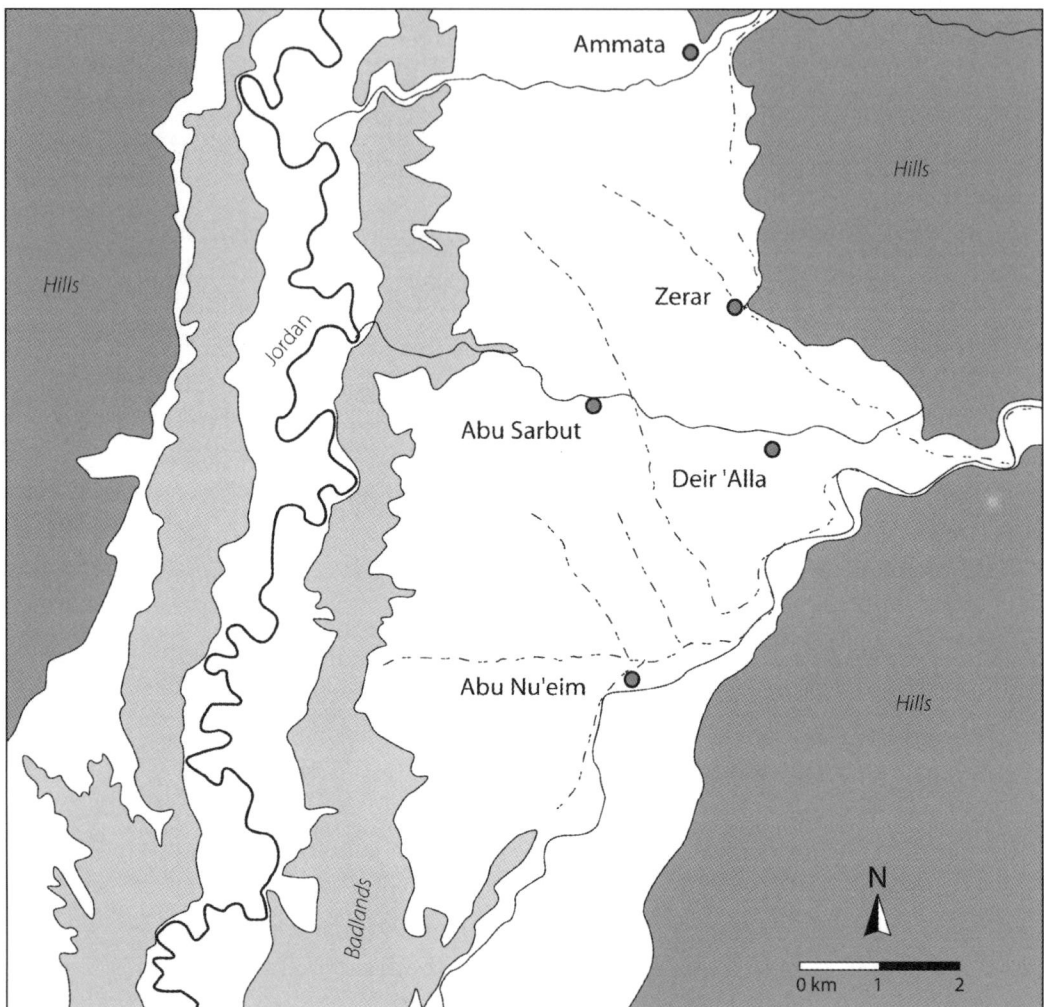

FIGURE 4. *Mamluk watermills and ethnohistorical irrigation canals.*

the mills are located along the routes of the three main irrigation channels and the Wadi al-Ghor (fig. 4). Two of the mills contained actual remains of where they had been connected to the canals. It seems likely, therefore, that the ethnohistorical irrigation system dates back to the Mamluk period.

Except for Zerar and Deir 'Alla, which shared a canal, all the sugar mills were supplied by a different main irrigation channel. Judging from the different types of pottery discovered in the survey, the villages were mostly at the same locations as the sugar refineries. The villagers probably worked as employees on the plantations, while subsistence crops were grown on fields that lay fallow from sugar cane. The people in the Zerqa Triangle were, therefore, small self-sufficient farmers on the one hand, and labourers in a capitalist economy on the other.

Although there is no direct evidence in the form of excavated canals, it is likely that a similar system of canal irrigation existed in the Iron Age IIa/b period (from 1000 to 725 BC). From the number of IA tells it is clear that the Zerqa triangle was rather densely populated during this period (Ibrahim et al. 1988; Petit 2009). Although these tells are only small, it is unlikely that so many small contemporary villages could exist in this dry area without irrigation. Moreover, several tells were located in the middle of the plain away from natural water sources. Archaeobotanical investigations show crops were grown which could not have been cultivated without irrigation. Large quantities of flax were grown, for example (van Zeist and Heeres 1973: table 1; Neef 1989: table 2). There is no direct evidence as to precisely how this irrigation was practised, but the presence of tells in the middle of the plain, in combination with the topography, suggests that canals were used to provide the water. Features in the landscape limit possibilities for canal routing, and channels may have been located along similar routes as in Mamluk and pre-modern times.

In the IA IIa/b, most tells were only 0.4 hectares or less. In the material culture, little differentiation between the sites is visible; they are small rural sites. Only Deir 'Alla, Mazar, Hammeh, Ammata and Damiyah are twice that size or larger. Tell al-Hammeh and Ammata are both situated at important locations in the irrigation system; both are located at a place where locating the intake of canals would have made most sense if a large area was to be reached. Damiyah, on the other hand, was located in the Zor or the actual stream bed of the Jordan, where it may have benefited from groundwater and floods. The locations of Deir 'Alla and Mazar, however, are less logical. The importance of Deir 'Alla may relate to its relatively long history and its position along the Wadi al-Ghor. Mazar is located downstream and is therefore in a dependent position. In the pre-modern system, however, this region was fed by canals stemming from both the Rajib and the Zerqa.

These have different drainage systems, resulting in differences in the timing of discharge. This may have contributed to the ability of Mazar to become larger than average. In general, it seems that the many small IA villages in the plain had little inequality and differentiation although the prerequisites were present.

A completely different system of cultivation was probably practised in the Late Chalcolithic and Early Bronze Age (from 3600 to 2300 BC). During the Late Chalcolithic and Early Bronze (EB) I periods, settlements were predominantly located along natural water courses like the Zerqa and the Wadi al-Ghor. Climatic proxy data suggest that during the EBA the climate was slightly moister than today (Rosen 2007). However, this does not mean that the high evaporation was met. The rivers and wadis were not as deeply incised at that time as they are today and would overflow regularly (Cordova 2007: 189; Rosen 2007: 88). It is likely that EBA communities in this area practised farming in the regularly overflowing floodplains, possibly using dams to retain the water for longer (Kapteijn 2009).

The EBA system using floodplain irrigation was different from the canal systems used in the other three periods. In floodplain irrigation, all areas close to the river have equal possibilities. In canal irrigation, the area downstream is potentially at the mercy of the area upstream. Canal irrigation is potentially hierarchical in nature, although this is not as straightforward as it seems (Ertsen and Van Nooijen 2009). The interplay between canals and cultures in the three periods resulted in different socio-economic structures, as discussed above. In our remaining discussions, we focus on the periods in which canal irrigation was likely to be the main provider of water for crops. For all three periods, we assume a similar system of irrigation using small open channels distributing water under gravity from the Zerqa throughout the valley. Furthermore, the agricultural techniques open to these communities were probably comparable. For example, until recently, farmers in this region did not use chemical fertilisers and ploughs did not differ from those used in the IA.

Crops, climate and carrying capacities

The cropping pattern documented for the 1950s in this part of the valley shows first of all that as much as 30% of the land lay fallow. Another 30% was taken up by cereals – mainly wheat – while fruit and vegetables took up small areas (Nedeco 1969). For the Mamluk and IA periods, estimations on cropping patterns were based on archaeobotanic data from excavations, ethnographical analogies and the few ancient texts available. For the IA, the considerable number of analysed archaeobotanical samples shows more or less the same pattern: lots of cereal, both in quantity of grains per sample, and in presence in number of samples (van Zeist and Heeres 1973; Neef 1989). Sesame was discovered in IA remains and may have grown during late spring. Different from the pre-modern system is the lack of summer crops like sorghum, which were introduced during the Islamic period. The Mamluk period is in some respects considerably different from the other two periods. While crop evidence in the other periods points to subsistence farming, the Mamluk period saw large-scale sugar cane farming in the valley. Sugar cane may have taken as much land as possible, even if it came at the expense of the subsistence of the local population as sugar cane crops degrade soil fertility rapidly. Mamluk writers describing sugar cane production state that two crops taking a year each can be obtained from a plot, after which it has to lay fallow for four years (Tsugitaka 1997).

Taking modern climate and crop data, crop water requirements can be calculated. Calculations for wheat and some of the other common regional crops confirm that dry farming agriculture is impossible in this area. As a result of the short winter season – which is too short for any of the crops to mature, especially when rains are late – irrigation is needed. Hence, even while the annual rainfall is above the 250 mm boundary of dry farming, local circumstances make such an absolute number meaningless (Kapteijn 2009: 339ff).

The amount of water needed per day over the season appears to be very different for the different periods (fig. 5). The pre-modern system has clearly the highest peak with its winter crops maturing while the summer crops have already been planted. After harvesting winter crops, water demand drops, but not as sharply as during the IA when only a few fruit trees and possibly some sesame needed water. The Mamluk sugar cane needed water throughout the summer, and water demand only decreased when precipitation began at the end of November. Comparing the amount of cereals potentially produced and the amount needed per person allows for an estimate of the number of people inhabiting the region in the different periods. These figures suggest quite huge numbers during the 1950s and some periods within the IA – around 4,500 people in a normal year – while the

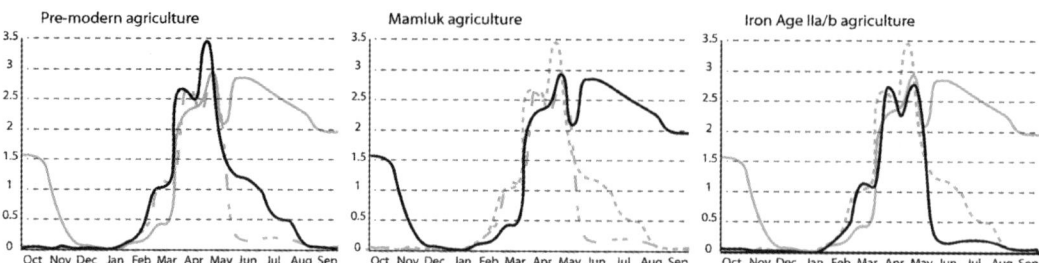

FIGURE 5. *Water demand trend lines per period in millimetres per day.*

Mamluk figure was significantly lower with an absolute maximum of 2,500 persons (Kapteijn 2009: 375–376).

What can we say about irrigation and climate in the Zerqa triangle? Firstly, we can state firmly that due to the climatic conditions in the region, crop growth requires irrigation to be consistent. Furthermore, the vulnerabilities of the cropping systems in terms of higher water demands, and thus of human societies, depend on the food and cash crops produced. However, the procedure we have applied so far has two important yet usually implicit simplifications, which are often ignored in discussions on irrigation in ancient societies. Apart from the well-known assumptions one needs to make on climatic and crop data, these two simplifications refer to how irrigation is conceptualised. Firstly, in building water balances for seasons, as we did, it is assumed that the water demand of crops needs to be maintained at optimal levels throughout the growing cycle in order to achieve a reasonable harvest. Furthermore, it is assumed that all the water needed is spread over the entire irrigated surface. In other words, it is assumed that irrigation water distribution through the canals is equal. Both these assumptions need to be challenged, as they link to the societal arrangements applied to the practice of irrigation at different levels – from the individual farmer to the larger society – and show that irrigation as a response to climate is actually a complex issue because societal arrangements, including the distribution of wealth or the lack of it, appear to be vital in explaining the successes and failures of responses to climate.

Irrigation to secure starting conditions

In many irrigation settings with comparable climatic conditions to the Zerqa triangle, irrigation is not the single source of water but is used to supplement the rainfall which will be available in the growing season in most years. The crop is sown when the rain begins, with irrigation starting once the river flow has increased due to the rains. This means that the crops in their first stage would need to survive on rainfall. For farmers, determining the optimal starting moments can be difficult. If one is too early, because of good rains that are subsequently not sustained, the young crop may not prosper because of drought conditions in the early stage. In the case of late sowing, the water availability is probably fine, but the crop may continue to grow too long into the dry season. The results of the difference between a good choice and a bad one may be as concrete as having a harvest or not. Young crops in particular, with their underdeveloped root system, are vulnerable to water stress. When water stress occurs early in the growing season, crops may not develop at all.

To illustrate the importance of water stress, we simulated evapotranspiration – which is a sign of crop growth – of two wheat crops. The first one started in dry soil, at wilting point (wp) when crops cannot extract the water in the soil anymore. We allowed another wheat crop to start with the soil being at field capacity (fc), when the soil contains the maximum amount of water for crops to be extracted. The crop starting out in wet conditions managed to develop fully, even on rain alone. Our poor dry crop did produce some biomass, as there was transpiration in the early season, but this biomass never managed to produce a yield (fig. 6). We also ran a simulation providing a single irrigation gift relatively late in the season, after one hundred days. This actually confirmed our findings: for dry starting conditions, this rescue irrigation gift is too late as the crop was already dead.

This simple example shows the importance of starting conditions when studying irrigation and yields but also suggests that several crops can be grown on the Zerqa plain on rainfall only, once the soil has been brought to field capacity. When farmers would start sowing their crops once the soil contained enough water, most crops would mature and give yields. However, how can we determine whether these starting conditions were reached? This issue is illustrated with an example about possible starting dates for wheat. Assume two possible scenarios to start the season: one is defined as

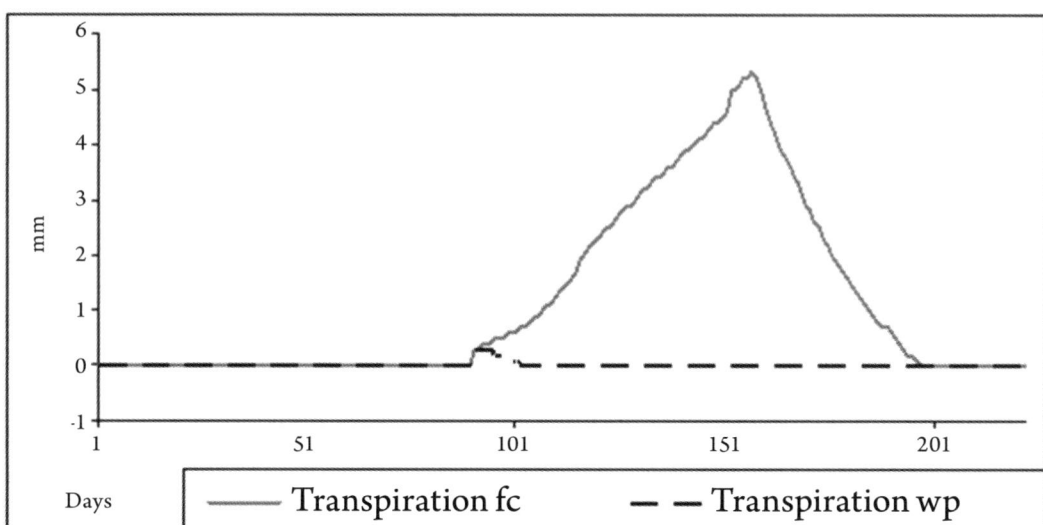

FIGURE 6. *Simulated crop water consumption.*

the moment when a total of 30 mm of rain has fallen, whereas in a second scenario, those same 30 mm must have fallen within a period of ten days before sowing is allowed to start. This allows for the fact that 30 mm in a longer period is more prone to evaporate before becoming effectively available for crops. Although both scenarios work with average rainfall, obviously starting dates will differ between these scenarios and between years. Taking into account more realistic sketchy rainfall patterns only increases the differences between years. This uncertainty brings irrigation back on the agenda. In a climate like that in the Zerqa triangle, securing starting conditions through one pre-sowing irrigation event – as is quite common in many irrigation systems in the world – would make a huge difference for crops such as wheat. It is also possible that such an event is arranged through flooding, whether facilitated or natural. Although we did not pursue this on this occasion, such water management might be typical for Bronze Age agriculture in the area.

Assuming that climatic conditions such as those in the Zerqa triangle would typically give irrigation a supplemental character, the issue is not any longer to match the crop water requirements exactly, but to give the crop a boost every now and then during the season to avoid water stress. The pre-sowing event especially would have been essential to allow crops to start growing, thus securing a successful crop. A second irrigation event should not take place too soon, as this could damage the young plants. Periods between gifts would have lasted for weeks or even months. Such irrigation strategies would have worked to support crops like wheat, which can sustain itself on soil moisture alone as soon as it has passed a certain threshold value in its growth and its roots reach deeper layers. For crops such as vegetables, whose roots go less deep and which need more frequent watering, irrigation may have been more vital during the growing cycle, but they would also have required less water per gift. Sugar cane is also a crop that would have needed regular short-term irrigation events.

Even in our supplementary irrigation scenario, there would have been a need to coordinate irrigation events. It makes little sense to shift irrigation turns from one end of the system to the other and back; it seems to be more rational to allow groups of farmers to share water for a certain period of time before moving to the next group. Allowing for comparable growing conditions, in terms of rainy season and temperatures, it may have been feasible to ensure farmers could irrigate their field within a reasonably short period. We assumed a pre-sowing event of 75 mm as reasonable, with an irrigation period of one month. Although not all the area might have been under irrigation, we assumed that this water would be available to the area covered by the canals. Given the three canals and their associated differences, both in terms of the groups (clans) that used them and their infrastructure and different intakes, we did the calculations for each canal separately. The upstream canal has by far the largest potentially irrigable area (fig. 2). Taking this area as the limiting area for seventy-five millimetres within thirty days, we came up with a needed continuous flow for the canal system of 0.22 m^3 per second. The middle and downstream canal areas are considerably smaller, but together they cover about the same area as the upstream canal.

This suggests two potential scenarios for diverting water from the Zerqa River. The upstream canal would need to be operated continuously during the irrigation period of thirty days. The other two canals, however, could be operated in at least two different ways. In one scenario, the canals could be operated at the same time. In another scenario, the canals could be operated one after the other, as their total operation time is also about thirty days. For both scenarios, we calculated how much of the Zerqa base flow – with daily values derived from averaging monthly base flows – would be required for the irrigation system. Whenever the three canals together required more than 50% of the available flow, we defined this as a potential problem. This 50% limit is obviously arbitrary, but would include such considerations since, in practice, more water would need to be diverted to give farmers seventy-five millimetres as water infiltrates, and since water cannot always be diverted from the river or downstream users may need extra water, etc. We did this basic calculation for three different starting dates of the growing season, and thus three different dates for the first gift. We counted the days with problems for the two scenarios and their different starting dates. Again, results are indicative at most, but two observations deserve to be made. First, there seems to be something akin to optimal timing. The early dates – gifts in December, thus sowing in November – and the late dates both yield more potential problems than the middle timing. Second, cooperation between the downstream canals pays off, as the system as a whole seems to be under much less stress, with fewer periods of water shortages. Obviously, how the relative positions of the canals come into play is not straightforward, even when it is clear that the upstream user has potentially less problems. Physical upstream positions do not necessarily coincide with social upstream positions (Ertsen and Van Nooijen 2009).

Concluding remarks

In the climatic conditions of the Jordan Valley and the Zerqa Triangle, even though the average rainfall may be on the limit considered the minimum for dry-farming, irrigation as a strategy in response to these conditions makes sense. As annual variability in precipitation is very large, with droughts occurring every few years, sustaining a stable society of some size without irrigation would have been rather difficult. At the same time, it is clear that in this arid but potentially very fertile region, sufficient water available for crops secures good harvests. We have discussed how the different societies – for all of whom irrigation was of the utmost concern – have shaped irrigation. Through time the manner in which irrigation was practised changed, but it is likely that after the Zerqa had become incised – some time before the Iron Age – canal irrigation became the norm. Although the Zerqa Triangle was probably always a rural area based on crop cultivation on fertile alluvial soils using the same system of irrigation, the socio-cultural characteristics of each period resulted in different socio-economic systems for farming.

In terms of responding to drought conditions, we have shown that securing soil moisture early in the growing season would have been enough for crops such as wheat to have reasonable yields. Even then, it is clear that reaching an acceptable distribution of water over the entire area within a reasonable time would have been vital. Arrangements to distribute risks evenly, such as those discussed for the ethnohistorical situation, when irrigation was collectively practised with a rotation of fields among clan members every few years, are a response to this challenge. It is not unlikely that similar arrangements would have been made in the Iron Age, with the advantage that the cropping pattern seems to have missed summer crops that would demand high amounts of water. For the sugar cultivation of the Mamluk period, such relatively simple arrangements would not have worked. Sugar cane is a water-demanding crop, with irrigation gifts needed regularly. It is not a coincidence that there was a separate administrative centre in Damascus that managed the sugar cane plantations in the Jordan Valley and probably the Zerqa Triangle. There was pressure on available water and thus a need for stricter social arrangements in the dry climate. It should not be concluded that the dry climate in the Zerqa Triangle was the ultimate deciding factor in terms of social arrangements. Especially with sugar cane – which is not the first crop which comes to mind when thinking about adapting to a dry climate – it is reasonable to assume that the political-economic context was more determining than climate. There is ample evidence that societies are able to develop economic activities within diverse climates, including activities that have increased their vulnerability to climate change. We should not blame the climate for everything.

Bibliography

Cordova, C. E. 2007. *Millennial landscape change in Jordan: Geoarchaeology and cultural ecology.* University of Arizona Press. Tucson.

Ertsen M. W., and R. Van Nooijen. 2009. "The man swimming against the stream knows the strength of it: Hydraulics and social relations in an Argentinean irrigation system." *Physics and Chemistry of the Earth* 34(3): 2000–2008.

Ibrahim, M. M., J. A. Sauer and K. Yassine. 1988. "The East Jordan Valley Survey 1976 (Part two)." In K. Yassine (ed.), *Archaeology of Jordan: Essays and reports,* 189–207. Amman.

Kapteijn, E. 2009. *Life on the watershed: Reconstructing subsistence in a steppe region using archaeological survey: A diachronic perspective on habitation in the Jordan Valley.* Sidestone Press. Leiden.

Kooij, G. van der. 2007. "Irrigation systems at Dayr 'Alla." In F. al-Khraysheh (ed.), *Studies in the history and archaeology of Jordan IX,* 133–144. Amman.

Kosok, P. 1965. *Life, land, and water in ancient Peru.* Long Island University Press. New York.

LaGro, H. E. 2002. *An insight into Ayyubid-Mamluk pottery: Description and analysis of a corpus of mediaeval pottery from the cane sugar production and village occupation at Tell Abu Sarbut in Jordan.* Unpublished PhD dissertation, Faculty of Archaeology, University of Leiden. Leiden.

Merrill, S. 1881. *East of the Jordan: A record of travel and observation in the countries of Moab, Gilead, and Bashan.* R. Bentley & Son. London.

NEDECO and Dar Al-Handasah. 1969 *Jordan Valley project: Agro- and socio-economic study. Final report.* Beirut and The Hague.

Neef, R. 1989. "Plants." In G. van der Kooij and M. M. Ibrahim (eds.), *Picking up the threads... : A continuing review of excavations at Deir 'Alla, Jordan,* University of Leiden Archaeological Centre, 30–37. Leiden.

Netherly, P. J. 1984. "The management of late Andean irrigation systems on the north coast of Peru." *American Antiquity* 49(2): 227–254.

Petit, L. P. 2009. *Settlement dynamics in the Middle Jordan Valley during Iron II-III.* BAR International Series 2033. Archaeopress. Oxford.

Orlove, B. 2005. "Human adaptation to climate change: A review of three historical cases and some general perspectives." *Environmental Science and Policy* 8: 589–600.

Rosen, A. Miller. 2007. *Civilizing climate. Social responses to climate change in the ancient Near East.* Altamira Press. Lanham, Maryland.

Scarborough, V. L. 2003. *The flow of power: Ancient water systems and landscape.* SAR Press. Santa Fe.

Tarawneh, M. F. 1989. *Aspects of rural transformation in the Jordan Valley: The case of Deir 'Alla.* Unpublished MA thesis, Yarmouk University.

Tsugitaka, S. 1997. *State and rural society in medieval Islam: Sultans, muqta's and fallahun.* Islamic history and civilization: Studies and Texts 17. E. J. Brill. Leiden and New York.

Zeist, W. van, and J. A. H. Heeres. 1973. "Paleobotanical studies of Deir 'Alla, Jordan." *Paléorient* 1: 21–37.

The Late Bronze Age Collapse and the Early Iron Age in the Levant: The Role of Climate in Cultural Disruption

David Kaniewski, Elise Van Campo, Karel Van Lerberghe, Tom Boiy, Greta Jans, Joachim Bretschneider

Abstract

It is generally accepted from the historical sources that the fall of the city and kingdom of Ugarit was the result of a military invasion by the Sea Peoples in the first quarter of the 12th century BC. Here we present an advanced picture of cultural and landscape changes for the Late Bronze Age collapse and the ancient Dark Age of history. The Gibala data indicate that the collapse of Levantine countryside towns occurred during a c. 1175–825 calibrated yr BC severe drought event corresponding with the Dark Age and suggest a link between climate induced environmental changes and eastern Mediterranean cultural history. This key study examines the diachronic urban development of the ancient coastal site of Gibala-Tell Tweini. Urban collapse and urban change of Gibala was linked with the Late Bronze Age to Early Iron Age new social adaption, possibly stimulated by a climatic stress event in the northern Levant.

Introduction

By the Late Bronze Age (LBA), the eastern Mediterranean was at the centre of some of the most advanced civilisations of the world. The Mycenaean culture in the Aegean was flourishing; in Anatolia, the Hittites had carved out a vast empire; in the Levant, the Canaanite coastal cities were prospering through trade from Egypt to Mesopotamia; and in Egypt, the New Kingdom was at its height (Drews 1993). However, around 1200 BC, at the end of the LBA, the eastern Mediterranean civilisation declined or collapsed

(Carpenter 1966; Brinkman 1968; Neumann and Parpola 1987; Alpert and Neumann 1989; Weiss 1982; Beckman 2000).

The last written correspondences between Levantine, Hittite and Egyptian kings indicate that nomadic raiders, the Sea Peoples (Gilboa 2006–2007), were of great concern for the coastal towns as well as for the great empires and the vassal kingdoms at the end of the LBA (Singer 1999). The disintegration of the cities and states has traditionally been attributed to these invasions. However, in considering other possible causes for the LBA collapse, suggestions include climate shift, environmental or natural disasters, technological innovations, internal collapses and states of inequality between centre and periphery (Neumann and Parpola 1987; Weiss 1982; Bryce 2005; Killebrew 2005). Throughout the eastern Mediterranean, the 12th century BC ushered in a Dark Age, which was not to lift for more than 300–400 years (Drews 1993).

The nearly contemporaneous decline of highly organised and powerful states warrants consideration of possible environmental causes likely to operate over sizable areas (Weiss 1982). A climate shift occurring during the 13th to 9th centuries BC may be of major interest in Mediterranean and West Asian environments where dry farming agro-production and pastoral nomadism were the primary or secondary subsistence systems. Reduced precipitation may have strongly affected the outlying nomad habitats, and led rain-fed cereal agriculturalists to habitat-tracking when agro-innovations were not available (Lewis 1987; Staubwasser and Weiss 2006; Reuveny 2007).

There are no written sources for these periods with any direct information on climate or climate changes, except for Aristotle's statement about the Mycenaean drought around 1200 BC (Neumann 1985). There is, however, other useful information that is related to food production, grain shortages and famine. Historical data that document episodes of food shortage in the eastern Mediterranean are rare but highly informative. The clay tablet RS 34.152 from Emar is a vivid testimony to the deteriorating conditions in inner Syria and severe food shortage around 1190 BC. The Emar year-names bear witness to a staggering increase in grain prices in the "year of hardship/famine" (Singer 2000; Cohen and Singer 2006). The clay tablet RS 18.38, dated to the late 13th century BC, indicates grain shipments from Egypt to the Hittites, suggesting grain shortages in eastern Anatolia (Bryce 2005). A particular note of urgency occurs in a letter (clay tablet RS 20.212) sent from the Hittite court to the Ugaritic king, either Niqmaddu III or Hammurabi (1215–1194/1175 BC), demanding ship and crew for the transport of 2000 kor of grain (c. 450 tons) from the Syrian coastal district Mukish to Ura. The letter ended by stating that it is a matter of life or death (Nougayrol et al. 1968). In Egypt, a famine struck the country during the reign of Merneptah (1213–1203 BC) (Bryson et al. 1974). The drop of Nile discharges during the reign of Ramses III (1186–1153 BC) led to crop failures/low harvests (Butzer 1976) and riots (Faulkner 1975). Hatti had very probably come to rely on grain importation during the last century of the kingdom. Following the 1259 BC treaty between Ramses II and Hattusili III, grain was probably imported from Egypt into Anatolia on a regular basis (Bryce 2005). This indicates that the Hatti kingdom was no longer self-sustainable in food procurement and had to rely on food imports. At the end of the 13th century BC, Pharaoh Merneptah (1213–1203 BC) sent the earliest known shipment of grain in the form of famine aid to the Hittites (Warburton 2003; Bryce 2005). The Hittite king Amuwanda III described the terrible hunger suffered during his father's day in Anatolia, and he mentioned drought as the reason (Warburton 2003). In Mesopotamia, written sources from Babylon and Assyria describe crop failures, famine and the outbreak of plague (Brinkman 1968; Neumann and Parpola 1987).

The drought hypothesis was first proposed by Carpenter (1966) to explain the collapse of the Mycenaean civilisation, and was later developed further by Weiss (1982) in relation to the disappearance of the LBA palatial civilisation in the eastern Mediterranean. Recently, a thousand year-long pollen-climate record from alluvial deposits around the ancient coastal city of Gibala (Bretschneider and Van Lerberghe 2008), the southernmost town in the Ugarit kingdom situated near modern Jableh (Syria), indicates climate instability and a severe drought episode at c. 1175–825 cal yr BC (Kaniewski et al. 2008, 2010). The 1 sigma (1σ) probability distribution of the radiocarbon (^{14}C) dates obtained for the climatic event ranges between 1260–1130 cal yr BC for the onset of the drought, and 900–830 cal yr BC for the termination (table 1).

Here we present an advanced picture of cultural and landscape changes for the LBA collapse and the Dark Age for the coastal Gibala-Tell Tweini site. We use archaeological data for the period 1200–900 BC, a refined numerical-derived climatic proxy and a pollen-derived record of food availability based on cultivated plants. Shifts in the socio-cul-

Code	Material	AMS ^{14}C yr BP	Calibrated dates BC	
			1σ	2σ
Beta-261721	Charcoal	2640±40	830-790	865-775
Beta-261722	Charcoal	2720±40	900-830	935-805
Beta-229047	Charcoal	2750±40	925-835	980-815
Poz-28165	Charcoal	2810±40	1010-915	1055-840
Beta-229048	Charcoal	2970±40	1260-1130	1315-1050

TABLE 1. *Details of the ^{14}C age determinations. All ages have been calibrated with IntCal09-Calib Rev 6.0.1.*

tural pattern, urban development and stratification of the site are analysed to integrate the climate trends in the main changes affecting human life in the eastern Mediterranean and West Asia during the LBA and the Iron Age (IA).

The site: Gibala-Tell Tweini

Tell Tweini (fig. 1) is a larger multi-period site of c. 12 hectares on the Syrian coast, which has sequences from the Early Bronze Age (EBA) IV to the Iron Age III (c. 2400–500 BC). So far, eleven main phases with ten levels have been excavated in the centre of the Tell (Bretschneider and Van Lerberghe 2008). Since 1999, Bronze and Iron Age chronology, pottery typology and urban patterns have been studied at the site. Tell Tweini can most probably be identified with the Ugaritic city of Gibala, the southernmost harbour of the Ugaritic kingdom. The term "Gi$_5$-bá-la" appears in the Akkadian tablets PRU 4, 71–76 and PRU 5, 74 (Bretschneider and Van Lerberghe 2008). The written LBA sources or epigraphic finds for Gibala ceased as soon as Ugarit was destroyed. The city of Gibala is mentioned again during the IA II, in an inscription of Tiglatpileser III (744–727 BC).

Eleven seasons of excavations at the site have revealed the importance of the ancient city at the end of the second millennium – the period contemporary with Ugarit – as well as in earlier and later periods. The research focuses on two primary aims: the exploration of early social complexity and

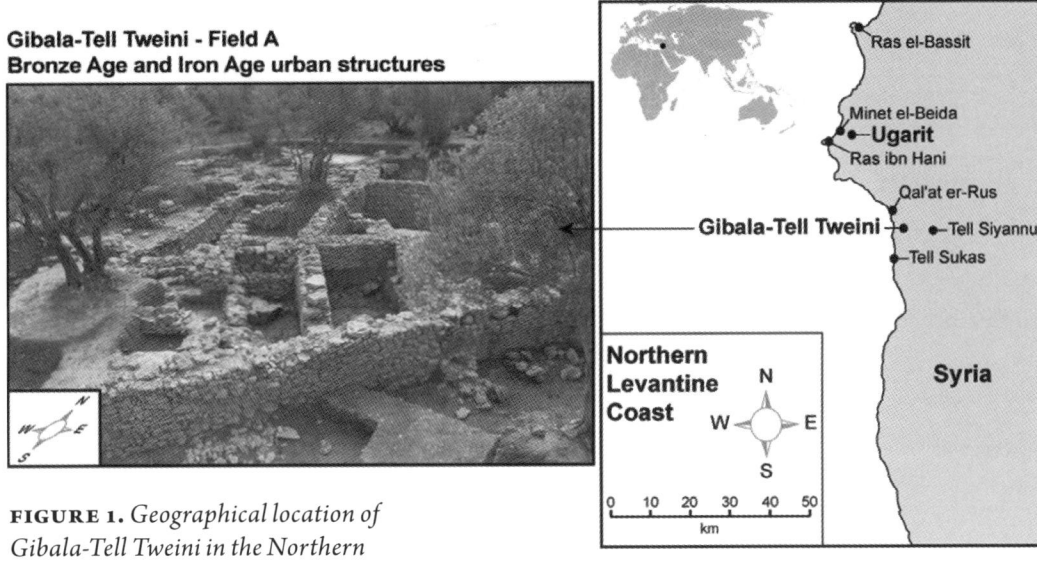

FIGURE 1. *Geographical location of Gibala-Tell Tweini in the Northern Levant.*

early urbanism in the late 2nd-early 1st millennium BC; and the investigation of the site's reaction to political change, from the collapse of the Ugarit kingdom through the growth of a new local site and the imposition of a new regional power.

Materials and methods

Core lithology, and chronology

The data presented in this paper are based on two cores (fig. 2) from the immediate vicinity of the pear-shaped Gibala-Tell Tweini. The TW-1 core (800 cm; 35°22'22.94" N, 35°56'12.49" E, 17 m a.s.l., 1.75 km from the Mediterranean) was extracted from the thick alluvial deposits of the Rumailiah River. The core chronology relies on four accelerator mass spectrometry (AMS) ^{14}C ages on charcoal (table 1). Datable plant remains are lacking

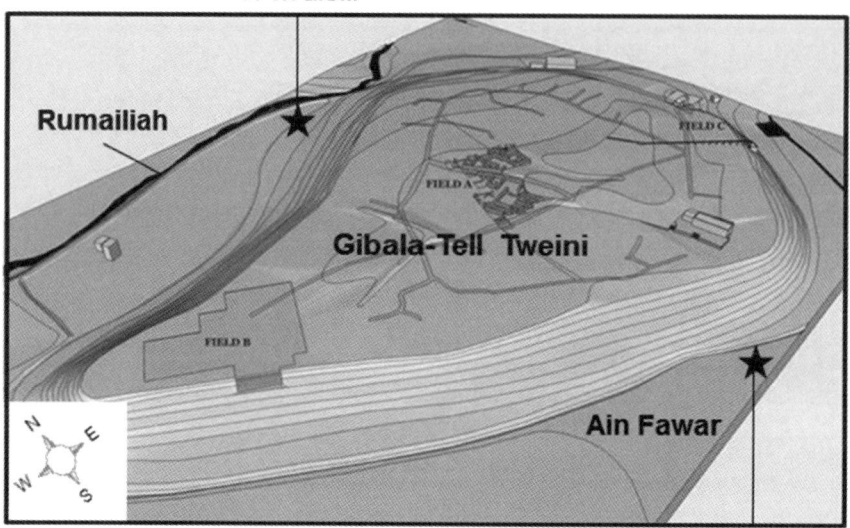

FIGURE 2. *Digital elevation model of Gibala-Tell Tweini showing the location of the cores TW-1 (alluvial plain of the Rumailiah River) and TW-2 (Ain Fawar spring complex).*

from the sediment column, above core depth 395 cm, which has the conventional age 2750±40 ^{14}C yr BP (table 1).

The TW-2 core (450 cm; 35°22'13.16"N, 35°56'11.36"E; 16.06 m a.s.l., 1.6 km from the Mediterranean) was sampled from the alluvial deposits of a small first-order, spring fed river valley bordering the Tell to the south (Ain Fawar). The core chronology is based on three AMS ^{14}C ages on charcoal (table 1).

Compaction corrected deposition rates have been computed between the intercepts of adjacent ^{14}C ages. The age of each sample was calculated by interpolation. The reliability of the age model was developed elsewhere (Kaniewski et al. 2010).

Pollen

A total of 83 samples from cores TW-1 and TW-2 were prepared for pollen analyses using standard palynological procedures. Pollen grains were counted under x400 and x1250 magnification using a Leitz microscope. Pollen frequencies (%) are based on the total pollen sum (400 pollen grains, on average), excluding local hygrophytes and spores of non-vascular cryptogams. Cultivated plants' and cereals' time-series have been plotted on the linear age scale.

Pollen data have been converted into Plant Functional Types (PFTs) and a pollen-derived biomisation of the PFTs has been elaborated (Prentice et al. 1996; Tarasov et al. 1998). Three semi-quantitative climatic indexes (SQCIs) have also been computed from pollen data (Kaniewski et al. 2008). The process used to convert environmental data into climatic proxy has been modified here and now includes the PdB and SQCI time-series in the principal components analysis (PCA) numerical matrix. The refined data are described using the computed age-scale model based on the AMS ^{14}C intercepts (fig. 3).

Results and discussion

Archaeological data

The final LBA occupation Level 7B at Tweini suffered from a massive destruction event. This destruction layer, termed Level 7A, corresponds to the second conflagration of the city (Kaniewski et al. 2011) with the ruins of the Late Bronze A houses. In Locus TWE-A-185 (Room 2/9), well-dated ceramic imports from the Greek mainland were preserved *in situ*. Most notable are a Late Helladic IIIB (Middle) kylix decorated with shell motifs, and a handmade and burnished vessel of North Syrian Cooking Ware, which are features of the last phase of the LBA at Ugarit and Ras Ibn Hani (Jung 2008; Vansteenhuyse 2010). Locus 185 was sealed by a fire destruction layer ending the LBA occupation in this part of the town.

An even more precise date for the LBA destruction was provided in the 2010 excavation of the western part of Field A in Locus TWE-A-07339, 07345, 7183 and 7169, where locally produced Late Helladic IIIC (Early) ceramics were found in the ruins of a LBA building (Jung 2010).

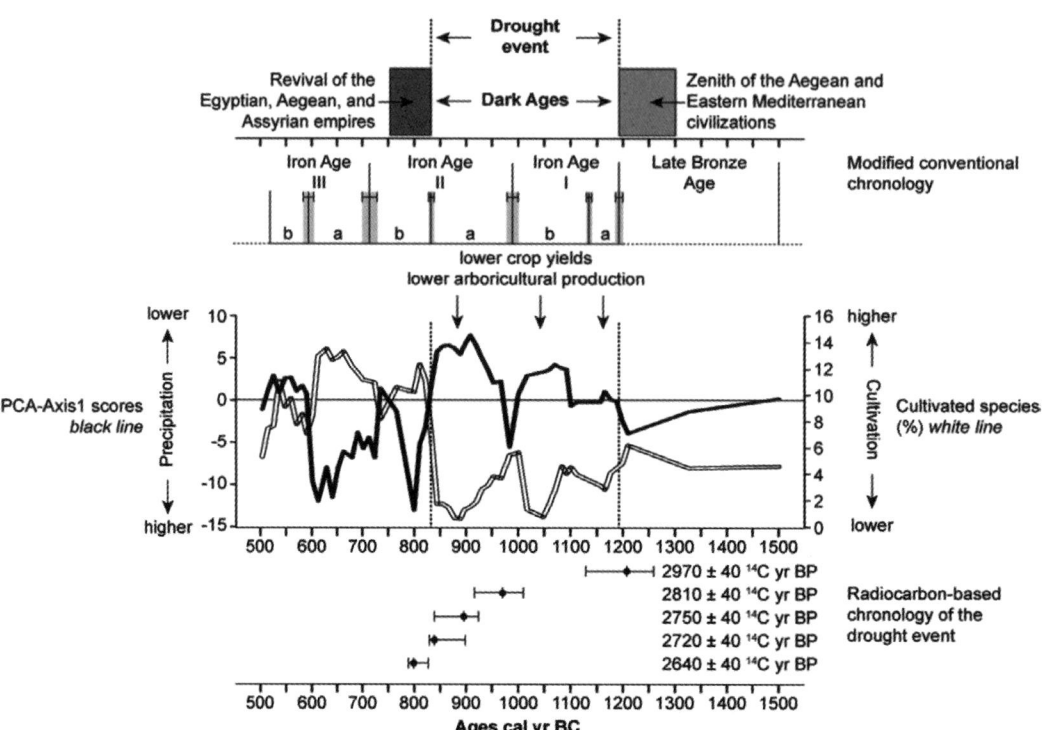

FIGURE 3. *The Late Bronze Age collapse and the Dark Ages from the viewpoint of climatology. Shown is the LBA-IA sequence from the alluvial deposits of the Rumailiah River, north of Gibala-Tell Tweini. The pollen-derived climatic proxy is drawn as PCA-Axis 1 scores. The Late Bronze Age and Iron Age modified conventional chronology is shown with the PCA-Axis 1 scores. Grey shades indicate cultural changes. Cultivated species time-series are plotted on linear age-scale. The main historical events are indicated at the top of the diagrams. Radiocarbon ages are displayed as 1 sigma calibration range. The black dots correspond to the intercepts with the calibration curve.*

The chronological sequence of the Late Helladic IIIC (Early) ceramics followed the fall of the Mycenaean palaces system in mainland Greece in the 12th century BC. Based on a new set of tree ring calibrated ^{14}C-data obtained from Kastanás (central Macedonia) and the latest Mycenaean-type pottery sequences from the Syrian coastal sites of Ugarit and Kazel (Monchambert 2004; Mountjoy 2004; Jung 2008), the Late Helladic IIIC (Early) period is now dated between 1210–1200 and 1170–1160 BC in the absolute chronology of the Aegean LBA (Weninger and Jung 2009).

The final LBA destruction of Tell Tweini (Level 7A) can be dated to the first three decades of the 12th century BC. This date matches the historical-archaeological dating of the Sea Peoples' invasions in the northern Levant with the fire destruction of Ugarit (Yon 2006) and other northern Levantine coastal sites such as Ras Ibn Hani, Ras el-Bassit, Tell Kazel and Tell Sukas (Bretschneider and Van Lerberghe 2008). Local Late Helladic IIIC (Early) ceramic is attested in Tell Tweini for the beginning of the 12th century BC (Jung 2010).

Near Eastern epigraphic and archaeological data document the invasions of the Sea Peoples (Gilboa 2006–2007; Yon 2006) and internal disintegration (Caubet 1989) as the proximate cause for the LBA collapse in the northern Levant. The chronology of the Sea People invasions is mainly based on letters just preceding the collapse of Ugarit (Singer 1999; Dietrich and Loretz 2002; Yon 1989, 2006) and on Egyptian sources (Singer 1999; Beckman 2000). The Sea People invasions were documented on the Ramses III's Medinet Habou Temple where they are illustrated with women and children suggesting movements of large kin-based units (Beckman 2000). The fall of Ugarit is currently dated between 1194 and 1176 BC, between the *terminus post quem* supplied by the letter of the Egyptian Beya (1194–1190 BC) and the *terminus ante quem* of Ramses III's eighth year (1179/1176 BC) (Singer 1999; Beckman 2000; Shaw 2000; Hornung et al. 2006). Freu (1988) concludes that tablet RS 86.2230 was sent to Ugarit during the reign of Pharaoh Siptah and Queen Tawsret (1194–1186 BC) and not during Sethnakht's reign, so that Ugarit has to have been destroyed after 1194 BC. A precise historical date of 1192–1185 BC is suggested by the combination of Ras Shamra clay tablet 86.2230 with the new dating of eclipse KTU 1.78 at 1192 BC (Dietrich and Loretz 2002). A new study, based on radiocarbon, historical, archaeological and astronomical data, suggests a date of 1192–1190 BC for the destruction of Tell Tweini by the Sea Peoples (Kaniewski et al. 2011).

The reuse of LBA ruins at Gibala-Tell Tweini and the construction of new buildings indicate a local reoccupation from the very beginning of the Early IA (Level 6G-H, around the second half of the 12th century BC) (Bretschneider et al. 2010), as was also the case for some other secondary coastal sites such as Tell Kazel (Capet 2003), Ras Ibn Hani and Ras el-Bassit (Caubet 1989). For the remainder of the kingdom, the survival of place names for both large and small villages from the LBA to the present pleads in favour of some continuity of occupation (Yon 1989). A second architectural phase is attested at Gibala-Tell Tweini during the end of the Early IA I (Level 6E-F). Level 6E (end of occupation), a 20–30 cm thick layer of powdery ashes, charcoal and charred seeds represents the second conflagration. This level is located between earlier IA structures (Level 6G-H) and is directly covered by foundation walls belonging to Early IA II structures (Levels 6D-C) (Bretschneider et al. 2010). The city was not fully regenerated until the 9th and 8th centuries BC (IA II, Levels 6D-A).

Environmental data

The PCA-Axis 1 ordination of the TW-1 data accounts for most of the variance (fig. 3), with +.749 of total inertia. Arid/saline SQCI-s (+.1107), PdB Hot desert (+.6188), and PdB Warm steppe (+.2705) are loaded in positive values, whereas negative values correspond to wet SQCI-s (-.5886), PdB Warm mixed forest (-.4067), and PdB Xerophytic woods/shrubs (-.047). Relative frequencies of pollen indicators of crop cultivation and arboriculture were considered as an indirect proxy of food availability. A straightforward relationship is evidenced between drought phases and periods of low crop production, which could induce famines (fig. 3).

The refined pollen-based climate record shows moist climate conditions at c. 1500–1200 cal yr BC, with a wetter pulse at c. 1210 cal yr BC. As a first approximation, the intercept age of 1210 cal yr BC (2970±40 ^{14}C yr BC) can be used to date the beginning of the climatic deterioration. This intercept age corresponds with the generally accepted age for the collapse of the LBA cultures in the eastern Mediterranean as being dated at c. 1200 BC, a conclusion based on a complex integration of archaeological data and literary sources coming mainly from Ugarit. Secure linkages between the LBA collapse and the onset of the drought event are particularly difficult to provide. The weak discrepancy between the written sources and the radiocarbon intercept may suggest that the drought event and the drought induced decline in crop production began in the late 13th/early 12th centuries BC. The climatic instability is characterised by increasing drought, peaking at c. 910 cal yr BC, but which is interrupted by a short wet pulse during c. 990–970 cal yr BC (fig. 3).

A dry and warm climate dated c. 1200–900 BC has also been suggested for the Middle East (Neumann and Parpola 1987). The major environmental shift, interpreted as a result of lower amounts of precipitation in the Syrian coastal area, is synchronous with a dry southern basin and a low lake level in the northern basin for the Dead Sea (Bookman et al. 2004). The lowest value of the northern lake was reached at 1400 BC, before the onset of the Syrian climatic shift, and the level stays low throughout the drought event. The onset of aridity is also reflected in the Dead Sea by a windblown sand layer at the lake bottom (Migowski et al. 2006). The change in rainfall inducing a shortage of water supply in coastal Syria is also correlated with *minima* in the Tigris and Euphrates river discharges from 1150 to 950 BC (Kay and Johnson 1981; Neumann and Parpola 1987; Alpert and Neumann 1989), and with higher $\delta^{18}O$ values in the Ashdod coast record (Schilman et al. 2001, 2002). The $\delta^{18}O$ study suggests that aridification, reflected by enhanced evaporation rates, started at 1050 BC, succeeding a 1650–1050 BC humid phase (Schilman et al. 2001, 2002). This climatic shift can be correlated with changes in atmospheric moisture over India (Singh et al. 1990) and low lake levels in Turkey (Lemcke and Sturm 1996). Peaking values of arid/saline Chenopodiaceae and a collapse of agriculture at c. 1050–850 BC in the Ze'elim and Ein Feshkha pollen sequences (Neumann et al. 2007) also reflect dry conditions. The onset of an aridification is recorded in the Arabian Sea somewhat earlier, at c. 1150 BC (Lückge et al. 2001).

Evidences for the crises during the late 13th/early 12th centuries BC in the eastern Mediterranean may serve as anchor points between the historical sources and the radiocarbon dated decline in crop production in coastal Syria. Agricultural activities during the Bronze Age correspond mainly to Poaceae cerealia (barley, wheat) and Fabaceae (fava bean, pea, chickpea, lentils). The two most drought resistant species are wheat (needs a minimum annual precipitation [MAP] of 300 mm yr^{-1}) and barley (MAP 200–300 mm. yr^{-1}). The potential MAP during the period with very low cultivation (1060–1000 and 925–850 BC; fig. 3) may be under or near 200–300 mm yr^{-1}.

The data suggest that the fall of Ugarit and the secondary cities has to be placed within the drought period which may have started at the end of the

13th century BC. Inhabitants of the destroyed and abandoned LBA cities probably sought refuge in the mountain villages which were somewhat protected by being located away from the coast (Caubet 1989; Yon 1989). The fact that certain village names have been preserved from the LBA to the present leads us to believe that the village communities managed to survive thanks to their inland location.

A causal process for the northern coastal Levant migration might also have been the transient ameliorating effect of moister conditions on crop and food resources concentrating population movement from the coast toward more fertile areas such as the riparian and adjacent karst aquifer-related settlements/cities of the Orontes River. The Sea Peoples may have induced the fall of the coastal settlements of Ugarit, Ras Ibn Hani, Ras el-Bassit, Tell Kazel, Tell Sukas and Gibala-Tell Tweini (Level 7A), followed by the destruction of several cities of the Hittite Empire such as Tarsus and Hattusas (Beckman 2000) and Alalakh, Tunip, Hamath and Qadesh near the Orontes River (Fugmann 1958; Woolley 1958; Bartl and Al-Maqdissi 2007; Whincop 2007), as well as others near the Euphrates such as Emar, Tell Bazi, Tell Faq'us, Tell Fray and Tell Suyuh (Beyer 2001; Adamthwaite 2001; Otto 2007; Cohen 2009).

The duration of the drought event in coastal Syria has been estimated by a series of AMS ^{14}C dates obtained in the two cores (table 1). Their ages range from the 13th/12th centuries BC to the 9th/8th centuries BC. The end of the drought event is enclosed in an interval between 2720±40 ^{14}C yr BP and 2640±30 ^{14}C yr BP. During this interval, defined by a 2σ confidence of, respectively, 935–805 cal yr BC (intercept at 840 cal yr BC) and 865–775 cal yr BC (intercept at 800 cal yr BC) (table 1), the drought suddenly ends. The TW-1 and TW-2 cores are consistent with a termination of the drought event during the 9th century BC, between 840 and 800 cal yr BC according to the intercepts. Archaeological data in coastal Syria show that dense occupation reappears during the end of the 9th or the beginning of the 8th century BC (Caubet 1989).

A pronounced wet peak at c. 825–800 cal yr BC marks the abrupt end of the c. 350 year drought event. A subsequent minor dry event, between c. 770 and 725 cal yr BC, is followed by a gradually increasing, c. 125 yr-long wet phase until c. 600 cal yr BC.

The Egyptian, Aegean and Assyrian empires recovered with diversified agro-production (manna ash, olive tree, vine tree, walnut tree), pastoral activities, and a sustained cultural revival (Weiss 1982). The archaeologically defined end of the Dark Age and the radiocarbon-dated end of the drought event are concordant in time.

Conclusion

Gibala is a rare settlement with Early IA I occupancy after the LBA collapse. The Rumailiah River and the Ain Fawar spring complex provided a stable water supply for resettlement on the surrounding alluvial plain, despite climate shifts and successive destructions during the following Dark Age. Gibala also demonstrates that there was no systematic, one way reaction of the people in regard to adverse environmental situations. At the late 13th/early 12th centuries BC period, the climate change may have induced cultural collapse; however, during IA I and II people were able to cope with the adverse situations.

Acknowledgements

This research is funded by het Fonds voor Wetenschappelijk Onderzoek, het Onderzoeksfonds Katholieke Universiteit Leuven, the Inter-university Attraction Poles Programme VI/34, Belgian Science Policy, Belgium, by the Paul Sabatier-Toulouse III University, and the INSU-CNRS Paleo2 MEDORIANT-MISTRAL program.

Bibliography

Adamthwaite, M. R. 2001. *Late Hittite Emar: The Chronology, Synchronisms, and Socio-Political Aspects of a Late Bronze Age Fortress Town*. Ancient Near Eastern Studies Supplement Series 8. Peeters Press. Leuven.

Alpert, P., and Neumann, J. 1989. "An ancient correlation between streamflow and distant rainfall in the Near East." *Journal of Near Eastern Studies* 48: 313–314.

Bartl, K., and M. al-Maqdissi. 2007. "Ancient settlements in the middle Orontes region between ar-Rastan and Qalcat Shayzar. First results of archaeological surface investigations 2003–2004." In D. Morandi Bonacossi (ed.), *Studi archeologici su Qatna*, Qatna 1, Udine 243–252.

Beckman, G. 2000. "Hittite chronology". *Akkadica* 119–120: 19–32.

Beyer, D. 2001. *Emar IV, Les sceaux*. OBO Series Archaeologica 20. Vandenhoeck & Ruprecht. Göttingen.

Bookman (Ken-Tor), R., Y. Enzel, A. Agnon and M. Stein. 2004. "Late Holocene lake levels of the Dead Sea." *Geological Society of America Bulletin* 116: 555–571.

Bretschneider, J., and K. van Lerberghe. 2008. "Tell Tweini, ancient Gibala, between 2600 BCE and 333 BC." In J. Bretschneider and K. van Lerberghe (eds.), *In Search of Gibala: An Archaeological and Historical Study Based on Eight Seasons of Excavations at Tell Tweini (1999–2007) in the A and C Fields*, Aula Orientalis, 12–66. Barcelona.

Bretschneider, J., G. Jans and A. S. van Vyve. 2010. "Les campagnes des Fouilles de 2009 et 2010." In M. Al-Maqdissi, K. van Lerberghe, J. Bretschneider and M. Badawi (eds.), *Tell Tweini: onze campagnes de fouilles syro-belges (1999–2010)*, Documents d'Archéologie Syrienne, Ministère de la Culture, Direction Générale des Antiquités et des Musées. Damascus.

Brinkman, J. A. 1968. *A Political History of Post-Kassite Babylonia, 1158–722 BC*. Analecta Orientalia. Rome.

Bryce, T. 2005. *The Kingdom of the Hittites*. Oxford University Press. Oxford.

Bryson, R. A., H. H. Lamb and D. Donley. 1974. "Drought and the decline of the Mycenae." *Antiquity* 43: 46–50.

Butzer, K. W. 1976. *Early Hydraulic Civilization in Egypt*. University of Chicago Press. Chicago and London.

Capet, E. 2003. "Tell Kazel (Syrie), Rapport préliminaire sur les 9ème –17ème campagnes de fouilles (1993–2001) du Musée de l'Université Américaine de Beyrouth, Chantier II." *Berytus* 47: 63–121.

Carpenter, R. 1966. *Discontinuity in Greek Civilization*. Cambridge University Press. Cambridge.

Caubet, A. 1989. "Reoccupation of the Syrian coast after the destruction of the 'Crisis Years'." In W. A. Ward and M. Sharp Joukowsky (eds.), *The Crisis Years: The 12th century BC. From Beyond the Danube to the Tigris*, Kendall/Hunt Publishing Company, 123–131. Dubuque, Iowa.

Cohen, Y. 2009. *The Scribes and Scholars of the City of Emar in the Late Bronze Age*. Harvard Semitic Studies 59. Eisenbrauns. Winona Lake, Indiana.

Cohen, Y., and I. Singer. 2006. "Late Synchronism between Ugarit and Emar." In Y. Amit, E. B. Zvi, I. Finkelstein and O. Lipschits (eds.), *Essays on Ancient Israel in its Near Eastern Context: A Tribute to Nadav Na'ama*, Eisenbrauns, 123–139. Winona Lake, Indiana.

Dietrich, M., and O. Loretz. 2002. "Der Untergang von Ugarit am 21. Januar 1192 v. Chn? Der astronomisch-hepatoskopische Bericht KTU 1.78 (RS 12.061)." *Ugarit-Forschungen* 34: 53–74.

Drews, R. 1993. *The End of the Bronze Age: Changes in Warfare and the Catastrophe ca. 1200 BC*. Princeton University Press. Princeton, New Jersey.

Faulkner, R. O. 1975. *Egypt, From the Inception of the Nineteenth Dynasty to the Death of Ramesses III*. Cambridge Ancient History. Cambridge University Press. Cambridge.

Freu, J. 1988. "La tablette RS 86.2233 et la phase finale du Royaume d'Ugarit." *Syria* 65: 395–398.

Fugmann, E. 1958. *Hama. Fouilles et recherches, 1931–1938*. Carlsbergfondet. Copenhagen.

Gilboa, A. 2006–2007. "Fragmenting the Sea Peoples, with an Emphasis on Cyprus, Syria and Egypt: A Tel Dor Perspective." *Scripta Mediterranea* 27–28: 209–244.

Hornung, E., R. Krauss and D. A. Warburton. 2006. *Ancient Egyptian Chronology*. Brill. Leiden and Boston.

Jung, R. 2008. "Die mykenische Keramik von Tell Kazel (Syrien)." *Damaszener Mitteilungen* 15: 147–218.

Jung, R. 2010. "La céramique de typologie mycénienne de Tell Tweini." In M. Al-Maqdissi, K. van Lerberghe, J. Bretschneider and M. Badawi (eds.),

Tell Tweini: onze campagnes de fouilles syro-belges (1999–2010), 115–121, Documents d'Archéologie Syrienne, Ministère de la Culture, Direction Générale des Antiquités et des Musées. Damascus.

Kaniewski, D., E. Paulissen, E. van Campo, M. Al-Maqdissi, J. Bretschneider and K. van Lerberghe. 2008. "Middle East coastal ecosystem response to middle-to-late Holocene abrupt climate changes." *Proceedings of the National Academy of Sciences (PNAS)* 105: 13941–13946.

Kaniewski, D., E. Paulissen, E. van Campo, H. Weiss, T. Otto, J. Bretschneider and K. van Lerberghe. 2010. "Late Second-Early First Millennium BC abrupt climate changes in coastal Syria and their possible significance for the history of the Eastern Mediterranean." *Quaternary Research* 74: 207–215.

Kaniewski, D., E. van Campo, K. van Lerberghe, T. Boiy, K. Vaansteenhuyse, G. Jans, K. Nys, H. Weiss, C. Morhange, T. Otto and J. Bretschneider, J. 2011. "The Sea Peoples, from Cuneiform Tablets to Carbon Dating." *PLoS ONE* 6(6): e20232, doi:10.1371.

Kay, P. A., and D. L. Johnson. 1981. "Estimation of the Tigris-Euphrates streamflow from regional palaeoenvironmental proxy data." *Climatic Change* 3: 251–263.

Killebrew, A. E. 2005. *Biblical Peoples and Ethnicity. An Archaeological Study of Egyptians, Canaanites, Philistines and Early Israel 1300–1100 BCE*. Society of Biblical Literature, Archaeology, and Biblical Studies. Atlanta.

Lemcke, G., and M. Sturm. 1996. "^{18}O and trace elements measurements as proxy for reconstruction of climate change at Lake Van (Turkey)." In H. N. Dalfes, G. H. Kukla and H. Weiss (eds.), *Third Millennium BC: Climate Change and Old World Collapse*, Springer Verlag, 653–678. Berlin.

Lewis, N. N. 1987. *Nomads and Settlers in Syria and Jordan, 1800–1980*. Cambridge University Press. Cambridge.

Lückge, A., H. Doose-Rolinski, A. A. Khan, H. Schulz and U. von Rad. 2001. "Monsoonal variability in the northeastern Arabian Sea during the past 5000 yr: Geochemical evidence from laminated sediments." *Palaeogeography, Palaeoclimatology, Palaeoecology* 167: 273–286.

Migowski, C., M. Stein, S. Prasad, J. F. W. Negendank and A. Agnon. 2006. "Holocene climate variability and cultural evolution in the Near East from the Dead Sea sedimentary record." *Quaternary Research* 66: 421–431.

Monchambert, J. Y. 2004. *La céramique d'Ougarit. Campagnes de fouilles*

1975 et 1976. Ras Shamra-Ougarit Series 15. Editions recherche sur les civilisations. Paris.

Mountjoy, P. A. 2004. "Miletos: A Note." The Annual of the British School at Athens 99: 189–200.

Neumann, J. 1985. "Climate change as a topic in the classical Greek and Roman literature." *Climate Change* 7: 441–454.

Neumann, J., and S. Parpola. 1987. "Climatic change and the eleventh-tenth-century eclipse of Assyria and Babylonia." *Journal of Near Eastern Studies* 46: 161–182.

Neumann, F. H., E. J. Kagan, M. J. Schwab and M. Stein. 2007. "Palynology, sedimentology and palaeoecology of the late Holocene Dead Sea." *Quaternary Science Reviews* 26: 1476–1498.

Nougayrol, J., E. Laroche, C. Virolleaud and C. Schaeffer. 1968. *Ugaritica V, Nouveaux textes accadiens, hourrites et ugaritiques des archives et bibliothèques privées d'Ugarit.* Mission de Ras Shamra 16. Librairie Orientaliste Paul Geuthner. Paris.

Otto, A. 2007. *Alltag und Gesellschaft zur Spaetbronzezeit: eine Fallstudie aus Tall Bazi.* Subartu XIX. Brepols. Turnhout.

Prentice, I. C., J. Guiot, B. Huntley, D. Jolly and R. Cheddadi. 1996. "Reconstructing biomes from palaeoecological data: a general method and its application to European pollen data at 0 and 6 ka." *Climate Dynamics* 12: 185–194.

Reuveny, R. 2007. "Climate change-induced migration and violent conflict." *Political Geography* 26: 656–673.

Schilman, B., M. Bar-Matthews, A. Almogi-Labin and B. Luz. 2001. "Global climate instability reflected by Eastern Mediterranean marine records during the Late Holocene." *Palaeogeography, Palaeoclimatology, Palaeoecology* 176: 157–176.

Schilman, B., A. Ayalon, M. Bar-Matthews, E. J. Kagan and A. Almogi-Labin. 2002. "Sea-Land paleoclimate correlation in the Eastern Mediterranean region during the Late Holocene." *Israel Journal of Earth Sciences* 51: 181–190.

Shaw, I. 2000. *The Oxford History of Ancient Egypt.* Oxford University Press. Oxford.

Singer, I. 1999. "A Political History of Ugarit." In W. G. E. Watson and N. Wyatt (eds.), *Handbook of Ugaritic Studies, Handbuch der Orientalistik, Erste Abteilung*, 603–733. Leiden.

Singer, I. 2000. "New evidence on the end of the Hittite Empire." In E. D. Oren (ed.), *The Sea Peoples and their World: A Reassessment*, University of Pennsylvania Museum, 21–33. Philadelphia.

Singh, G., R. J. Wasson and D. P. Agrawal. 1990 "Vegetational and seasonal climatic changes since the last full glacial in the Thar Desert, northwestern India." *Review of Palaeobotany and Palynology* 64: 351–358.

Staubwasser, M., and H. Weiss. 2006. "Holocene climate and cultural evolution in late prehistoric-early historic West Asia." *Quaternary Research* 66: 372–387.

Tarasov, P. E., R. Cheddadi, J. Guiot, S. Bottema, O. Peyron, J. Belmonte, V. Ruiz-Sanchez, F. Saadi and S. Brewer. 1998. "A method to determine warm and cool steppe biomes from pollen data: Application to the Mediterranean and Kazakhstan regions." *Journal of Quaternary Science* 13: 335–344.

Vansteenhuyse, K. 2010. "La céramique du chantier A." In M. Al Maqdissi, K. van Lerberghe, J. Bretschneider and M. Badawi (eds.), *Tell Tweini: onze campagnes de fouilles syro-belges (1999–2010)*, 115–121, Documents d'Archéologie Syrienne, Ministère de la Culture, Direction Générale des Antiquités et des Musées. Damascus.

Warburton, D. 2003. "Love and War in the Later Bronze Age: Egypt and Hatti." In R. Matthews and C. Roemer (eds.), *Ancient Perspectives on Egypt*, UCL Press, 75–100. London.

Weninger, B., and R. Jung. 2009. "Absolute Chronology of the End of the Aegean Bronze Age." In S. Deger-Jalkotzy and A. Baechle (eds.), *LH IIIC Chronology and Synchronisms III: LH IIIC Late and the Transition to the Early Iron Age*, Proceedings of the International Workshop at the Austrian Academy of Sciences, February 23–24, 2007, Austrian Academy of Sciences Press (VÖAW), 373–416. Vienna.

Weiss, B. 1982. "The decline of the Late Bronze Age civilization as a possible response to climate change." *Climatic Change* 4: 173–198.

Whincop, M. 2007. "The Iron Age II at Tell Nebi Mend: Towards an explanation of ceramic regions." *Levant* 39: 185–212.

Woolley, G. L. 1958. *Alalakh: An Account of the Excavations at Tell Atchana in the Hatay*. Oxford University Press. Oxford.

Yon, M. 1989. "The End of the Kingdom of Ugarit." In W. A. Ward and M. Sharp Joukowsky (eds.), *The Crisis Years: The 12th Century BC. From*

Beyond the Danube to the Tigris, Kendall/Hunt Publishing Company, 111–122. Dubuque, Iowa.

Yon, M. 2006. *The City of Ugarit at Tell Ras Shamra*. Eisenbrauns. Winona Lake, Indiana.

Long Term or Short Term? Climate Change and the Demise of the Old Kingdom

Miroslav Bárta

Abstract

There has been a long-standing debate within Egyptology whether the decline of the Old Kingdom (dated to about 2200 BCE) was due to internal and/or external factors or a combination of both of them. Equally important has been the question of the duration of this process. The present study shall make the case that the external environmental factors which contributed to the demise of the Old Kingdom state can be discerned to have happened much earlier than traditionally supposed. There is now sufficient evidence which demonstrates a continual adaptation of the ancient Egyptian society to a changing/worsening environment over a period which lasted almost two centuries. This climate change went hand in hand with several deep internal processes in the society that contributed significantly to the eventual demise of the Egyptian empire. The analysis will use iconographic material from Old Kingdom tombs, evidence from archaeological excavations, interpretation of unique finds of beetles in Abusir and a model describing the functioning of the administrative system of the Egyptian state. The emerging picture will show that the collapse of the Old Kingdom was a long process triggered by several internal crises in combination with a change in climate.

The ultimate demise of the Egyptian Old Kingdom around 2150 BC has been attracting the attention of many scholars for quite some time. It was by no means a unique phenomenon: societies and cultures of the same period experienced an economic decline and social crisis due to a climate change that affected the entire northern hemisphere – Central and Western Europe, the Mediterranean, Mesopotamia, North Africa and India (for an overview,

FIGURE 1. *Desert hunt in the mastaba of Nyankhkhnum and Khnumhotep in Saqqara (Courtesy Oxford Expedition to Egypt, P. J. Scremin).*

see Dalfes, Kukla, Weiss 1997). Given the global impact and similar character of development, a detailed comparative study of individual cases may prove useful.

I shall make the case here that the external environmental factors which contributed to the demise of the Old Kingdom state can be discerned much earlier than traditionally supposed. Earlier authors have ascribed the Old Kingdom's collapse to an abrupt climate change and to a rapid decline in the annual Nile floods (Bell 1970, 1971; Schenkel 1978; Staubwasser and Weiss 2006). I shall argue that there is now sufficient evidence which demonstrates the continual adaptation of the ancient Egyptian society to a changing/worsening environment over a period which lasted almost two centuries. In doing so, I will be relying on independent sets of evidence provided by the interpretation of iconographic material from Old Kingdom tombs (1), archaeological excavations and a study of settlement drift on the island of Elephantine (2), recent unique finds of beetles in Abusir (3), and an analysis of the administrative system of the state (4).

It is crucially important that we establish the moment when climate change actually began to interfere with the ancient Egyptian society and in what ways precisely.

Nyuserra, introduction of the desert theme and the end of the Early Bronze Age III period

Ancient Egyptian tombs feature a large array of scenes relating to the daily life of the ancient Egyptians. They depict principal agricultural activities, religious scenes, family and leisure activities and settings for the nobility of the day (Harpur 1987). Many scenes are relevant to the reconstruction of the natural environment. One of the first scientists who paid attention to the iconography of animals and their significance for the reconstruction of climate changes in the past was Karl Butzer in his 1959 study. Most significant among these are scenes depicting desert motifs, specifically hunting in the desert. This group of scenes is rather limited in both occurrence and date. All but two of the scenes listed below (nos. 1 and 2) are attested from the mid-Fifth Dynasty onwards and can be found in the residential cemeteries of ancient Memphis (see also Herb and Föster 2009; McFarlane and Mourad 2013). The list is rather brief and includes the following non-royal tombs:

1. Mastaba of Nefermaat and Atet (reign of Sneferu) (Petrie 1892: pl. IX; Harpur 2001: 77, 189–191, pls. 11–12);
2. Mastaba of Rahotep and Nofret (reign of Sneferu) (Petrie 1892: pls. XVI–XVIII);
3. Mastaba of Nyankhkhnum and Khnumhotep (reign of Nyuserra) (Mousa and Altenmüller 1977: 109–110, pl. 40);
4. Mastaba of Ti (reign of Nyuserra) (Épron 1939: pls. XCV and CXXVII);
5. Mastaba of Pehenuika (reign of Djedkara) (Lepsius 1897: vol. II, pl. 46);
6. Mastaba of Ptahhotep II called Fefi (late Fifth Dynasty) (Davies 1898, pl. XXXII);
7. Mastaba of Raemkai (late Fifth Dynasty) (Hayes 1953: 98ff, fig. 56);
8. Mastaba of Tjefu Ptahhotep (late Fifth Dynasty) (Hassan 1975b: 112–113, fig. 59, pl. LXXXVIc);
9. Mastaba of Fetekti (late Fifth Dynasty) (Bárta 1999: 103–104, fig. 3.17);
10. Mastaba of Isesimerynetjer Nimaatra (late Fifth Dynasty) (Roth 1995: 132–133, pls. 95–97, 189);
11. Mastaba of Akhetmerunesut (late Fifth Dynasty) (Smith 1946: 198–199);

12. Mastaba of Seshemnefer (late Fifth Dynasty) (Junker 1953: 153–157, fig. 63);
13. Mastaba of Mereruka Mereri (reign of Pepy I) (Duell 1938: pl. 25);
14. Mastaba of Meryteti Meri (reign of Pepy I) (Kanawati and Abder-Raziq 2004: 24–26, pl. 46);
15. Mastaba of Sankhuptah (Sixth Dynasty)(Kanawati and Abder-Raziq 1998: 51–57, pls. 1b, 26–30, 71–73);
16. Mastaba of Nebkauhor Idu (reign of Pepy I) (Hassan 1975a: 28–32, figs. 12–14, pls. XIX–XX).

Most of the mastabas featuring motifs of desert hunting are located in Saqqara; some come from the nearby site of Abusir and some from Giza and Meidum. The motif becomes more prominent during the reign of Nyuserra and is used continuously through the reign of Pepy I. From the time prior to the Fifth Dynasty, the only known examples of the theme occur in two tombs in Meidum, dating to the early Fourth Dynasty. The reason for these scenes appearing in the mastabas in Meidum is most likely that these tombs belonged to blood members of the royal family (unlike the later examples) and as such contained depictions of the tomb owners' leisure activities. In the royal sphere, the earliest attestations of the desert motif originate from the mortuary complexes of Userkaf from the beginning of the Fifth Dynasty (Ćwiek 2003; el-Awady 2009).

The analysis of the content of these scenes indicates a rather uniform approach to their composition. Most of them feature schematic depictions of desert landscapes with undulating sandy terrain painted in hues of yellow colour. We can observe hunters with hunting dogs attacking and tearing apart desert game. The hunters are also lassoing desert animals. Sometimes there is a lion attacking a bull, biting at its muzzle to suffocate it. In the background there are minor depictions of animals such as hedgehogs and juvenile desert mammals, and low shrubs indicating sparse desert vegetation. Thanks to the fact that the desert animals are represented in a quite realistic manner, it is possible to identify them with a high degree of precision. They are as follows: the lion (*Panthera Leo*), wild bull (*Bos africanus*), hartebeest (*Alcelaphus buselaphus*), Isabela gazelle (*Gazela Isabella*), oryx antelope (*Oryx algazel dammah*), addax (*Addax nasomaculata*), dorcas gazelle (*Gazella dorcas*), capricorn (*Capra nubiana*), barbary sheep (*Ammotragus lervia*), hyena (*Hyaena hyaena*), wild dog (*Canis lupaster*), domestic hunting dog (*Canis lupaster domesticus*), hare (*Lepus aegyptius*), Saharan striped polecat (*Ictonyx libyca*), desert jerboa (*Jaculus jaculus*), fennec fox (*Vulpes zerda*) and hedgehog (*Erinaceus auritus*).

One may wonder if the scenes were socially exclusive, but this is most likely not the case after the end of the Fourth Dynasty. Amongst the tomb owners we find not only viziers and members of their families (nos. 1, 3, 5, 13), but also lower ranking officials (compared to the social milieus of the specific cemeteries) of the central administration, scribes and priests. This makes it possible to suppose that the desert hunt motif was a theme accessible to the members of several layers of society who had the potential to build and decorate tombs. At the same time, it is not far-fetched to suggest that the desert was a phenomenon observed on a daily basis and thus a subject familiar to the Egyptians. Evidently, the desert motif became an element of ever increasing importance from the time of Nyuserra.

Elephantine settlement evidence and the Palermo Stone Nile flood heights

Interesting evidence of the fluctuations of the Nile was provided by the settlement drift observed during archaeological excavations on the island of Elephantine at Aswan. There was settlement on the island for the better part of the third millennium BC and thus it may be used as an indicator of the oscillation of the Nile over the centuries. From the data available we know that there was a significant drop in the height of the Nile during the second half of the third millennium BC. Regarding settlement dated to the first half of the First Dynasty, we know that the building line lies at a height of 96 m above sea level. The Second Dynasty fortress occupies a location of 94.50–94.00 m above sea level; and during the Fourth and Fifth Dynasties the flood line dropped below 94 m above sea level, whereas the Sixth Dynasty city houses are located some 90.80–91.30 m above sea level (Ziermann 1995: 138–140 and fig. 15; Seidlmayer 2001: 81 and 90, table 7). This evidence indicates that from the beginning to the end of the Old Kingdom period, the Nile flood became less and less substantial, which made it possible for permanent settlements to be established at relatively very low locations with respect to the Nile.

Similar evidence is provided by the Palermo Stone and its related fragments, originally representing a stone block containing annals of the Old Kingdom kings. Seventy-two entries can be found for the annual height of the Nile flood, beginning with the reign of Djer of the First Dynasty up to the first half of the Fifth Dynasty (reign of Neferirkara) (for a summary of the Palermo Stone and the history of associated fragments, see Wilkinson 2000). It is interesting to observe that there were reasonably high Nile floods down to the Fourth Dynasty, whereas during the latter half of the Fourth and during the Fifth Dynasty, the floods diminished in height.

With regard to the Elephantine evidence, it is worth mentioning that the Egyptians experienced some exceptional flood levels during the First Dynasty: for example, in the reigns of Djer (3.15 m and 3.20 m), Den (2.60 m and 2.71 m) and Andjib (up to 4.22 m). Some exceptionally high floods are also attested from the reign of Sneferu and Khufu from the beginning of the Fourth Dynasty, which may be important for our understanding of why the largest Old Kingdom royal tombs were built precisely at this time.

Beetles from Abusir South

The Abusir beetles represent a unique set of palaeoecological data from the Old Kingdom. They were discovered in two different archaeological contexts: in the burial chamber of the vizier Qar, dated to the reign of Teti, and in a ritual deposit shaft located within the tomb complex of one of Qar's sons, Inti (Shaft E), dated to the early reign of Pepy II (Bárta and Bezděk 2008).

From the burial chamber of Qar originates, together with some other species, an example from the family Carabidae: *Scarites* (sbg. *Scallophorites*) sp. the genus *Scarites* (Fabricius 1775), which inhabits sandy soils. This type of beetle is common on seashores and the banks of salt lakes. In the material from the tomb of the vizier Qar, only one incomplete specimen was found. Members of the family Carabidae *Poecilus* (sbg. *Ancholeus*) *pharao* (Lutshnik 1916) were found in the mummification deposit from the complex of Inti. *Poecilus pharao* is a species endemic to Egypt (e.g. Bousquet 2003). Like other members of the subgenus *Ancholeus* (Dejean 1828), this species is associated strictly with saline habitats (Alfieri 1976; Schatzmayr 1936).

In our specific case it is interesting to note that there is evidence already from the early Sixth Dynasty that indicates that there was desert to the west of the Lake of Abusir. This lake existed on the eastern edge of the Abusir South cemeteries and constituted until the Fifth Dynasty a major access route connecting the settlement of Memphis in the east with the Abusir and North and Central Saqqara necropoleis.

The genus *Scarites* identified in the burial chamber of Qar has a preference for sandy soils and is common on seashores and the banks of salt lakes. A similar conclusion may be drawn from Shaft E in the complex of Inti. It is certain that the pottery found ritually buried in Shaft E was primarily used for mummification. The exclusive saline habitat of *Poecilus* (*Ancholeus*) *pharao* makes it clear that the mummification was carried out in a dry salty environment, most likely in the desert and very close to the present tomb location. Thus it is possible to suppose that Inti's body mummification was carried out in a desert environment. What follows is that the process of mummification was in both cases (Qar and Inti) carried out near a lake (a sufficient amount of water was understandably one of the prerequisites for any mummification installation) and at the same time in a salt-rich biotope. Thus we posses rather reliable evidence that by this time (the reign of Teti) the desert encroached quite closely on the Nile valley.

I wish to emphasise that this evidence seems to appear in the reign of Teti. This may indicate that a major climate decline began much earlier than previously thought, i.e. at least 180 years before the official end of the Old Kingdom. The beetle finds of Abusir fully support the notion of a continuous worsening of the climate conditions at the end of the Old Kingdom.

Administration

During the third millennium BC, Egypt experienced the rule of several personalities who profoundly changed their time and society. To name but a few: Hor Aha, Den, Netjerikhet (known as Djoser), Sneferu, Sahura, Nyuserra, and Djedkara or Unas. Of these, Nyuserra, a ruler of the Fifth Dynasty, was undoubtedly one of the most outstanding.

The reign of Nyuserra marks a profound change in the history of Old Kingdom Egypt. According to the latest relative scheme, he reigned between 2402 and 2374 BC, most likely for 11 to 31 years (Verner 2006: 139). His successor to the throne, King Menkauhor, reigned only briefly and the next king, Djedkara Isesi, took the throne around 2365 BC and died around 2322 BC. We may note that the date for Djedkara's ascension to the throne is very close to the date proposed for the end of the Early Bronze Age III and the beginning of the Early Bronze Age IV in Syria and Palestine, i.e. around 2350 BC.

From Nyuserra's reign onwards we posses many indications that there were continuous attempts on the part of the government to readjust the state's approach to the rapidly changing situation within the society. There may have been various reasons for this sudden dynamic development. A closer look at this development and its characteristics shows, however, that the factors are quite similar to those considered currently to be critical for the demise of the Old Kingdom. Moreover, as suggested by iconographic evidence and the entomological evidence from Abusir, the beginning of this development more or less coincided with a significant climate change.

During the reign of Nyuserra, the wealth of individual officials increased enormously, as indicated, among other things, by the high number of private landholdings. We may also discern a strong tendency in the society towards nepotism; for example, many offices became hereditary in principle. Nyuserra (and his successors to the throne) tried hard, on the one hand, to limit the increasing independence of wealthy officials with large private landholdings (Helck 1994) and, on the other, to further empower a few persons it was felt they could trust. Nyuserra and his successors followed a line leading to a more centralised rule that it was hoped would curb centrifugal tendencies in the state administration. This is demonstrated by a reduced number of high and trusted officials, a trend discernible in the long series of titles of the viziers and the highest officials of the time and indicating that these officials were accumulating many functions and offices within the state administration. It is also worth noting that from Nyuserra onwards, the kings often pursued a policy of political marriages of their daughters to these officials, thus ensuring their loyalty (Bárta 2005 and 2013a).

Interestingly, contradictory centralist and centrifugal tendencies found their almost immediate reflection in the material culture. As a direct result of centralist policy, the development of wealthy and lavishly decorated multi-room tombs for the highest officials of the state becomes strikingly noticeable. These tombs represent the unique status of their owners. So-called family tombs also developed during this period, built by officials of significantly lesser status. These modest mud brick tombs usually comprised a simple cult chapel and several shafts for the burials of the family members.

The obvious purpose of these tombs was to indicate family ties and inherited rights to certain positions within the central administration.

At the same time, the king created the office of the 'Overseer of Upper Egypt', probably because he needed to practise more efficient control over both the remoter regions of the country and the officials there who experienced the centrifugal tendencies most of all (Pardey 1976: 152; Strudwick 1985: 142–144 for the oldest office holder). The rapidly worsening situation in the administration of the country and the central government's decreasing ability to enforce the application of central decisions in the provinces and the remoter regions, and to collect taxes and control professional careers in the administration, led in essence to every king after Nyuserra having to impose a new set of administrative reforms. These were to counteract the contemporary development during each respective reign, but we can only speculate as to what extent the reforms really impacted on the natural flow of things (Kanawati 1980).

Thus, we can identify and examine two opposing trends during Nyuserra's reign: one towards the usurpation of more and more power and influence by officials and their families, and another towards a limiting of the power of these officials by the king. That this development was effectively initiated under the reign of Nyuserra is corroborated by R. Müller-Wollermann's study. According to her research, these trends represent a crisis of participation that appeared precisely at this time (Müller-Wollermann 1986: 69–70 and 96–97).

From the reign of Nyuserra and the tomb of his favourite official, Ptahshepses, originates a unique relief which depicts the transportation of a Syro-Palestinian amphora with two handles (Bárta 2010b: 120, fig. 10). These vessels were frequently imported to Egypt as containers for liquids such as wine and oils. They are typical of the Early Bronze Age II and III periods and do not seem to occur at later times. The iconographic attestation of this artefact from Abusir is probably the last of its kind as metallic ware disappears by the end of the Early Bronze Age III, which means very shortly after the reign of Nyuserra or Djedkara (Greenberg and Porat 1996; Bárta 2009). For possibly the closest date for this event, we may use the date of ascension of Djedkara or even the end of Nyuserra's rule. It has traditionally been supposed that the Early Bronze Age III was roughly parallel with the Fourth and Fifth Egyptian Dynasties. It can be presumed that the Early

Bronze Age III end was parallel to the reign of Djedkara, mainly because of historical similarity and context of the dates both in Egypt and in Syria and Palestine.

Around this date, Syria and Palestine entered troubled times that were dominated by a vital change in the subsistence economy. The culture of city states disappeared rather quickly, the process of urbanisation came to a halt and most of the area became populated with nomads (Weiss 2001). There are three alternative explanations for this development: a sudden drop in annual precipitation, external influences in the form of a wave of Amorite warriors coming from the north, and the destruction of individual city states by invaders. This major setback was used by the Egyptians who despite the economic decline of their homeland seem to have organised an unusually high number of military campaigns to Palestine. This is documented in the reliefs which depict besieged cities in some late Fifth Dynasty tombs at Saqqara and Deshasha, and in the military account of Weni (Ben-Tor 1992: 122–125; Bárta 2010a).

To counteract the slow disintegration of the centralised state and the ever increasing power of the high officials, Djedkara embarked on a series of reforms focusing on administration in the provinces, which continued throughout the Sixth Dynasty. The character of these reforms is such that it is difficult to escape the conclusion that their primary purpose was to curb the increasing independence of the far away regions south of Memphis. From the late Fifth Dynasty onwards, these regions became places where local wealthy families extended their influence, such as the famous Weni family of Abydos (Richards 2002). It is interesting to see that every king following Djedkara considered it important to come up with new reforms which always attempted to respond in specific ways to the current development in the south.

The policies of the kings regarding the remote yet economically important nomes provide evidence for the generally accepted assumption that the power and authority of the central government were gradually being eroded. For example, the creation of the office of the Overseer of Upper Egypt by Djedkara, if not earlier; the installation of two viziers, one in Memphis and one in Upper Egypt by Teti, and the political marriages of Pepy I to the daughters of an Abydos monarch demonstrate that the king was trying to re-establish his influence over the remote provinces.

Conclusions

The above observations have been made on the basis of iconographic evidence from tombs, archaeological research, and a survey of the administration, economy and political development of the ancient Egyptian state beginning in the reign of Nyuserra in the Fifth Dynasty and continuing during the Sixth Dynasty. All this evidence points towards a series of single historical events combined with gradual processes which inevitably led to the decline of the Old Kingdom state. In light of this information, we should see this decline (but not necessarily the collapse) of the Old Kingdom as a continual process inherent in the development of the society. This process lasted almost two centuries and demonstrated itself through several endogenous processes that were accelerated by a major climatic change at the end (for this climate change, see most recently Hassan 1997; Issar and Zohar 2004). Based on the evidence from Abusir South, it may be suggested that the worsening of the climate can be observed as early as the early Sixth Dynasty, i.e. the reign of Teti. Given the iconographic evidence of the desert hunt scenes we may suppose that this deterioration began even earlier, as early as during the reign of Nyuserra (Cílek, Bárta, Lisá Pokorná et al. 2012; Bárta 2013b). Most recent study of the Giza depositional processes indicates that climate instability characterised by devastating spells of rain occurred already during the second half of the Fourth Dynasty (Butzer, Butzer and Love 2013).

Compared to the previous, more traditional approach which related the demise of the Old Kingdom to low Nile floods towards the period near the end of the Old Kingdom, we now know that the climatic change was not rapid but continuous, lasting for the greater part of the Sixth Dynasty. The ^{18}O of Lake Van as well as ^{18}O and ^{13}C of the speleothems of the Galilee caves from about 4500 BP show clearly that they tend to become heavier and indicate that the climate was getting warmer and drier, thus causing a major setback in the Syropalestinian region. This eventually led to the abandonment of cities and a return to a semi-nomadic way of life. At the same time, it allowed for larger-scale migrations across large distances. It has been observed that this major climatic decline lasted several centuries, began perhaps as early as 2400 BP and had a significant impact on almost all cultures in the northern hemisphere (Issar and Zohar 2004: 131ff). A recent study by Bar-Matthews and Ayalon (2011) demonstrates clearly the frequency of climate fluctuations in the eastern Mediterranean during the Holocene. Their study, based on a careful examination of speleothems from Soreq Cave, Israel, demonstrates the existence of cycles, each lasting of about 1500 years, with changes of about 400 mm in annual rainfall. From our perspective the most interesting period of drought established by this research is the one around 4200–4050 BP which corresponds very well with the decline of the Old Kingdom. Thus it is tempting to establish a causal relationship between the administrative development of the state and the encroaching depredation of the climate, at least in ancient Egypt. The isotope curve of the Lake of Van indicates that after 4000 BP the climate began to get cooler and more humid again (Issar 1998: 120), a trend which roughly coincides with the beginning of the Middle Kingdom state.

The large number of beetle species discovered in the tomb complexes of Qar and Inti in Abusir can be associated with a dry and salty environment. This calls for some specific comments, especially with regard to the decline of the Old Kingdom. The reasons for this decline have been attracting closer attention since the 1980s when a stimulating publication by R. Müller-Wollerman was made available (Müller-Wollermann 1986). Given the time of its compilation, it is no wonder that she dealt exclusively with the social and political phenomena of the Old Kingdom demise during the 23rd century BC. Egyptology in those days paid only limited attention to environmental issues which, in general, have not been studied systematically during the 1980s. Moreover, from the current perspective, most of the critical factors identified by Müller-Wollermann can be detected considerably earlier, i.e. as early as the second half of the Fifth Dynasty (Bárta 2005). The major factors involved in the gradual erosion of the centralised state may be enumerated as follows:

- A crisis of identity: the way that the ruling group is accepted
- A crisis of participation: who takes part in the state administration and in what form
- A crisis of the ability of the executive to control the state's administration and economy
- A crisis of legitimacy: the authority and ability to enforce decision making

- A crisis of distribution: the inability to manage the redistribution of economic resources
- A crisis of land-ownership, due to the fact that more and more land was being transferred from the state to the funerary, non-taxable domains, whose only raison d'être was to provide economic means for the upkeep of both royal and non-royal cults, revenues from which an ever increasing number of officials participating in them were sustained.

On a general level, we could also call this economic and social decline a *distribution of power and rule* crisis and the general failure of the state machine to set and keep the *norms and standards in every possible walk of life*. To these aspects we may now add one more critical factor: the significant and long-term worsening of the climate.

Over the past few years it has become clear that the decline of the Old Kingdom had not only its internal reasons brought about by the genesis of this early primary state, but also several other factors, climate change being the principal one. Comparative evidence shows convincingly that not only Egypt but also large areas of the Middle East, and even most of the European cultures of the time, were affected by the sudden climate change which meant a significant drop in the annual precipitation and the desiccation of vast areas. This trend lasted for almost two centuries and had a profound impact on the subsistence strategies of the people of those generations, as well as on the political systems in the areas indicated above.

It is self-evident that throughout the history of civilisations, individual states and cultures had to cope with the impact of natural elements. At the same time, there was always an increasing inner weakness as a consequence of major social changes, the disintegration of the principal administrative bodies and the decline of the societies, and their particularisation and inability to provide general solutions for the large and global problems that surrounded them. When looking at some of the global environmental changes we witness today, maybe this is just one small lesson that we may take from history. What is at stake? Not only the hectares of agricultural land and villages and cities being swallowed up by the rising sea. Above all, at least to me, this is a development that can potentially lead to a major crisis of civilisations and of individual societies.

Chronological note

For the purposes of absolute dates, the following dating system has been applied (Hornung, Krauss and Warburton 2006: 490–91): First Dynasty 2900–2730; Second Dynasty 2730–2590; Third Dynasty 2592–2544; Fourth Dynasty 2543–2436; Fifth Dynasty 2435–2306; Sixth Dynasty 2305–2118.

The work on this project was supported by a grant provided by the Grant Agency of the Czech Republic no. P405/11/1873.

Bibliography

Alfieri, A. 1976. "The Coleoptera of Egypt." *Mémoires de la Société Entomologique d'Egypte* 5: 1–361.

el-Awady, T. 2009. *Abusir XVI. Sahure. The pyramid causeway: History and decoration program in the Old Kingdom*. Prague: Charles University in Prague.

Bar-Matthews, M., and A. Ayalon. 2011. "Mid-Holocene climate variations revealed by high-resolution speleothem records from Soreq Cave, Israel and their correlation with cultural changes." *The Holocene* 21(1): 163–171.

Bárta, M. 1999. *Abusir V: The cemeteries at Abusir South I*. Set Out. Prague

Bárta, M. 2005. "Architectural Innovations in the development of the non-royal tomb during the reign of Nyuserra." In P. Jánosi (ed.), *Structure and Significance: Thoughts on ancient Egyptian Architecture*, Akademie Verlag, 2005, 105–130; Vienna.

Bárta, M. and A. Bezděk. 2008. "Beetles and the decline of the Old Kingdom: Climate change in ancient Egypt." In M. Bárta and H. Vymazalová (eds.), *Chronology and archaeology in ancient Egypt (The third millennium B.C.). Proceedings of the Conference Held in Prague (June 11–14, 2007)*, Czech Institute of Egyptology, 214–222. Prague.

Bárta, M. 2009. "A mistake for the afterlife?" In T. I. Rzeuska and A. Wodzinska (eds.), *Studies on Old Kingdom Pottery*, Neriton, 43–48. Warsaw.

Bárta, M. 2010a. "Borderland Dynamics in the Era of the Pyramid Builders in Egypt." In W. I. Zartman (ed.), *Understanding Life in the Borderlands. Boundaries in Depth and in Motion*, University of Georgia Press, 21–39. Athens, Georgia.

Bárta. M. 2010b. "Biblical Archaeology and Egypt During the EB III and EB IV: New Connections" In T. E. Levy (ed.), *Historical Biblical Archaeology and the Future. The New Pragmatism*, Equinox, 103–125. London and Oakville, Connecticut.

Bárta, M. 2013a. "Kings, viziers and courtiers: Executive power in the third millenniun B.C." In J. C. Moreno García (ed.), *Ancient Egyptian Administration* (Handbuch der Orientalistik), Leuven: Peeters, 153–175.

Bárta, M. 2013b. "In mud forgotten: Old Kingdom palaeoecological evidence from Abusir." *Studia Quaternaria* 30, *forthcoming*.

Bell, B. 1970. "The Oldest Records of the Nile Floods." *Geographical Journal* 136: 569–573.

Bell, B. 1971. "The Dark Ages in Ancient History. I. The First Dark Age in Egypt." *American Journal of Archaeology* 75(1): 1–26.

Ben-Tor, Amnon. 1992. *The archaeology of ancient Israel*. New Haven [Tel Aviv]: Yale University Press; Open University of Israel.

Bousquet, Y. 2003. "Tribe Pterostichini Bonelli, 1810". In I. Löbl and A. Smetana (eds.), *Catalogue of Palaearctic Coleoptera. Volume 1. Archostemata – Myxophaga – Adephaga*, Stenstrup: Apollo Books, 469–521.

Butzer, K. 1959. *Studien zum vor-und frühgeschichtlichen Landschaftswandel der Sahara. III. Die Naturlandschaft Ägyptens während der Vorgeschichte und der Dynastischen Zeit*. Akademie der Wissenschaften und der Literatur. Bonn.

Butzer, K., E. Butzer and S. Love 2013. "Urban geoarchaeology and environmental history at the Lost City of the Pyramids, Giza: Synthesis and review." *Journal of Archaeological Science* 40 (8): 3340–3366.

Cílek, V., M. Bárta, L. Lisá, A. Pokorná, L. Juřičková, Vl. Brůna, Abdel Moneim, A. Mahmoud, A. Bajer, J. Novák and J. Beneš, 2012. "Diachronic development of the Lake of Abusir during the third millennium BC, Cairo, Egypt." *Quaternary International* 266 (2012): 14–24.

Ćwiek, A. 2003. *Relief decoration in the royal funerary complexes of the Old Kingdom: Studies in the development, scene content and iconography*, PhD. Dissertation, Institute of Archaeology, Faculty of History, Warsaw University.

Dalfes, H. N., G. Kukla and H. Weiss. 1997. *Third Millennium BC Climate Change and Old World Collapse*. NATO ASI Series. Springer Verlag. Berlin and Heidelberg.

Davies, Norman de Garis 1898. *The mastaba of Ptahhetep and Akhethetep at Saqqareh*. 2 vols, *Memoir / Archaeological survey of Egypt*. London.

Duell, P. 1938. *The mastaba of Mereruka*. The University of Chicago Press. Chicago.

Épron, L. 1939. *Le Tombeau de Ti*. Impr. de l'Institut français d'archéologie orientale. Cairo.

Greenberg, R., and N. Porat. 1996. "A Third Millennium Levantine Pottery Production Center: Typology, Petrography, and Provenance of the Metallic Ware of Northern Israel and Adjacent Regions." *Bulletin of the American Schools of Oriental Research* 301: 5–24.

Harpur, Y. 1987. *Decoration in Egyptian tombs of the Old Kingdom: Studies in orientation and scene content*. KPI. London.

Harpur, Y. and P. J. Scremin. 2001. *The tombs of Nefermaat and Rahotep at Maidum: Discovery, destruction and reconstruction*. Oxford Expedition to Egypt. Prestbury, Cheltenham.

Hassan, F. 1997. "Climate, Famine, and Chaos. Nile Floods and Political Disorder in Early Egypt." In H. N. Dalfes, G. Kukla and H. Weiss (eds.), *Third millennium BC climate change and old world collapse*. Springer. Berlin and London.

Hassan, S. 1975a. *The Mastaba of Neb-Kaw-Her. Excavations at Saqqara, 1937–1938, Volume I*. General Organisation for Government Printing Office. Cairo.

Hassan, S. 1975b. *Excavations at Saqqara, 1937–1938, vol. 2, Mastabas of Nyankh-Pepy and Others*. General Organisation for Government Printing Office. Cairo.

Hayes, W. Ch. 1953. *The scepter of Egypt: A background for the study of the Egyptian antiquities in The Metropolitan Museum of Art*. Harper in co-operation with the Metropolitan Museum of Art. New York.

Helck, W. 1986. *Politische Gegensätze im alten Ägypten, Hildesheimer Ägyptologische Beiträge* 23, Gerstenberg Verlag. Hildesheim.

Helck, W. 1994. "Wege zum Eigentum und Boden im Alten Reich." In S. Allam (ed.), *Grund und Boden in Altägypten*, Im Selbstverlag des Herausgebers. Tübingen.

Herb, M. and F. Föster. 2009. "From desert to town: The economic role of desert game in the Pyramid Ages of ancient Egypt as inferred from historical sources: (c. 2600–1800 BC). An outline of the workshop's inspiration and objectives." In H. Riemer, F. Föster, M. Herb and N. Pöllath (eds.), *Desert animals in the eastern Sahara: Status, economic significance, and cultural reflection in antiquity. Proceedings of an Interdisciplinary ACACIA Workshop held at the University of Cologne, December 14–15, 2007*, Heinrich Barth Institut, 17–44. Cologne.

Hornung, E., R. Krauss and D. Warburton. 2006. *Ancient Egyptian Chronology*. Brill. Leiden.

Issar, A. S. 1998. "Climate change and history during the Holocene in the eastern Mediterranean region." In A. S. Issar and N. Brown (eds.), *Water, environment and society in times of climatic change*, Kluwer Academic Publishers, 113–28. Dordrecht, Boston and London.

Issar, A. I., and M. Zohar. 2004. *Climate Change – Environment and Civilisation in the Middle East*. Springer. Berlin and Heidelberg.

Junker, H. 1953. *Gîza XI. Gîza. Bericht über die von der Akademie der Wissenschaften in Wien auf gemeinsame Kosten mit Dr. Wilhelm Pelizaeus unternommenen Grabungen auf dem Friedhof des Alten Reiches bei den Pyramiden von Gîza. Band XI. Der Friedhof südlich der Cheopspyramide. Ostteil.* Hölder-Pichler-Tempsky A.G.: Vienna.

Kanawati, N. 1980. *Governmental Reforms in Old Kingdom Egypt.* Aris & Phillips. Warminster.

Kanawati, N., and M. Abder-Raziq. 1998. *The Teti Cemetery at Saqqara. Volume III. The Tombs of Neferseshemre and Seankhuiptah.* Aris & Phillips. Warminster.

Kanawati, N., M. Abd El-Raziq, and E. Alexakis. 2004. *Mereruka and his family. Part 1. The tomb of Meryteti,* The Australian Centre for Egyptology: Reports 21. Aris and Phillips. Oxford.

Lepsius, K. R. 1897. *Denkmaeler aus Aegypten und Aethiopien nach den Zeichnungen der von Seiner Majestaet dem Koenige von Preussen Friedrich Wilhelm IV nach diesen Laendern gesendeten und in den Jahren 1842–1845 ausgefuehrten wissenschaftlichen expedition.* J. C. Hinrichs. Leipzig.

McFarlane, A., and A. L. Mourad (eds.). 2013. *Behind the Scenes: Daily Life in Old Kingdom Egypt,* Australian Centre for Egyptology Studies 10, Aris and Phillips. Oxford.

Moussa, A. M., and H. Altenmüller. 1977. *Das Grab des Nianchchnum und Chnumhotep.* P. v. Zabern. Mainz.

Müller-Wollermann, R. 1986. *Krisenfaktoren im ägyptischen Staat des ausgehenden Alten Reichs,* Tübingen: Tübingen, published dissertation.

Pardey, E. 1976. *Untersuchungen zur ägyptischen Provinzialverwaltung bis zum Ende des Alten Reiches.* Gerstenberg. Hildesheim.

Petrie, W. M. Flinders, and F. Ll Griffith. 1892. *Medum.* London: D. Nutt.

Richards, J. 2002. "Text and Context in late Old Kingdom Egypt: The Archaeology and Historiography of Weni the Elder." *JARCE* 39: 75–102.

Roth, A. M., P. Der Manuelian and W. K. Simpson, 1995. *A Cemetery of Palace Attendants: Including G 2084-2099, G 2230+2231, and G 2240.* Dept. of Ancient Egyptian, Nubian, and Near Eastern Art, Museum of Fine Arts, Boston. Boston, Massachusetts.

Schatzmayr, A. 1936. *Risultati scientifici della spedizione entomologica di S.A.S. il Principe Alessandro della Torre e Tasso in Egitto e nella penisola del Sinai. XII. Catalogo ragionato dei Carabidi finora noti d'Egitto e del Sinai.* Pubblicazioni del Museo Entomologico "Pietro Rossi" 1: 1–114.

Schenkel, W. 1978. *Die Bewässerungsrevolution im alten Ägypten.* Zabern. Mainz.

Seidlmayer, S. J. 2001. *Historische und moderne Nilstände: Untersuchungen zu den Pegelablesungen des Nils from der Frühzeit bis in die Gegenwart.* Achet Verlag. Berlin.

Smith, W. S. 1946. *A history of Egyptian sculpture and painting in the Old Kingdom.* Oxford University Press. London.

Staubwasser, M., and H. Weiss 2006. "Holocene climate and cultural evolution in late prehistoric-early historic West Asia." *Quaternary Research* 66, 372–387.

Strudwick, N. 1985. *The Administration of Egypt in the Old Kingdom: The Highest Titles and Their Holders.* KPI. London and Boston.

Tainter, J. A. 1988. *The Collapse of Complex Societies.* Cambridge University Press. Cambridge and New York.

Verner, M. 2006. "Contemporaneous evidence for the relative chronology of Dyns. 4 and 5." In E. Hornung, R. Krauss and D. Warburton (eds.), *Ancient Egyptian chronology.* Leiden: Brill.

Wilkinson, T. A. H. 2000. *Royal Annals of Ancient Egypt: The Palermo Stone and its Associated Fragments.* Kegan Paul International. London.

Ziermann, M. 1995. "Rekonstruierte Hochwasserstände von der Frühzeit bis zum Mittleren Reich." In W. Kaiser, P. Becker, M. Bommas, E. Hoffmann, H. Jaritz, S. Müntel, J.-P. Pätznick and M. Ziermann (eds.), "Stadt und Tempel von Elephantine. 21./22. Grabungsbericht", *MDAIK* 51, 138–140.

Archaeological Evidence for Pollution and its Ecological Implications

New Data on Animal Exploitation from the Mesolithic to the Neolithic Periods in Northern Sudan

Louis Chaix and Matthieu Honegger

Abstract

Northern Sudan and southern Egypt are important regions for the understanding of climatic variations at the beginning of the Holocene period and of the first steps towards the adoption of domesticated cattle in Africa. However, insufficient research has been conducted to propose a precise scenario of these phenomena, and the archaeological sites considered are often very eroded. This is particularly the case with the well known sites of the Nabta Playa area (Wendorf and Schild 2001). There is still much discussion, therefore, about the dating of these sites, about the chronological homogeneity of the remains collected and the high degree of fragmentation of animal bones (e.g. Wengrow 2003). The scenario of the first steps in local cattle domestication in southern Egypt at the beginning of the Holocene remains a hypothesis more than a demonstration.

Introduction

The Holocene chronology of human occupations is relatively well known in two regions of the Nile valley (fig. 1). In Lower Nubia, the sequence of Nabta Playa and Bir Kiseiba, 200 km west of the Nile, is the most complete, from 8800 to 3000 BC. This sequence shows very early, and controversial, indications of pottery manufacture and cattle domestication (Wendorf and Schild 2001). In the south, central Sudan is rich in sites from the Mesolithic and Neolithic periods (from 7000 to 3500 BC), but the introduction of domesticated cattle occurs more than 3,000 years later than in Lower Nubia. Between these two regions more than 700 km apart, we have tried to build a new chronological framework in the area of Kerma, south of the Third

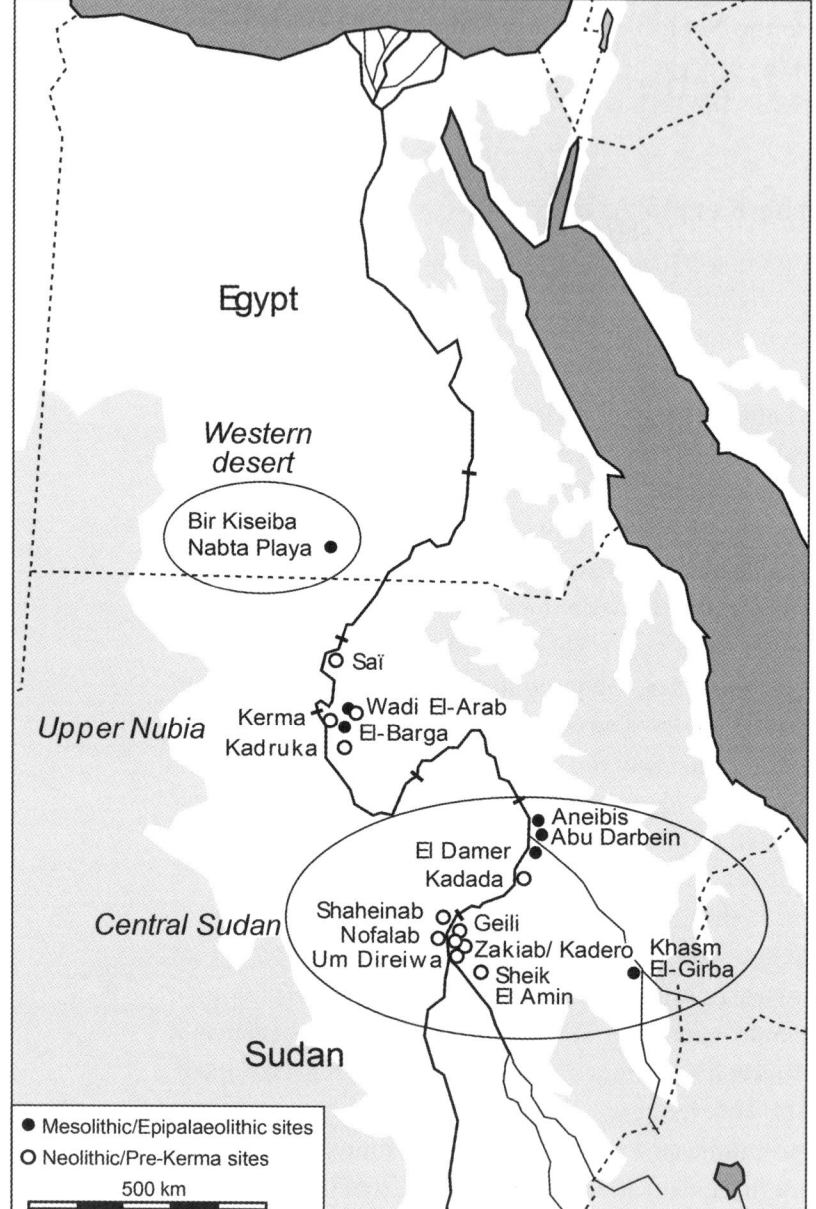

FIGURE 1. Map of Egypt and Nubia with the localisation of the main areas where a chronological framework was established for the Holocene prehistory. Kerma, the area from which the new data in this paper comes, is located between them. Other sites mentioned in the text are indicated.

Cataract, to follow the evolution of human societies during the Holocene period. We want to present here the main results of animal exploitation from the Mesolithic to the Neolithic, and even later periods. We will then propose a synthesis of the process of cattle adoption in northeastern Africa, including its relation with climate fluctuations and cultural influences.

Archaeology in the Kerma Area

In the last ten years, survey work and excavations undertaken in the Kerma region have brought to light remains from several periods. Four hundred and forty sites have been identified (Honegger 2007; Honegger et al. 2009). A great number of them are eroded and partially destroyed by agricultural fields, but others are better preserved and are, occasionally, of significant archaeological interest. The spatial distribution of sites shows a distinctive split between the occupations located on the alluvial plain and those outside, along the desert edge.

The sites of the first half of the Holocene period are located outside the alluvial plain and correspond to an older and more humid climatic phase (fig. 2). Access to the alluvial plain might have been difficult and human groups thus preferred settling on small mounds near its edge, safe from the Nile flood. They also settled around a wide depression, which was filled by an ancient swamp fed by rainwater and Nile floods. The most important sites excavated are Wadi el-Arab and el-Barga (Honegger 2006; Honegger et al. 2009). The first contains stratified layers with settlements and graves dated from 8300 to 6500 BC. The second has revealed a habitation structure dug into the sandstone bedrock and includes an important archaeological assemblage dated to about 7400 BC. Two cemeteries were found nearby that were in use from 7800 to 5500 BC.

The sites of the second half of the Holocene period correspond to a more arid climate and, logically, they are found within the alluvial plain nearer to then-extant Nile channels. These sites belong to different archaeological periods such as Middle Neolithic, Pre-Kerma, Kerma, Napatan and Meroitic periods as well as later ones. The most important are eroded habitation sites partially washed away by Nile floods. We have also included in our study cemetery sites from the region of Kadruka, often located at the same zone as settlements, from which faunal remains were collected.

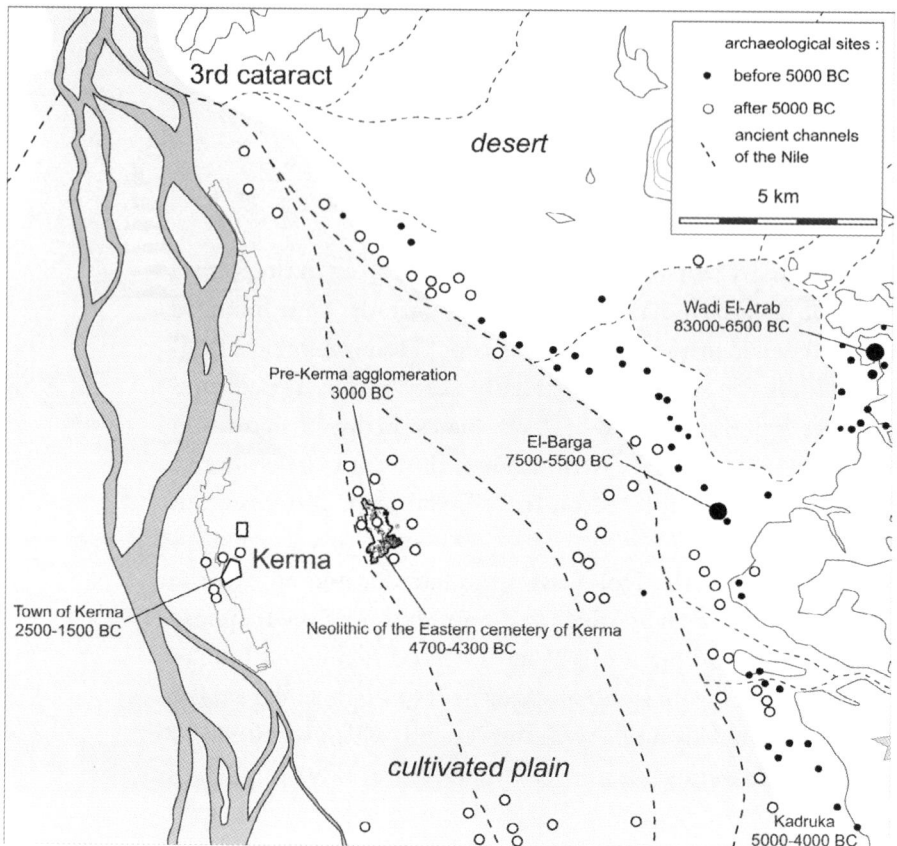

FIGURE 2. *Map of the Kerma area showing the distribution of the sites during the Holocene period. The most important of them are indicated and were excavated during these last 15 years or are still being excavated today.*

According to a recent study on climatic changes in the eastern desert (Kuper and Kröpelin 2006), there is a change in the location of human installations during the Holocene. This occurred around 5500 and 5000 BC and is signalled by the shifting of sites closer to the Nile River. In Kerma, we observe the same phenomena but at a local scale. The difference with Kuper and Kröpelin's study is that the sites at Kerma before 5500 BC are much closer to the alluvial plain than in Egypt, where they are located further in the desert. This situation is due to the fact that the authors did not include

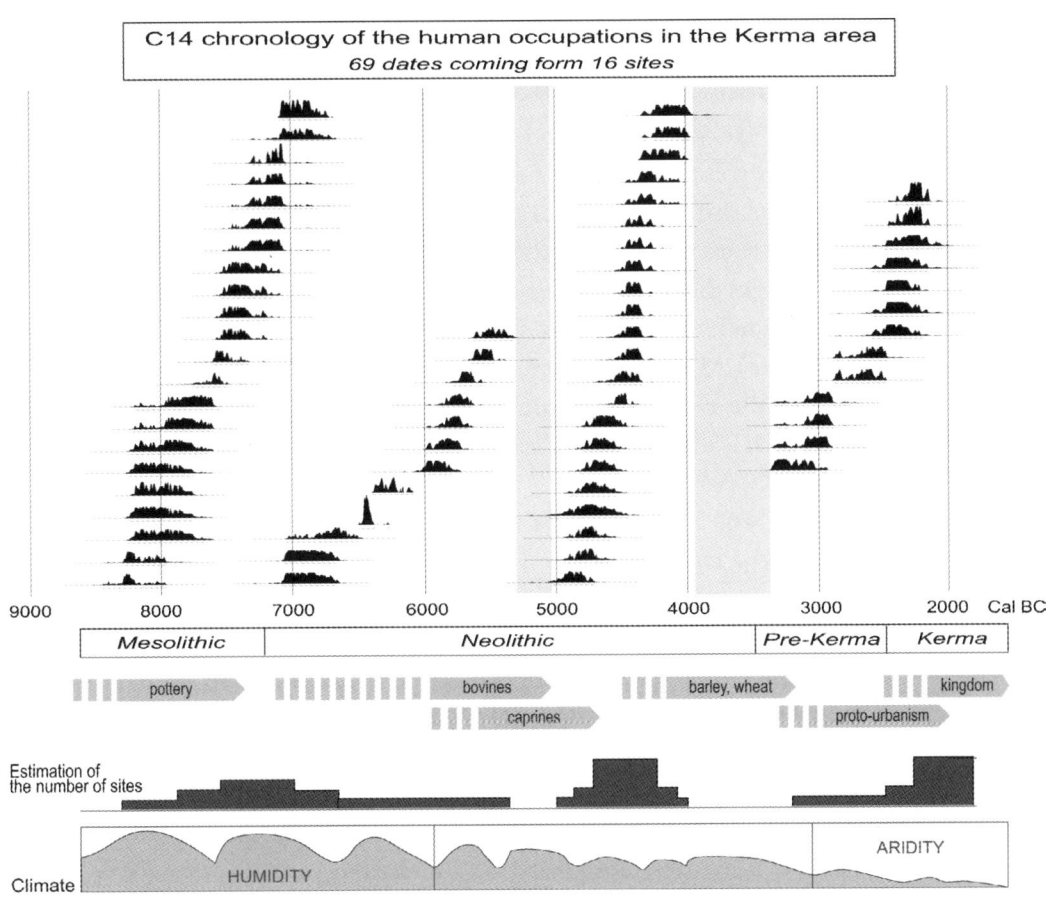

FIGURE 3. *Chronological framework of the Kerma area obtained with 69 ^{14}C dates made between 8300 BC and 2000 BC. The principal innovations are indicated as well as an estimation of the number of sites for each period and the schematic climatic fluctuation (after Hassan 2002).*

in their study all the dated sites located between the second and the fifth cataract, which means that their appreciation of the situation in the Sudanese Nile valley during the Holocene is not really representative.

In Kerma, the chronological and cultural frameworks are known from 8300 BC to the present with exceptional continuity (all the dates in BC are calibrated). More than 60 ^{14}C dates were obtained from sites that show original cultural features from Epipalaeolithic times to the ancient Kerma period (Honegger 2009: fig. 3). They include the results obtained by our colleague Jacques Reinold on the cemeteries of Kadruka (Reinold 2000, 2001 and 2004). The chronology is interrupted by two gaps and we do not know if the absence of sites during these periods is linked to climatic changes, demographic decrease or if it results from the current state of research. Trial or long-term excavations of important sites were made in order to understand the major evolutionary stages of Nubian societies, such as the invention of pottery, the shift to a sedentary way of life, the transition to stockbreeding and agriculture, and urbanisation and state formation (Honegger 2009). An estimation of the number of sites grouped according to periods shows a typical situation for the Sudanese Nile valley: a high density of Epipalaeolithic (Mesolithic), Middle Neolithic and Kerma sites. The most problematic gap is that of the fourth millennium. It corresponds to Predynastic times in Egypt and it is difficult to understand why there are so few sites dated to this period in Upper Nubia.

The analysis of pottery style and technique offers the possibility of reconstructing the cultural influences between the different regions (Honegger and Gatto, in prep.):

— The first pottery found in Kerma, dated to around 8300 BC, consists primarily of sherds decorated using the return technique. No comparison is known except in Acacus (Libya), where this decoration is present at a later period.
— The el-Barga style pottery (7800–7200 BC) is characterised by a monotonous decoration composed of alternating pivoting stamps. We find similar decorations between the Second and Third cataracts, and in the Nubian Desert. It appears to be a regional cultural group distinctive from that of Nabta Playa and Central Sudan.
— The next phases were identified in Wadi el-Arab. They developed between 7200 and 6200 BC and can be related with the sequence of Nabta Playa, with the Nabta and the el-Jerar phases. We will see that the first introduction of domesticated cattle dates to this period.
— Finally, the Middle Neolithic phase (5000–4000 BC) is characterised by burnished pottery, with the first evidence of red and black top ware, and ripple ware decoration. Today, there is no attempt to define regional cultural variations, but such characteristics can be found on Badarian pottery as well as on Abkan and central Sudan Neolithic ceramics. This is the period of fully fledged pastoral societies.

Analysis of faunal remains

Mesolithic (8300–7200 BC)

In the area of Kerma, the Mesolithic period is known from sites such as el-Barga and Wadi el-Arab. The study of the faunal remains attests to a broad spectrum economy with the exploitation of a multitude of environments, characterised by large amounts of aquatic resources in terms of number of specimens (Peters 1995). However, if we speak of the quantity of meat available, this vision must be corrected as mammals, particularly bovines, represent, by weight, the main source of proteins (fig. 4). Domesticated cattle are not known at this time in Sudan.

The Mesolithic sites in our area and also in central Sudan are clearly linked to the Nile or to other permanent rivers like the Atbara. The proportions of fish and molluscs are high in terms of NISP (number of identified specimens) and the rest of the spectrum shows the presence of various animals from the deciduous savannah, which had a pluviometry of between 300 and 400 cubic cm per year (Peters 1991 and 1995). Amongst the mammals, the presence of giraffes and elephants is indicative of large trees, and different bovids are clearly linked to riverine forests. Medium and small antelopes form the bulk of the game species. Meanwhile, some gazelles (such as *Gazella dorcas*) indicate hunting activities in more arid zones. Reptiles, tortoises and snails are not rare but they represent, like birds, marginal activities. In contrast to the Egyptian Nile valley, wild cattle are unknown in the Kerma area, as in central Sudan, with the exception of Khashm el-Girba in the Atbara region, a site dated to around 10,000 BC (Marks et al. 1987; Peters 1986; Linseele 2004). This last site is problematic because among all the sites of the early Holocene period in Sudan, it is the only one that contains wild cattle. In our view, there are two possible explanations for this: either it belongs to an isolated ecological area disconnected from the other sites and their environments, or there is a problem of connection between the ^{14}C sample, the lithic industry and the faunal remains, which is a relatively common situation in north African archaeology. As we have seen above, fish are abundant and varied. Some specimens are quite large: amongst the Clariidae at el-Barga, two individuals measure more than one metre and two Nile perch exceed 1.50 metres.

The analysis of the faunal remains from the different sites in our area and

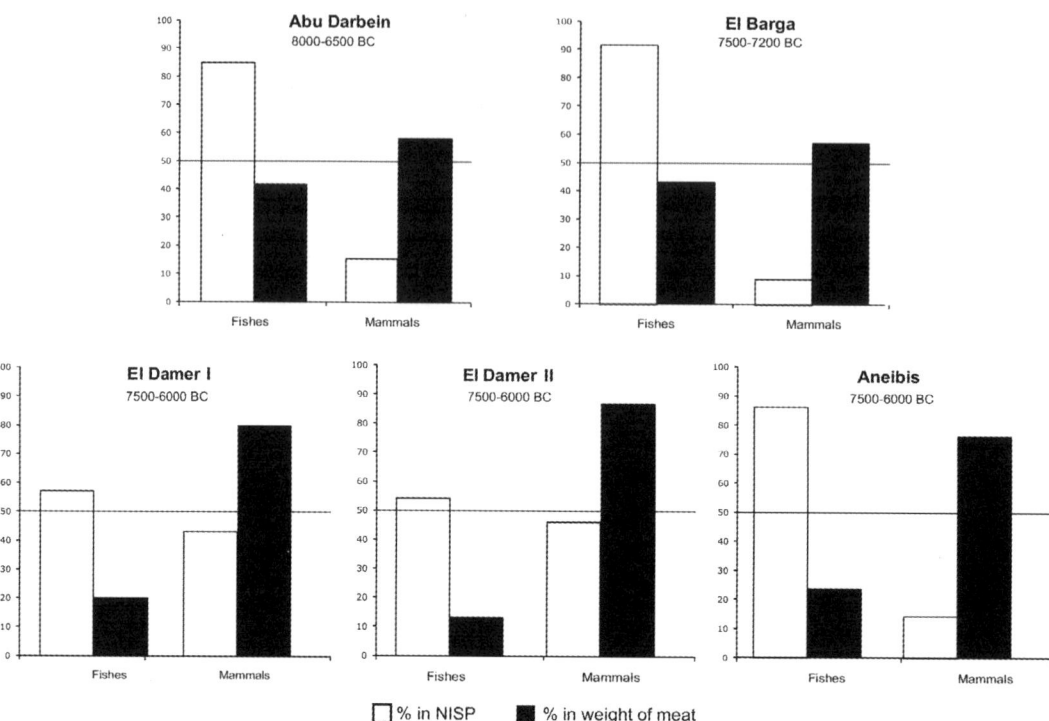

FIGURE 4. *Proportion in different Mesolithic sites of the number of mammals bones and fishes (NISP) compared to the weight of meat calculated on the same samples (after Peters 1995). If the fish remains are much more numerous, they represent often a small part of the alimentation.*

in central Sudan indicates that Mesolithic people had a sedentary way of life linked to the presence of a permanent water source and within a limited area. The presence of cemeteries is another indication of a non-mobile population.

Ancient Neolithic (7200–5500 BC)

Few sites from this period are known in our area and the faunal remains are scarce and badly preserved. The majority come from surface sites with the well-known problem of differential preservation, except for the site of Wadi el-Arab where we can observe stratified layers of 40 to 70 cm. Despite the scarcity of the recovered remains, the few finds indicate a deep change:

the appearance of domestic cattle in very low proportions (fig. 5; Honegger and Chaix, in prep.). With the exception of these rare remains of domestic bovines, the majority of the fauna consists of game animals (such as elephant, various bovids and warthog). It is important to note that caprines, originating from the Middle East, are not present in the layers of this site. Fish and molluscs were also exploited, but in many cases the preservation of the site can explain their paucity or even their absence.

At el-Barga, a grave dated about 5750 BC revealed a cattle skull (bucranium), deposited on the body of a child (Honegger 2005). This deposit is proof of the importance of cattle in funerary practices and in society and it means that this cemetery of el-Barga (6000–5500 BC), with more than 100 graves, is the most ancient cemetery known in Africa from Neolithic times. It is possible that the economic importance of bovines was growing during the sixth millennium and that the numbers of this animal increased, but the few sites of this period known in the Kerma area do not present enough evidence to confirm this. Cattle graves are known in other places in the Sahara at the same period, such as the grave found in Nabta Playa and dated to about 5400 BC (Paris 2000; Gautier 2007b; Wendorf and Schild 2001).

Middle Neolithic (5000–4000 BC)

This period is attested in the area of the Third Cataract by some sites, mainly found on the surface. Furthermore, we have added two sites located south of Kerma – Kadruka 5 and 29 – which also date to the fifth millennium. Three of these sites each delivered more than 200 bones, a number large enough to be statistically interesting and upon which some conclusions may be drawn. Despite the problems of preservation, one can observe that in all these settlements cattle are clearly dominant, followed by caprines where sheep and goats are attested. Wild animals are very rare and hunting activities do not represent a substantial part of the economy. Fish remains are generally not preserved, except for some large vertebrae. Many graves of the cemeteries at Kadruka contain cattle skulls located close to the body (Reinold 2004, 2006). In central Sudan the sites from the Middle Neolithic are numerous and relatively well studied (Sobocinski 1977; El Mahi 1982; Gautier 1983; Chaix 2003; Gautier 2007a). Even if the samples vary – from a few dozen determined bones to nearly a thousand specimens – the trend is the same with a substantial dominance of cattle and caprines compared with the disappearance of the wild animals.

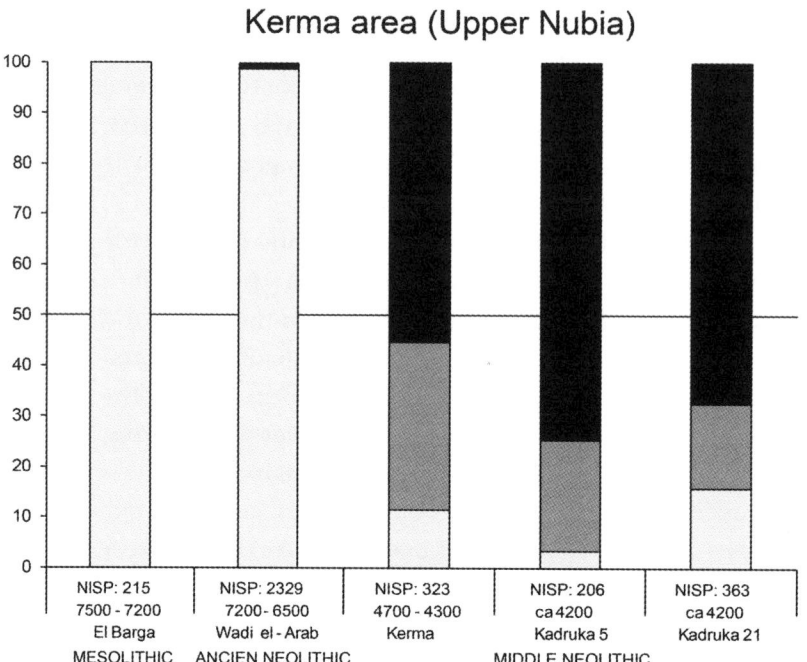

FIGURE 5. *Comparisons of the faunal spectrum between different sites from Kerma area.*

To us, this period shows the most important change, namely pastoralism becoming the principal way of life and announcing the first complex pastoral societies. Rare measurements and observations show that the cattle were strongly built animals with long horns, very similar to those later found at Kerma (Chaix 2007a and forthcoming b).

Late Neolithic/Pre-Kerma (4000–2500 BC)

Pre-Kerma culture, which is contemporary with the A-Group in Lower Nubia (Predynastic period), is known from a few sites located between Kerma and the Second Cataract. At Saï Island (2900–2600 BC; Gueus 2004), many storage pits have revealed faunal remains where caprines (sheep and goats) are clearly dominant and cattle represent only a small part of the total (Chaix, forthcoming a). On Site No. 21 at Kerma (Honegger 2004), cattle top the list with seven bones from a total of eleven discovered.

This suggests that the Pre-Kerma culture (3500–2500 BC; Honegger, 2004), with one of the first agglomerations known in Africa (fortified and comprising several cattle enclosures), is the forbearer of the Kingdom of Kerma (2500–1500 BC). As at the capital of Kerma (Bonnet and Valbelle 2004) or the rural settlement of Gism el-Arba (Gratien et al. 2003), the exploitation of the animal world is entirely based on stockbreeding of cattle and caprines, forming more than 90% of the fauna (Chaix 2007b). Since Middle Neolithic times, cattle played an important role in the beliefs and ideology of the communities (Chaix and Grant 1992). It is regularly implied in the funerary tradition with bucrania placed in or near the graves. The most spectacular example is the deposit of 5,600 skulls close to a Middle Kerma grave dated c. 1900 BC (Chaix 2001; Chaix and Hansen 2003).

Conclusion

The faunal remains of the Kerma area confirm and clarify the scenario of pastoralism spreading from the western desert in southern Egypt to central Sudan. The evidence at Nabta Playa and Bir Kiseiba shows the presence of few cattle – probably domesticated or in the process of becoming domesticated – from 8800 to 6000 BC (fig. 6). Even if these data are controversial, our research in Kerma tends to confirm this local process of cattle domestication which happened before the introduction of caprines, and probably of bovines, from the Middle East around 6000 BC (Blench and MacDonald 2000; Hanotte et al. 2002; Pérez-Pradal et al. 2010). In Kerma, the first occupations analysed reveal a typical Mesolithic economy such as that known in central Sudan for the eighth millennium. Wild or domesticated cattle are not present. The next phase, which was called Ancient Neolithic, is characterised by an economy very similar to the Mesolithic one and is mainly based on hunting and fishing, but with the presence of a few cattle bones. This phase begins around 7200 BC at Wadi el-Arab and ends in the 6th millennium. The situation is different in central Sudan where this phase is not yet represented and where we see a quick transition from a Mesolithic way of life to full pastoral societies. The pastoral economy appears in our area only with the Middle Neolithic, between 5500 and 5000 BC, where the faunal spectrum is clearly dominated by bovines and caprines. It is in this period

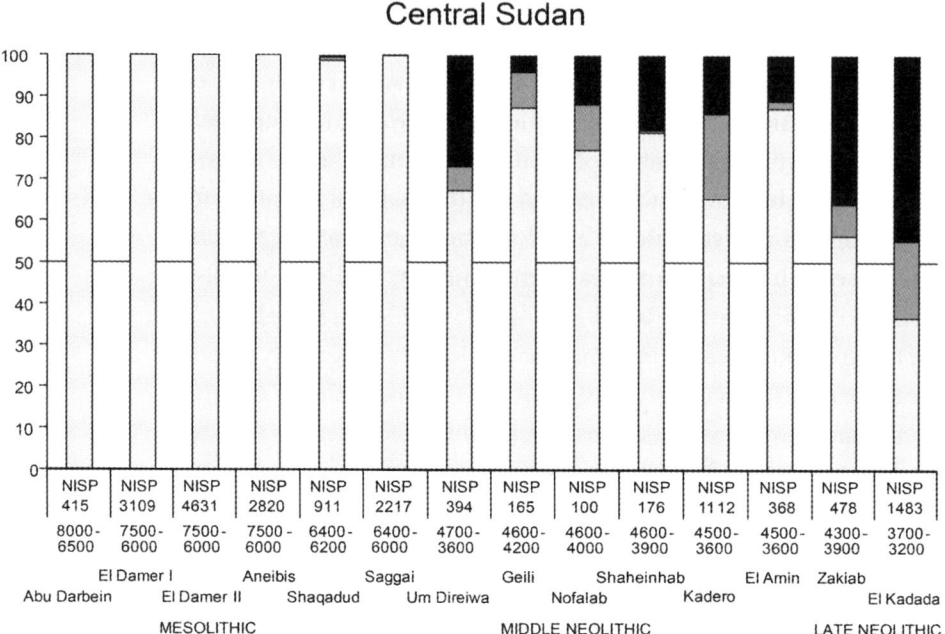

FIGURE 6. *Spread of the pastoral economy from the North (Nabta Playa) to the South (Central Sudan) between 8800 BC and 5000 BC (after El-Mahi 1982; Gautier, 1983, 1986, 2007; Chaix 2003).*

that the first dogs appear in cemeteries and that the first attempts at the cultivation of barley followed by wheat are made (Reinold 2000).

The terminology used in Sudanese Holocene prehistory has been criticised on many occasions for being too influenced by European traditions. The term Mesolithic introduced by Arkell (1949) could be replaced by Epipalaeolithic or Pre-pastoral phase (Garcea 2004) where it would be important to distinguish if the pottery is present or not. The term Ancient Neolithic, that we use according to Wendorf's terminology, is characterised by an economy based essentially on game, gathering and fishing and is not very different from the term Mesolithic. It could be called the Proto-pastoral phase. Finally, with the Middle Neolithic there is a shift towards the emergence of a full pastoral economy.

If the diffusion of the Ancient Neolithic (or Proto-pastoral phase) from Nabta Playa to central Sudan takes time, we see a quick spread of the full pastoral economy from north to south at a moment that corresponds to the main climatic change, if we consider the work of Kuper and Kröpelin (2006). With more arid conditions, populations settle close to the Nile River and in other humid ecosystems. As mentioned in different works (see for instance Hassan 2002), drier climatic conditions and the necessity of adaptation to these new conditions can stimulate the adoption of innovation, which is probably the case with the quick spread of pastoral way of life in Northern Africa.

Addendum

This paper was written in November 2010. Since this date, our research in the area of Kerma does not confirm the presence of cattle or of *aurochs* before 6000 BC. This means that the long process of local domestication supposed by the discoveries of Nabta Playa and Bir Kiseiba is not confirmed by our results.

Bibliography

Arkell, A.J. 1949. *Early Khartoum: An account of the excavation of an early occupation site carried out by the Sudan Government Antiquities Service in 1944–5.* Oxford University. Press. London.

Bonnet, C., and D. Valbelle. 2004. *Le temple principal de la ville de Kerma et son quartier religieux.* Errance. Paris.

Blench, R. M., and K. MacDonald (eds.). 2000. *The origins and development of African livestock: Archaeology, genetics, linguistics and ethnography.* UCL Press. London and New York.

Caneva, I. 1987. "Pottery decoration in prehistoric Sahara & Upper Nile: A new perspective." In Barbara E. Barich (ed.), *Archaeology and Environment in the Libyan Sahara: The excavations in the Tadrart Acacus, 1978–1983*, British Archaeological Reports International Series 368, 231–254. Oxford.

Chaix, L. 2001. "Animals as symbols: The bucrania of the grave KN 24 (Kerma, Northern Sudan)." In H. Buitenhuis and W. Prummel (eds.), *Animals and man in the past: Essays in honour of Dr. A. T. Clason emeritus professor of archaeozoology Rijkuniversiteit Groningen, the Netherlands*, ARC Publicatie 41, 364–370. Groningen.

Chaix, L. 2003. "La faune des sites mésolithiques et néolithique de la zone du Nil Bleu (Soudan central)." In V. M. Fernandez (ed.), *The Blue Nile Project: Holocene archaeology in Central Sudan*, Complutum 14, Universidad Complutense, 273–396. Madrid.

Chaix, L. 2007a. "Contribution to the knowledge of domestic cattle in Africa: The osteometry of fossil *Bos taurus* L. from Kerma, Sudan (2050–1750 BC)." In G. Grupe and J. Peters (eds.), *Skeletal series and their socio-economic context*, Documenta Archaeobiologiae 5, M. Leidorf, 170–249. Rahden, Westphalia.

Chaix, L. 2007b. "New data about rural economy in the Kerma culture: The site of Gism-el-Arba (Sudan)." *Studies in African Archaeology* 9: 25–38.

Chaix, L. 2010. "Le 'bucrane' dans les cultures préhistoriques de la vallée du Nil." In O. Larroque and M. Nougarede (eds.), *Vaca Bruto, du quotiden au sacré, taureaux d'Afrique noire*, exhibition catalogue, Editions de Musée des Cultures Taurines, 16–29. Nîmes.

Chaix, L., and A. Grant. 1992. "Cattle in Ancient Nubia." *Anthropozoologica* 16: 61–66.

Chaix, L., and J. W. Hansen. 2003. "Cattle with 'forward-pointing horns': Archaeozoological and cultural aspects." In L. Krzyzaniak, K. Kroeper and M. Kobusiewicz (eds.), *Cultural Markers in the Later Prehistory of Northeastern Africa and Recent Research*, Studies in African Archaeology 8, Poznan Archaeological Museum, 269–281. Poznan.

Chaix, L. (Forthcoming, a). "Contribution à l'étude de l'économie de la période Pré-Kerma. Premiers résultats sur la faune du site 8B-10A sur l'Ile de Saï (Nord Soudan)." In *Hommages à Patrice Lenoble*. Presses Universitaires de la Sorbonne. Paris.

Chaix, L. (Forthcoming, b). "Holocene cattle (*Bos taurus* L.) in the Sudan." *Reports in African Archaeology, supplement to the Journal of African Archaeology*.

El-Mahi, T. 1982. *Fauna, ecology and socio-economic conditions in the Khartoum Nile environment*. PhD dissertation, University of Bergen.

Garcea, E. A. A. 2004. "An alternative way towards food production: The perspective from the Libyan Sahara." *Journal of World Prehistory* 18(2): 107–154.

Gatto, M. C. 2006. "The Khartoum Variant Pottery in Context: Rethinking the Early and Middle Holocene Nubian Sequence." *Archéologie du Nil Moyen* 10: 57–72.

Gautier, A. 1983. "Animal life along the prehistoric Nile: The evidence from Saggai I & Geili (Sudan)." *Origini. Preistoria e protostoria delle civilta antiche* 12: 50–115.

Gautier, A. 1986. "La faune de l'occupation néolithique d'El-Kadada (secteurs 12-22-32) au Soudan Central." *Archéologie du Nil Moyen* 1: 59–111.

Gautier, A. 2007a. "The faunal remains of the Early Neolithic site Kadero, Central Sudan." In K. Kroeper, M. Chlodnicki and M. Kobusiewics (eds.), *Archaeology of Early Northeastern Africa. In memory of Lech Krzyzaniak*, Studies in African Archaeology 9, Poznan Archaeological Museum, 113–117. Poznan.

Gautier, A. 2007b. "Animal domestication in North Africa." In M. Bollig, O. Bubenzer, R. Vogelsang and H.-P. Wotzka (eds.), *Aridity, Change and Conflict in Africa: Proceedings of an International ACACIA Conference held at Königswinter, Germany, October 1–3, 2003*, Colloquium Africanum, 2, Heinrich Barth Institut, 75–89. Cologne.

Gratien, B., S. Marchi, O. Thuriot and J.-M. Willot. 2003. "L'habitat 2 de

Gism el-Arba. Rapport préliminaire sur un centre de stockage Kerma." *CRIPEL* 23: 29–43.

Gueus, F. 2004. "Pre-Kerma storage pits on Saï island." In T. Kendall (ed.), *Nubian Studies 1998: Proceedings of the Ninth Conference of the International Society for Nubian Studies, Boston, Massachusetts, 21–26 August 1998*, Northeastern University, 76–51. Boston, Massachusetts.

Hanotte, O., D. G. Bradley, J. W. Ochieng, Y. Verjee, E. W. Hill and J. E. O. Rege. 2002. "African pastoralism: Genetic imprints of origins and migrations." *Science* 296: 336–339.

Hassan, F. A. 2002. *Droughts, food and culture: Ecological change and food security in Africa's later prehistory*. Kluwer Academic/Plenum Publishers. New York.

Honegger, M. 2004. "The Pre-Kerma: A cultural group from upper Nubia prior to the Kerma civilisation." *Sudan & Nubia* 8: 38–46. http://www.kerma.ch/index.php?option=com_wrapper&Itemid=155

Honegger, M. 2005. "Kerma et les débuts du Néolithique africain." *Genava n.s.* 53: 239–249. http://www.kerma.ch/index.php?option=com_wrapper&Itemid=120

Honegger, M. 2006. "El-Barga: un site clé pour la compréhension du Mésolithique et du début du Néolithique en Nubie." *Revue de paléobiologie*, vol. spécial 10, Hommage à Louis Chaix: 95–104.

Honegger, M. 2007. "Aux origines de Kerma." *Genava n.s.* 55: 201–212. http://www.kerma.ch/index.php?option=com_wrapper&Itemid=127

Honegger, M. 2009. "Kerma (Soudan) : origine et développement du premier royaume d'Afrique noire." *Archéologie suisse* 32(1): 2–13. http://www.kerma.ch/index.php?option=com_wrapper&Itemid=216

Honegger, M., C. Bonnet and collaborators. 2009. *Kerma (Soudan), report to the 2008–2009 season*. Documents de la mission archéologique suisse au Soudan 1, Université de Neuchâtel. http://www.kerma.ch/index.php?option=com_wrapper&Itemid=201

Honegger, M., and L. Chaix. in prep. New data on the beginning of cattle domestication in Africa.

Honegger, M., and M. Gatto. in prep. Early Holocene occupations in the Kerma Area (Upper Nubia, Sudan): Chronology and cultural affinities.

Kuper, R., and S. Kröpelin. 2006. "Climate-controlled Holocene occupation in the Sahara: Motor of Africa's evolution." *Science* 313: 803–807.

Linselee, V. 2004. "Size and size change of the African aurochs during the Pleistocene and Holocene." *Journal of African Archaeology* 2(2): 165–186.

Marks, A. E., J. Peters and W. van Nerr. 1987. "Late Pleistocene and Early Holocene Occupations in the Upper Atbara River Valley. Sudan." In A. Close (ed.), *Prehistory of Arid North Africa*, Southern Methodist University Press, 137–161. Dallas.

Paris, F. 2000. "African livestock remains from Saharan mortuary contexts." In R. M. Blench and K. MacDonald (eds), *The origins and development of African livestock: Archaeology, genetics, linguistics and ethnography*, UCL Press, 111–126. London.

Pérez-Pradal, L., L. J. Royo, A. Beja-Pereira, S. Chen, R. J. C. Cantet, A. Traoré, I. Curik, J. Solkner, R. Bozzi, I. Fernandez, I. Alvarez, J.-P. Gutiérrez, E. Gómez, F. A. Ponce De León and F. Goyache. 2010. "Multiple paternal origins of domestic cattle revealed by Y-specific interspersed multilocus microsatellites." *Heredity* 105: 511–519.

Peters, J. 1986. *Bijdrage tot de Archeozoölogie van Soedan en Egypte*. PhD dissertation, Ghent University.

Peters, J. 1991. "Mesolithic fishing along the Central Sudanese Nile and the Lower Atbara." *Sahara* 4: 33–40.

Peters, J. 1995. "Mesolithic subsistence between the 5th and the 6th Nile cataract: The archaeofaunas from Abu Darbein, El Damer and Aneibis (Sudan)." In R. Haaland and A. Abdul Magid (eds.), *Aqualithic sites along the Rivers Nile and Atbara, Sudan*, Alma Mater Forlag, 178–244. Bergen.

Reinold, J. 2000. *Archéologie au Soudan. Les civilisations de Nubie*. Errance. Paris.

Reinold, J. 2001. "Kadruka and the Neolithic in the Northern Dongola Reach." *Sudan & Nubia* 55: 2–10.

Reinold, J. 2004. "Kadruka." In D. E. Welsby and J. R. Anderson (eds.), *Sudan, Ancient treasures: An exhibition of recent discoveries from the Sudan National Museum*, The British Museum Press, 20–28. London.

Reinold, J. 2006. "Les cimetières préhistoriques au Soudan: coutumes funéraires et systèmes sociaux." In I. Caneva and A. Roccatti (eds.), *Tenth International Conference of the Society for Nubian Studies, Rome, 9–14 September 2002*, Libreria delle Stato, 139–162. Rome.

Sobocinski, M. 1977. "Szczatki Zwierzece z Osady Neolitycznej w Kadero (Sudan)." *Roczniki Akademii Rolniczej w Poznaniu* 93: 49–61.

Wendorf, F. and R. Schild. 2001. *Holocene Settlement of the Egyptian Sahara. Volume 1: The Archaeology of Nabta Playa.* Kluwer Academic/Plenum Publishers. New York.

Wengrow, D. 2003. "On desert origins for the ancient Egyptians." Review of F. Wendorf and R. Schild (2001), *Holocene Settlement of the Egyptian Sahara, Vol. 1: The Archaeology of Nabta Playa. Antiquity* 77(297): 597–601.

Large Game Depression and the Process of Animal Domestication in the Near East

Benjamin S. Arbuckle

Abstract

In this chapter the role of large game depression in the process of animal domestication in the Neolithic Near East is discussed. Using a large sample of published faunal data from sites dating from the Paleolithic to early Neolithic periods, I explore geographic variability in large game depression, with a focus on sheep and goats, the first domesticated food animals. In general, the results do not support the idea that depression of game resources was the causal factor behind the initial domestication of livestock in the Near East and the implications of these findings are discussed.

Introduction

One of the most common explanations for the domestication of animals and plants in the Near East focuses on this transition as a rational economic response to long-term declines in wild resources around increasingly sedentary communities in the late Epipaleolithic and early Neolithic (Alvard and Kuznar 2001; Munro 2004b; Tchernov 1991). It is argued that sedentism along with post-glacial population growth and increased packing led to a significant increase in the spectrum of resources exploited by foragers. This 'broad spectrum revolution' represents a major decline in foraging efficiency as foragers, in response to declining populations of big game, increasingly turned to a range of lower ranked, and more costly to exploit, animal and plant resources (Flannery 1969; McCorriston and Hole 1991; Munro 2004b; Stutz et al. 2009; Tchernov 1991). This broad spectrum revolution, and the pressure on large game populations that it represents, has been documented in great detail in the southern Levant where it is seen as the primary engine driving the development of agro-pastoral economies in the Pre-Pottery Neolithic A and B (PPNA and PPNB) cultural periods c. 11500–8500 BP calibrated (e.g. McCorriston and Hole 1991).

Although the depression of large game in late Epipaleolithic and PPNA sedentary communities is commonly cited as a main factor in the process of animal domestication (Munro 2003; Tchernov 1993) there have been relatively few efforts to test this model outside of the southern Levant in regions where current research suggests the process of animal domestication was most precocious (Peters et al. 2005; Zeder 2008).

In this chapter I address the role of large game depression in the process of animal domestication across the Near East. Using a large sample of published faunal data from sites dating from the Paleolithic to early Neolithic periods, I explore the scale of, and geographic variability in, large game depression, with a focus on sheep and goats, the first domesticated food animals. Chronologically, I focus attention on the PPNA, the period just prior to the emergence of domestic animals. Although differences in the recovery and reporting of faunal assemblages makes inter-regional comparisons difficult, clear differences are beginning to emerge in the degree to which large game depression was an issue just prior to the emergence of agro-pastoral economies in the Near East.

Large game depression model of domestication

Discussions of the relationship between large game depression and domestication have been dominated for decades by Flannery's (1969) concept of the "broad spectrum revolution". This long-term, pre-agricultural change in subsistence practices represents a shift away from big game hunting in favour of a diverse array of smaller and lower-return taxa. More recently, researchers working within the conceptual frameworks of foraging theory have interpreted this transition in terms of long-term human and prey population dynamics (Munro 2004a, b; Stiner et al. 2000; Stutz et al. 2009). These studies have interpreted increases in low-return small game as a symptom of overall declines in high-return, large game driven by a combination of hunting pressure, post-glacial growth in human populations, and sedentism.

A causal relationship has often been proposed between sedentism, resource depression and domestication. This position has been most clearly expressed by Tchernov (1991, 1993, 1994). Starting with the assumption that sedentary foragers "will consume themselves out of even the richest environments" (Simmons et al. 1988: 39), Tchernov argued that the intensive use of small game such as hares, birds and tortoises by Natufian and PPNA foragers indicated significant depression of large game populations (e.g. ungulates such as gazelle and deer) in the hunting territories directly around long-term settlements. Increases in the frequencies of juvenile gazelles identified in PPNA assemblages was seen as another sign of intensive hunting pressure and a shift towards the exploitation of lower-return, immature animals with low meat and fat yields (Davis et al. 1994; Munro 2004b).

According to this paradigm, "Sedentism [...] led to over-hunting, reduction in meat resources and consequently to domestication" (Tchernov 1997: 220), firstly of sheep and goats and followed by cattle and pigs. Thus several decades of intensive work on faunal assemblages in the southern Levant have resulted in a picture in which severe depression of large game resulting from sedentism and post-glacial population growth in the Late Epipaleolithic and PPNA is causally linked to the eventual domestication of livestock in the PPNB (Binford 1968; Flannery 1973; McCorriston and Hole 1991; Reed and Perkins 1984).

Other responses to game depression

It is well documented, both archaeologically and ethnographically, that resource depression does occur around long-term residential settlements, particularly with respect to large mammals. But a fairly robust literature on this issue suggests that there are alternate responses to large game depression that do not include domestication. The concept of resource stress and its association with early sedentary communities has been extensively addressed in North American archaeology, especially in the American Southwest (Badenhorst and Driver 2009; Broughton 1994; Cannon 2000; James 2004; Janetski 1997; Speth and Scott 1989). A review of this literature, in a region where local ungulate populations were heavily exploited but never domesticated, suggests several alternate human responses to the depression of large game populations in addition to the Near Eastern domestication model.

Many studies in the American West have shown that sedentism is associated with depression of large game and subsequent, often dramatic, increases in the use of small game (Badenhorst and Driver 2009; Cannon 2000; Garfinkel et al. 2009; James 2004; Janetski 1997). In regions where large game populations around early sedentary communities were severely depressed, systems of 'garden hunting' developed, which focused almost exclusively on exploiting small game with high reproductive rates – especially hares and jackrabbits – encountered in and around agricultural fields (Badenhorst and Driver 2009; Linares 1976).

Despite predictions that sedentism should result in a decline in large game and an increase in diet breadth, this pattern is not always evident. Szuter and Bayham (1989) show that local environmental productivity as well as the size of the human population are major factors in mitigating the scale of local large game depression. They report that large sedentary Hohokam settlements located in arid lowland regions of the American Southwest exhibit severe evidence for large game depression, whereas large game remains abundant in smaller farming settlements located in more productive upland regions. Thus, in productive environments, even with sedentary populations, the scale of resource depression may be so small that it does not seriously impact human behaviour.

In some instances, the frequency of large game has been shown to increase with the development of large sedentary communities. Speth and Scott (1989) describe increases in the abundance of ungulates in some of the largest farming societies in the American Southwest, including Chaco and Snaketown. These increases in high-return large mammals are argued to represent the development of specialised, logistically organised hunting strategies (also Cannon 2000). In response to declines in local game populations, hunting became increasingly specialised with hunters regularly travelling to territories far from settlements to target high-return large game.

Furthermore, in societies where the plant-based diet is predictable and stable, hunters may invest considerable energy in risky, but potentially highly rewarding, large mammal hunting (Speth and Scott 1989). If success in big game hunting is linked with social status and prestige, these specialised hunting systems may develop a high degree of ritual elaboration, especially in cases where communal hunting and feasting are mechanisms for social competition (Dean 2001; Hildebrandt and McGuire 2002). This may lead to an overall decline in foraging efficiency as the social benefits of a successful hunt may come to outweigh the considerable economic costs of locating and transporting large game from distant hunting territories. Although resorting to specialised hunting may not seem to fit with the predictions of optimal foraging models, it emphasises instead the social aspect of hunters' decision making. Hunters may be willing to incur high costs associated with locating large game if it confers a strong social benefit in the form of status, prestige, increased access to marriage partners, etc. (Hawkes 1991).

In sum, the literature suggests that there are multiple responses to sedentism-related game depression other than domestication. The first of these involves turning increasingly towards lower ranked prey such as hares and birds in the context of "garden hunting". In cases where environmental productivity is high and human populations modest, the impact on game populations may be minimal, requiring no significant changes in hunting strategies. Finally, game depression may lead to the development of elaborate, specialised and logistically organised hunting strategies in which hunters target resource rich areas far from sedentary settlements.

Testing the Big Game Depression Model using Near Eastern Archaeofaunas

The hypothesis that large game resource depression led to the process of animal domestication can be examined using several indices drawn from published faunal reports from Near Eastern archaeological sites. Firstly, I use the ungulate index, which measures the frequency of ungulates compared to small vertebrates in archaeofaunal assemblages. Near Eastern faunal assemblages are dominated by ungulate taxa including wild/domestic sheep (*Ovis orientalis* and *Ovis aries*); wild/domestic goats (*Capra aegagrus* and *Capra hircus*); Nubian ibex (*Capra nubiana*); aurochs/domestic cattle (*Bos primigenius* and *B. taurus*); cervids (*Cervus elaphus*; *Dama dama* and *D. mesopotamica*; *Capreolus capreolus*); and boar/domestic pig (*Sus scrofa*). The most common small taxa include hare (*Lepus capensis*), tortoise (*Testudo graeca*) and a variety of edible birds including pheasants, ducks and bustard (*Otis tarda*) (following Munro 2004b; Stiner et al. 2000). This index provides a very general measure of the abundance of the highest ranked game resources and provides a proxy for monitoring any depression in those resources through time and space.

Secondly, I calculated a caprine index, which looks specifically at the frequency of *Ovis* and *Capra* (including *O. aries*, *O. orientalis*, *C. hircus*, *C. aegagrus* and *C. nubiana*) compared to all other ungulates. This measure provides a way to identify any declines in the availability of sheep and goats over time, which is of particular interest since these were the first successfully domesticated livestock.

I also calculated the small game index, which compares the ratio of "fast" versus "slow" small game, with fast game represented by hares and edible birds while slow game is represented primarily by tortoises (Munro 2004b; Stiner et al. 2000). Due to differences in the reproductive rates and acquisition costs of these two types of small prey, this index has been used as a measure of occupational intensity with high frequencies of "fast" small game indicating more intensive sedentism and higher frequencies of "slow" small game suggesting lower degrees of occupational intensity.

Finally, prey demographics, or mortality profiles, are also examined in order to monitor for evidence of increasingly selective or intensive culling of caprines in early sedentary communities, which might indicate increasing hunting pressure over time (Davis et al. 1994; Munro 2004b). Juvenile status is defined based on both dental eruption and long bone fusion: the former based on the presence or absence of deciduous fourth premolars, and the latter on the fusion status of distal metapodial epiphyses.

In order to test the role of resource depression in the process of caprine domestication, these indices were calculated for 110 faunal assemblages across the Near East, representing time periods from the Middle Paleolithic to the PPNB (table 1). Chronological control is maintained through the use of time horizons defined based on the traditional Levantine cultural history framework (Bar-Yosef and Bar-Yosef Mayer 2002) (table 2). Although the use of these chronological divisions is somewhat problematic, especially in the northern Levant where there is no clear boundary between the Epipaleolithic and PPNA (e.g. Peasnall 2000), they are necessary for looking at diachronic patterns on a macro-regional scale. Geographically, sites were divided into five groupings including the southern Levant, the Negev, lowland Mesopotamia, upland northern Mesopotamia and Anatolia, and the Zagros region. Broadly speaking, these regions show a degree of internal homogeneity in terms of archaeological cultures, geography and subsistence adaptations.

TABLE 1

	Southern Levant		Negev/southern Jordan		lowland Mesopotamia	
Paleol.	Amud	1			Dederiyeh	27
	Misliya Cave	2			Umm el-Tlel	27
	Hayonim Cave	3				
	Kebara Cave	4				
	Ohalo II	5				
	Ksar Akil	6				
Epipal	Ksar Akil	6	Rosh Horesha	19	Mureybet	28
	Hayonim Cave	3	Wadi Mataha	20		
	Nahal Hadera	7				
	Neve David	7				
	Hefsibah	7				
	Kebara Cave	4				
	Nahal Oren	8				
	Mallaha	9				
	Hatoula	10				
	El Wad Cave	11				
	El Wad Terrace	7				
	Hilazon Tachtit	11				
	Wadi Judayid	12				
PPNA	Netiv Hagdud	9	Abu Salem	21	Göbekli Tepe	29
	Hatoula	10	Wadi Faynan 16	22	Abu Hureyra	30
	Nahal Oren	9	El Khiam	23	Mureybet	31
	Jericho	9			Nemrik	32
	Gilgal	13			M'Lefaat	33
					Qermez Dere	34
PPNB	Aswad	14	Wadi Tbeik	24	Halula	35
	Ghoraife	15	Ujrat El-Mehed	25	Nevali Çori	36
	Motza	16	Beidha	26	Gürcütepe II	37
	Beisamoun	17	El Khiam	23	Gritille	38
	Ain Jammam	18			Sabi Abyad	39
					Abu Hureyra	30
					Mureybet	31

TABLE 1 *(continued).*

	Upland Mesopotamia/ Anatolia		Zagros	
Paleol.	Üçağızlı	40	Bisitun	48
			Tamtama	48
			Kobeh Cave	49
			Shanidar	50
			Warwasi	51
			Ghar-i-Khar	52
			Yafteh Cave	53
Epipal.	Üçağızlı	40	Palegawra	51, 54
	Karain B	41	Warwasi	51
	Öküzini	41	Ghar-i-Khar	52
	Direkli	42	Zawi Chemi	55
			Karim Shahir	56
PPNA	Hallan Çemi	43	Asiab	54, 57
	Körtik Tepe	44		
	Çayönü	45		
	Pinarbasi A	46		
PPNB	Cafer	47	Ganj Dareh	58
	Çayönü	45	Ali Kosh	59

List of faunal assemblages used in this study.

1=(Horwitz and Hongo 2008); 2=(Yeshurun et al. 2007); 3=(Stiner 2006); 4=(Saxon 1974); 5=(Rabinovich and Nadel 2005); 6=(Kersten 1989, 1991); 7=(Bar-Oz and Dayan 2005); 8=(Noy et al. 1973); 9=(Tchernov 1993, 1994); 10=(Davis et al. 1994); 11=(Munro 2004); 12=(Henry et al. 1985); 13=(Noy et al. 1980); 14= (Helmer and Gourichon 2008); 15= (Ducos 1995); 16=(Sapir-Hen et al. 2009); 17=(Davis 1982); 18=(Makarewicz 2009); 19=(Horwitz and Goring-Morris 2000); 20=(Janetski and Baadsgaard 2005, Whitcher et al. 2000); 21=(Butler et al. 1977); 22=(Carruthers 2002); 23=(Ducos 1997); 24=(Tchernov and Bar-Yosef 1982); 25=(Dayan et al. 1986); 26=(Hecker 1975); 27=(Griggo 1998, Griggo 2004); 28=(Helmer 1991); 29=(Peters and Schmidt 2004); 30=(Legge and Rowley-Conwy 2000); 31=(Gourichon and Helmer 2008); 32=(Lasota-Moskalewska 1994); 33=(Lasota-Moskalewska 1998); 34=(Dobney et al. 1999); 35=(Sana Segui 2000); 36=(Peters et al. 2005); 37=(von den Driesch and Peters 1999); 38=(Monahan 2000); 39=(Cavallo 2000). 40=(Kuhn et al. 2009); 41=(Atici 2009); 42=(Arbuckle and Erek 2012); 43=(Starkovich and Stiner 2009); 44=(Arbuckle and Özkaya 2007); 45=(Hongo and Meadow 2000); 46=(Carruthers 2005); 47=(Helmer 2008); 48=(Coon 1951); 49=(Marean and Kim 1998); 50=(Evins 1982); 51=(Turnbull and Reed 1974); 52=(Hesse 1989); 53=(Mashkour et al. 2009); 54=(Munro 2009); 55=(Perkins 1964); 56=(Stampfli 1983); 57=(Bökönyi 1977); 58=(Hesse 1978); 59=(Hole et al. 1969).

Cultural periods	Approximate dates (BP calibrated)
Middle Paleolithic	>100,000–45,000
Upper Paleolithic	45,000–18,000
Epipaleolithic	18,000–11,500
PPNA	11,500–10,500
PPNB	10,500–8500

TABLE 2. *Time horizons used in this study.*

Expectations

The game depression model for animal domestication predicts that the ungulate index should decline step-wise from the Upper Paleolithic to the Epipaleolithic and from the Epipaleolithic to the PPNA in response to the spread of increasingly sedentary hunter-gatherer communities across the Fertile Crescent. The ungulate index should then increase dramatically in the PPNB when the domestication of sheep and goats, and later cattle and pigs, provided increased access to these animal resources. If animal domestication was driven by game depression then the caprine index should follow the same general pattern, reaching its lowest point in the PPNA just prior to the domestication of sheep and goats. The frequency of sheep and goats should then rebound following their domestication in the early PPNB.

For the small game index, a general increase in occupational intensity from the Epipaleolithic to the PPNA and continuing into the PPNB should result in a steady decline in "slow" small game, such as tortoises, and an increase in the frequency of hares and birds. Expectations for prey demographics are largely derived from faunal work in the southern Levant showing that an increase in the frequency of juvenile gazelles in the Epipaleolithic and PPNA represents increased hunting stress (Davis et al. 1994; Munro 2004b). An increase in the frequency of juvenile caprines in the PPNA would similarly suggest that these populations were under significant hunting pressure just prior to domestication.

Results

The ungulate index was calculated for eighty-seven assemblages throughout the Near East covering periods from the Middle Paleolithic through the Neolithic (fig. 1). This clearly shows that large ungulates are generally very well represented in the archaeofaunas of the region in all periods and in all regions. Big game frequencies decline from the Upper Paleolithic to the Epipaleolithic and remain constant in the PPNA. They increase in the PPNB and Pottery Neolithic corresponding to the advent of herding economies in the region.

Ungulate index values were also calculated for five geographic regions within the Near East in order to examine evidence for spatial variation in changes in the abundance of big game (fig. 2). In the upland regions of the Zagros and northern Mesopotamia (including Anatolia) ungulates are abundant in the Epipaleolithic and decline only slightly in the PPNA, while in the Negev and the Mesopotamian lowlands ungulates increase in frequency in PPNA assemblages. The clear outgroup is the southern Levant with the lowest index values of any region, and which shows a dramatic decrease in ungulates in the PPNA followed by a major increase in the PPNB.

The caprine index provides an important way to monitor for changes in access to sheep and goats, the first domesticated food animals (fig. 3). Caprine index values vary dramatically between regions reflecting the habitat preferences of wild sheep and goats whose home ranges include the mountain and piedmont regions of the Fertile Crescent including upland northern Mesopotamia, Anatolia and the Zagros. A third type of caprine, the Nubian ibex (*Capra nubiana*), was heavily exploited in the Negev in the Epipaleolithic and early Neolithic. Caprines were a minor part of the local fauna in the southern Levant and lowland Mesopotamia where gazelle and equids constituted the primary prey species until the PPNB. Caprine index values remain relatively stable through time in each region, suggesting only minor changes in the availability of wild caprines, with the exception that they increase in every region in the PPNB with the domestication and sheep and goats and the spread of pastoralism throughout the region.

Finally, the small game index is presented in figures 4–5 for sites in the southern Levant, and Anatolia and the Zagros regions. For sites in the southern Levant, fast game are abundant in the early Epipaleolithic including the Kebaran and Early Natufian. The use of fast game decreases in the Late

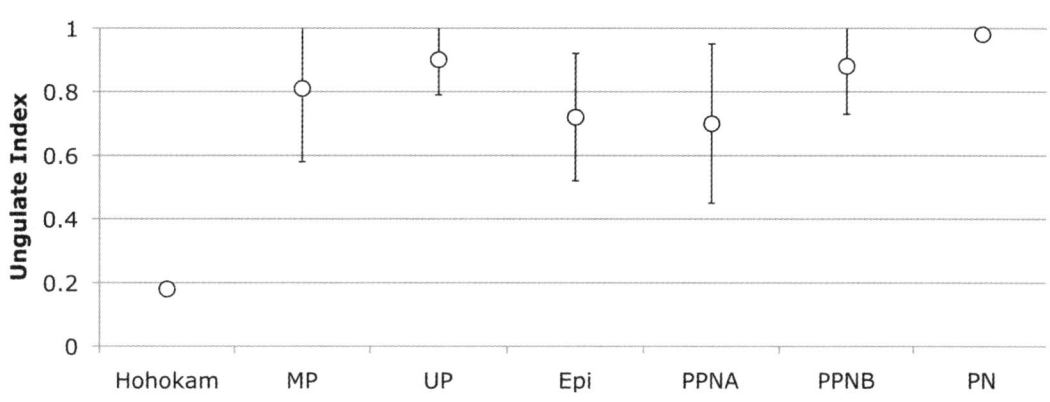

FIGURE 1. *Mean and one standard deviation ranges for ungulate index values calculated for time horizons. MP = Middle Paleolithic; UP = Upper Paleolithic; Epi = Epipaleolithic without Natufian values; PPNA = Pre-Pottery Neolithic A; PPNB = Pre-Pottery Neolithic B; PN = Pottery Neolithic. Hohokam (an early farming culture in the American Southwest) values are included for comparison (from James 2004; Szuter and Bayham 1989).*

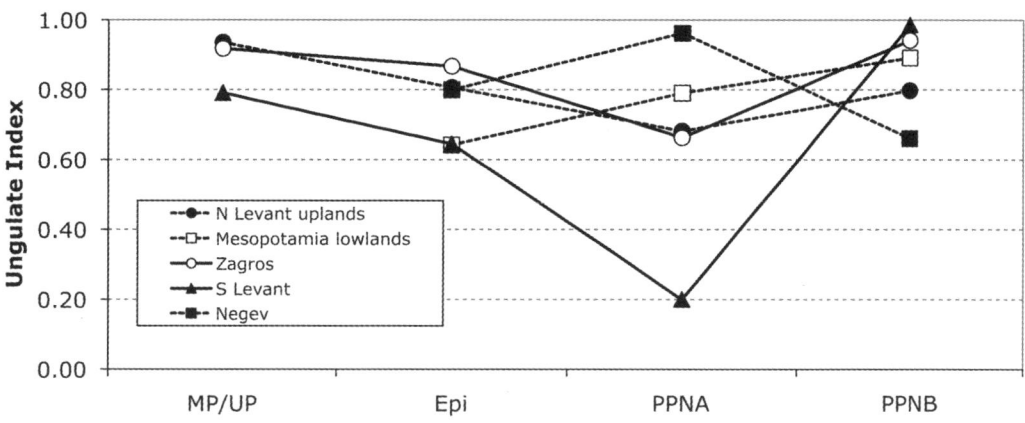

FIGURE 2. *Mean ungulate index values calculated for five geographic regions through time.*

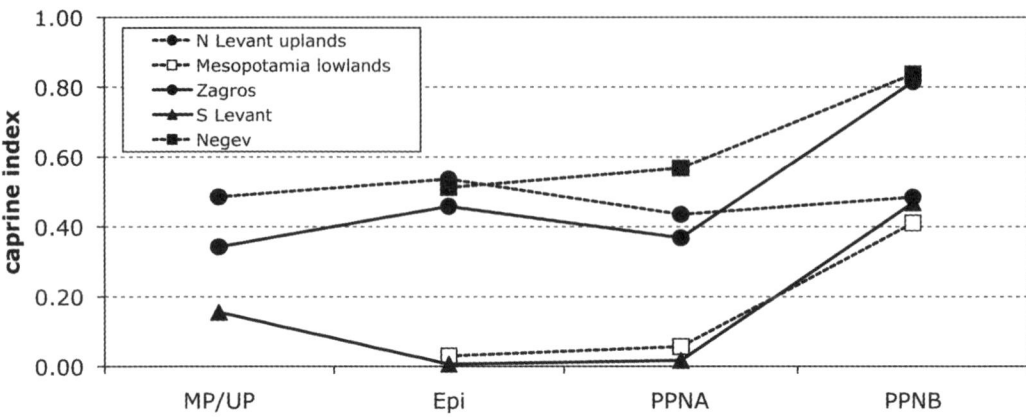

FIGURE 3. *Mean caprine index values for five geographic regions through time.*

Natufian, likely as a result of a reduction in occupational intensity as Late Natufian foragers adapted to the difficult conditions of the Younger Dryas with increased mobility (Munro 2003, 2004b, 2009). The frequency of fast game increases dramatically at sites such as Netiv Hagdud and Hatoula, indicating a major increase in occupational intensity and resource stress in the PPNA in the southern Levant.

Outside of the southern Levant, the small game index exhibits very different patterns. The frequency of fast game is relatively high in Upper Paleolithic and Epipaleolithic sites such as Üçağızlı, Karain B and Öküzini caves indicating intensive and multi-seasonal use of cave sites during the late Pleistocene on the Mediterranean coast (Atici 2009; Kuhn et al. 2009). Although there are no faunal assemblages from which to calculate the small game index for the latest Epipaleolithic – including the Younger Dryas – outside the southern Levant, fast game decline in the PPNA at the sedentary sites Hallan Çemi (southeastern Turkey) and Asiab (western Iran) and at Direkli Cave (southern Turkey) suggests relatively low occupational intensity at these sites. Fast game increase dramatically in PPNB villages in both Anatolia and the Zagros at sites such as Aşıklı Höyük, Ganj Dareh and Gritille Höyük.

The mean frequencies of juvenile caprines in Near Eastern assemblages are presented in figure 6. The frequency of juvenile caprines is generally low in Paleolithic assemblages but increases dramatically to 35% in the Epipaleolithic, perhaps associated with an increase in hunting pressure combined with the widespread adoption of bow and arrow technology. There is virtually no change in the frequency of juveniles culled in PPNA assemblages (38%) while a major increase (51% juveniles) occurs with the advent of herding in the PPNB.

Discussion

In general, the data presented above do not support the idea that depression of big game resources was the causal factor behind the domestication of livestock in the Near East. The ungulate index indicates that large game was abundant in Near Eastern assemblages in most regions and in most periods. While there is evidence for some depression in large game populations in the late Epipaleolithic and PPNA, outside of the southern Levant the scale of this decline is relatively small, and, importantly, we do not see consistent or severe declines in ungulates in the PPNA just prior to the domestication of livestock.

In the southern Levant, however, there is ample evidence for depression of ungulate populations around sedentary communities in the Epipaleolithic and especially the PPNA, when index values are comparable to those from the American Southwest (fig. 1). At sites like Netiv Hagdud, PPNA hunters responded to declines in big game by turning increasingly to the exploitation of hare and birds, perhaps utilising strategies of "garden hunting" to supplement their diets in an environment unable to support large ungulate populations.

Contrary to the predictions of the resource depression model there are no significant changes in the frequency of caprines, the first domesticated food animals, in any region in the PPNA. In the upland regions of Anatolia, the northern Levant and the Zagros wild caprines continue to remain the primary prey animal as they had in previous periods, and in the Negev Nubian ibex were increasingly the target of hunting. In the southern Levant and lowland Mesopotamia, where caprines were rare prior to their domestication in the PPNB, measurements of their relative abundance remain virtually identical in the Epipaleolithic and the PPNA signifying no major changes in availability.

Unlike the case in the southern Levant, where increasingly intensive gazelle hunting led to a focus on younger animals in the PPNA, mortality profiles from sites in Mesopotamia, Anatolia and the Zagros provide no evidence for increasingly intensive harvesting of young caprines in the PPNA. Finally, the small game index suggests that although proxy measures of occupational intensity are high for the PPNA of the southern Levant (fig. 4), the occupational intensity of PPNA communities outside of the southern Levant may have been highly variable (fig. 5). At key PPNA period sites in the northern uplands and Zagros – such as Hallan Çemi, Asiab and Çayönü – there is virtually no evidence that wild caprines, or any ungulates for that matter, were under any significant stress at all.

However, data from the American Southwest suggest that there are potentially multiple ways to interpret the relative stability in the ungulate and caprine indices seen in the Near Eastern dataset in the Epipaleolithic and PPNA. In the American Southwest, in response to local resource depression, sedentary hunters instituted systems of increasingly specialised, logistically organised hunting, which maintained access to big game (albeit at higher costs) despite declines in local big game populations (Speth and Scott 1989).

Although this remains a possible interpretation of the ungulate and caprine data for the Near East, there is currently no evidence to suggest that hunting practices became increasingly specialised or that hunters had to travel increasingly long distances to find game in the PPNA, particularly in the environmentally productive upland regions of the Fertile Crescent. In cases where hunters do move greater distances to find game, we expect to find changes in the representation of the skeletal elements brought back to residential settlements with those elements with the highest food utility being disproportionately selected for transport (e.g. Monahan 1998). In the few cases where such data are available for PPNA sites, there is no evidence to suggest such selective transport of caprine carcasses and therefore little support for the presence of specialised, logistical hunting strategies in the PPNA (Arbuckle and Özkaya 2007; Starkovich and Stiner 2009).

Together, the ungulate and caprine indices and demographic data do not suggest that there was a major change in the availability of big game or that animal exploitation systems changed dramatically in the PPNA just prior to the emergence of agro-pastoral economies. Instead, most indices suggest that major changes took place in the late Epipaleolithic with the appearance

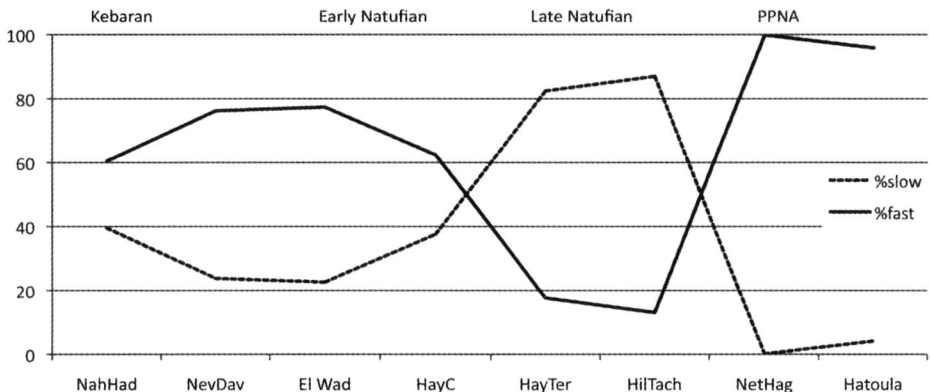

FIGURE 4. *Small game index values for sites in the southern Levant. NahHad = Nahal Hadera; NevDav = Neve David; HayC = Hayonim Cave; HayTer = Hayonim Terrace; HilTach = Hilazon Tachtit; NetHag = Netiv Hadud (Davis et al. 1994; Munro 2009).*

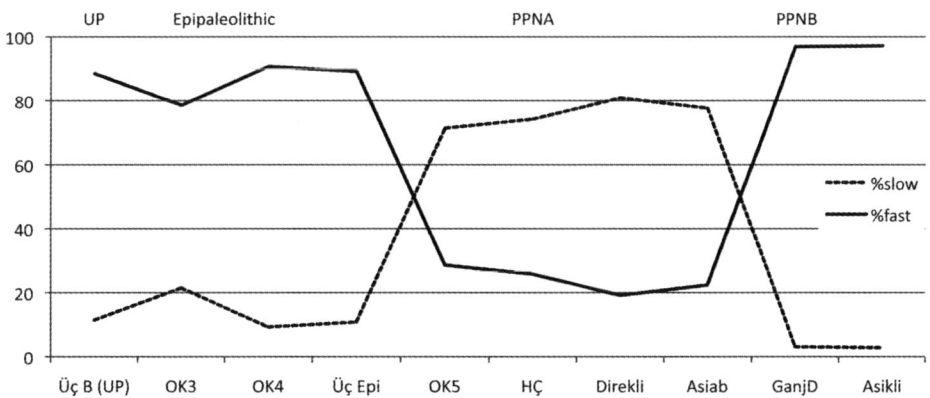

FIGURE 5. *Small game index values for sites outside of the southern Levant. Üç B = Üçağızlı layer B (Upper Paleolithic); OK3–5 = Öküzini Cave phases 3–5; Üç Epi = Üçağızlı Epipaleolithic; HÇ = Hallan Çemi; GanjD = Ganj Dareh (Arbuckle and Erek 2012; Atici 2009; Buitenhuis 1997; Kuhn et al. 2009; Munro 2004a; Starkovich and Stiner 2009).*

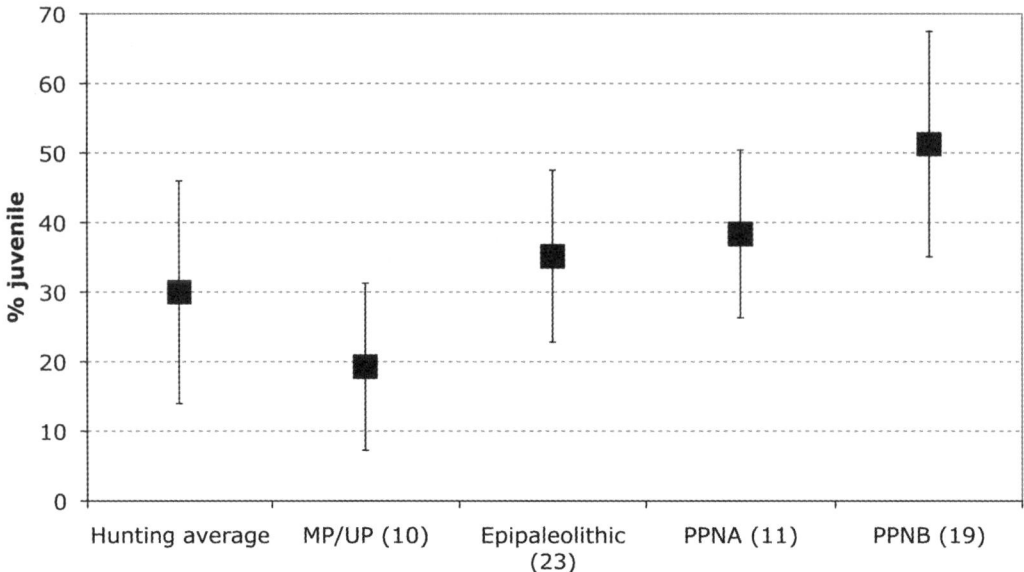

FIGURE 6. *Mean values and one standard deviation ranges for the frequency of juvenile caprines in faunal assemblages through time. Numbers in parentheses indicate number of assemblages used to calculate these values.*

of the first sedentary communities. In many ways, PPNA subsistence strategies are fundamentally Epipaleolithic in nature (Bar-Yosef and Meadow 1995; Colledge 2001; Horwitz 1996; Savard et al. 2006) and the data presented here show that the conditions of caprine, and more generally ungulate, resource depletion were basically identical in the late Epipaleolithic and PPNA for most regions outside of the southern Levant. This is significant because it suggests that the pressures that are widely thought to have fuelled the process of animal domestication at the PPNA/PPNB transition (i.e. sedentism-related resource depletion) seem to have been in place some 3,000 years before the domestication of the first livestock animals, i.e. sheep and goats, in the early PPNB.

So what, then, explains the lack of fit between the resource depression model and the Near Eastern faunal data? Part of the problem lies in the fact that this model was developed by researchers working in the southern

Levant, a region where ungulate depression was, in fact, very intense in the PPNA. However, it is likely that the southern Levant was not an early center of animal domestication, and the scale of game depression seen in that region is not evident in other, more environmentally productive parts of the Near East where the domestication process is thought to have begun (Peters et al. 2005; Zeder 2008). This suggests that intensive resource stress, as seen in the PPNA of the southern Levant, did not lead to the initial domestication of livestock but, rather, to the adoption of caprine management practices already in use in neighbouring regions.

Exploring the social side of hunting and the domestication process

The lack of zooarchaeological support for the resource depression model of animal domestication suggests that modelling the transition from hunting to herding in purely economic terms is problematic. The resource depletion model focuses on the diminishing returns of hunting in early sedentary communities but does not acknowledge the social context in which hunting took place in prehistoric communities (see Garfinkel et al. 2009; Hamilakis 2003). By focusing more on hunting as a social, rather than strictly as an economic behaviour, we may find an explanation for the failure of the resource depression model to explain the archaeological patterns.

When we look at the animal domestication process we can imagine two competing forces at work. On the one hand, there is the undeniable economic value of a managed herd and on the other there is the social value of hunting large game. Ethnographically, hunting large game often plays a central role in the development of male identity, acquisition of prestige and even access to marriage partners (Dean 2001; Garfinkel et al. 2009; Hawkes 1991). In order for the process of animal domestication to be initiated it is not enough that there be an economic problem (such as resource depression) that needs to be addressed. As we have seen in the data from the American Southwest, there are multiple possible responses to resource depression, and so the domestication option also has to outweigh the social value of continuing to hunt the very same animals.

Epipaleolithic and PPNA communities facing even mild forms of depression in large game may have responded to this problem in ways similar to many of their counterparts in the American Southwest and intensified their efforts to hunt large game with increasingly elaborate and ritualised systems of hunting (Garfinkel et al. 2009; James 2004; Szuter and Bayham 1989). This is made more likely by the presence of intensive plant processing technologies, which suggests that male hunters could devote more time to high risk and reward big game hunting (Speth and Scott 1989). These increasingly elaborate hunting systems – which seem to be reflected in Epipaleolithic and Neolithic art in the form of figurines, wall paintings, feasting events, etc. (Helmer et al. 2004; Peters and Schmidt 2004; Rosenberg 1999; Russell and Meece 2005) – may have pushed the social value of hunting so high that it stifled any attempts at innovation in the form of managing and domesticating animals.

As a result, the timing of the appearance of domestic caprines may have more to do with reaching a 'tipping point' where the economic benefits of herding finally overcame the social benefits of hunting, rather than being the direct result of a rapid increase in sedentism-related game depression. This would partially, at least, explain the time lag between the initiation of game depression in the late Epipaleolithic, as seen in the ungulate and caprine indices and demographics, and the initial emergence of livestock management in the early PPNB several millennia later.

In addition, it may be significant that sheep and goats were the first food animals to come under human control (Zeder 2008). This suggests that the point of inflection where the economic benefits of herding overcame the social value of hunting was reached first for these small stock. Caprines were the ungulates with the least symbolic importance in late Pleistocene hunting systems, as suggested by their low representation in Neolithic symbolism (e.g. caprines represent less than 4% of animal representations at Göbekli Tepe and Çatalhöyük) (Peters and Schmidt 2004; Russell and Meece 2005), and the combination of the high economic value of herding caprines coupled with the relatively low social value of hunting them may explain why sheep and goat herding was initiated before the management of the more symbolically potent aurochs and boar.

Conclusion

Despite being the most widely cited explanation for the initial domestication of livestock, Tchernov's resource depression model is not supported by a combination of zooarchaeological measures of access to big game from across the Near East. Although stress on big game was clearly a major issue in the southern Levant in the PPNA, it seems to have played only a minor role in the process of initial animal domestication. Instead, these datasets suggests that large game, including ungulates and especially caprines, remained abundant in Near Eastern faunal assemblages even in the PPNA, the period just prior to the domestication of livestock. Moreover, moderate game depression likely occurred in the late Epipaleolithic and continued relatively unchanged into the PPNA in most regions, suggesting that the processes leading to animal domestication were more complex than a simple decline in the availability of favoured prey species.

It is further suggested that the long delay between the initiation of sedentary communities in the late Epipaleolithic and PPNA and the domestication of sheep and goats in the early PPNB (c. 10,500 BP calibrated) may be understood as the result of competition between the economic value of herding domestic animals and the social value of hunting. This leads ultimately to the suggestion that one of the most effective economic responses to resource depletion for sedentary foragers/farmers, i.e. animal domestication, may not have been a viable social option for most Near Eastern hunters who functioned in a world where status, prestige and access to marriage partners may have been intertwined with hunting prowess.

These results have several interesting implications including the idea that the real engine behind the process of animal domestication may have been people on the fringes of society, those outside of traditional power structures, and those who had the least to gain from participating in the traditional systems of hunting-related social competition. This focus on the social context of hunting is an area that needs further exploration and may bring new insights into our understanding of the complex social processes involved in the domestication of animals.

Bibliography

Alvard, M. S., and L. Kuznar. 2001. "Deferred Harvests: The Transition from Hunting to Animal Husbandry." *American Anthropologist* 103: 295–311.

Arbuckle, B. S., and V. Özkaya. 2007. "Animal Exploitation at Körtik Tepe: An Early Aceramic Neolithic Site in Southeastern Turkey." *Paléorient* 32: 198–211.

Arbuckle, B. S., and C. M. Erek. 2012. "Late Epipaleolithic Hunters of the Central Taurus: Faunal Remains from Direkli Cave, Kahramanmaras, Turkey." *International Journal of Osteoarchaeology* 22: 694–707.

Atici, L. 2009. "Implications of Age Structures for Epipaleolithic Hunting Strategies in the Western Taurus Mountains, Southwest Turkey." *Anthropozoologica* 44: 13–40.

Badenhorst, S., and J. C. Driver. 2009. "Faunal Changes in Farming Communities from Basketmaker II to Pueblo III." *Journal of Archaeological Science* 36: 1832–1841.

Bar-Oz, G., and T. Dayan. 2005. "Zooarchaeological Diversity and Palaeoecological Reconstruction of the Epipalaeolithic Faunal Sequence in the Northern Coastal Plain and the Slopes of Mount Carmel, Israel." In H. Buitenhuis, A. M. Choyke, L. Martin, L. Bartosiewicz and M. Mashkour (eds.), *Archaeozoology of the Near East VI*, ARC Publication, 50–60. Groningen.

Bar-Yosef, O., and R. H. Meadow. 1995. "The Origins of Agriculture in the Near East." In T. D. Price and A. B. Gebauer (eds.), *Last Hunters, First Farmers: New Perspectives on the Prehistoric Transition to Agriculture*, School of American Research Press, 39–94. Santa Fe.

Bar-Yosef, O., and D. E. Bar-Yosef Mayer. 2002. "Early Neolithic Tribes in the Levant." In W. A. Parkinson (ed.), *The Archaeology of Tribal Societies*, International Monographs in Prehistory Archaeological 15, 340–371. Ann Arbor.

Binford, L. R. 1968. "Post-pleistocene Adaptations." In S. Binford and L. R. Binford (eds.), *New Perspectives in Archaeology*, Aldine, 313–341. Chicago.

Bökönyi, S. 1977. *Animal Remains from the Kermanshah Valley, Iran*. BAR Supplement Series 34. Oxford.

Broughton, J. M. 1994. "Declines in Mammalian Foraging Efficiency during the Late Holocene, San Francisco Bay, California." *Journal of Anthropological Archaeology* 13: 371–401.

Buitenhuis, H. 1997. "Aşıklı Höyük: A 'Protodomestication' Site." *Anthropozoologica* 25–26: 655–662.

Butler, B. H., E. Tchernov, H. Hietala and S. Davis. 1977. "Faunal Exploitation during the Late Epipaleolithic in the Har Harif." In A. Marks (ed.), *Prehistory and Paleoenvironments in the Central Negev, Israel*, vol. 2, Institute for the Study of Earth and Man, 327–346. Dallas.

Cannon, M. D. 2000. "Large Mammal Relative Abundance in Pithouse and Pueblo Period Archaeofaunas from Southwestern New Mexico: Resource Depression among the Mimbres-Mogollon?" *Journal of Anthropological Archaeology* 19: 317–347.

Carruthers, D. 2002. "The Dana-Faynan-Ghuwayr Early Prehistory Project Preliminary Animal Bone Report on Mammals from Wadi Faynan 16." In H. Buitenhuis, A. Choyke, M. Mashkour and A. H. Al-Shiyab (eds.), *Archaeozoology of the Near East V*, ARC Publication 62, 93–97. Groningen.

Carruthers, D. 2005. "Hunting and Herding in Central Anatolian Prehistory: The 9th and 7th Millennium Site at Pınarbası." In H. Buitenhuis, A. M. Choyke, L. Martin, L. Bartosiewicz and M. Mashkour (eds.), *Archaeozoology of the Near East VI*, ARC Publication 123, 85–95. Groningen.

Cavallo, C. 2000. *Animals in the Steppe: A Zooarchaeological Analysis of the Later Neolithic Tell Sabi Abyad, Syria*. BAR International Series 891. Oxford.

Colledge, S. 2001. *Plant Exploitation on Epipaleolithic and Early Neolithic Sites in the Levant*. British Archaeological Reports, International Series 986. Oxford.

Coon, C. S. 1951. *Cave Explorations in Iran, 1949*. University of Pennsylvania Museum. Philadelphia.

Davis, S. J. 1982. "Climatic Change and the Advent of Domestication: The Succession of Ruminant Artiodactyls in the Late Pleistocene-Holocene in the Israel Region." *Paléorient* 8: 5–15.

Davis, S. J., O. Lernau and J. Pichon. 1994. "The Animal Remains: New Light on the Origin of Animal Husbandry." In M. Lechevallier and A. Ronen (eds.), *Le site de Hatoula en Judée occidentale, Israel*, Mémoires et Travaux du Centre de Recherche Français de Jérusalem 8, 83–100. Jerusalem.

Dayan, T., E. Tchernov, O. Bar-Yosef and Y. Yom-Tov. 1986. "Animal Exploitation in Ujrat El-Mehed, a Neolithic Site in Southern Sinai." *Paléorient* 12: 105–116.

Dean, R. 2001. "Social Change and Hunting during the Pueblo III to Pueblo

IV Transition, East-central Arizona." *Journal of Field Archaeology* 28: 271–285.

Dobney, K., M. Beech and D. Jaques. 1999. "Hunting the Broad Spectrum Revolution: The Characterization of Early Neolithic Animal Exploitation in Qermez Dere, Northern Mesopotamia." In J. C. Driver (ed.), *Zooarchaeology of the Pleistocene/Holocene Boundary*, BAR International Series 800, 47–58. Oxford.

Ducos, P. 1995. "Note préliminaire dur les faunes d'Aswad et Ghoraifé." In H. de Contenson (ed.), *Aswad et Ghoraifé. Sites Néolithiques en Damascene (Syrie), aux IXeme et VIIIeme millénaires avant l'ere Chrétienne*, Bibliotheque Archéologique et Historique, 339–349. Beirut.

Ducos, P. 1997. "A Re-evaluation of the Fauna from the Neolithic Levels of El-Khiam." *Journal of the Israeli Prehistoric Society* 27: 75–81.

Evins, M. A. 1982. "The Fauna from Shanidar Cave: Mousterian Wild Goat Exploitation in Northeastern Iraq." *Paléorient* 8: 37–58.

Flannery, K. V. 1969. "Origins and Ecological Effects of Early Domestication in Iran and the Near East." In P. J. Ucko and G. W. Dimbleby (eds.), *The Domestication and Exploitation of Plants and Animals*, 73–100. Duckworth, London.

Flannery, K. V. 1973. "The Origins of Agriculture." *Annual Review of Anthropology* 2: 271–310.

Garfinkel, A. P., D. A. Young and R. M. Yohe. 2009. "Bighorn Hunting, Resource Depletion, and Rock Art in the Coso Range: A Computer Simulation." *Journal of Archaeological Science* 37: 42–51.

Gourichon, L., and D. Helmer. 2008. "Étude archéozoologique de Mureybet." In J. J. Ibáñez (ed.), *Le site néolithique de Tell Mureybet (Syrie du Nord)*, BAR International Series 1843, 115–228. Oxford.

Griggo, C. 1998. "La fauna Moustérienne du site d'Umm el Tlel (Syrie). Étude préliminaire. Fouilles 1991–1993." *Cahiers de l'Euphrate* 8: 11–26.

Griggo, C. 2004. "Mousterian Fauna from Dederiyeh Cave and Comparisons with Fauna from Umm El Tlel and Douara Cave." *Paléorient* 30: 149–162.

Hamilakis, Y. 2003. "The Sacred Geography of Hunting: Wild Animals, Social Power and Gender in Early Farming Societies." In E. Kotjabopoulou, Y. Hamilakis, P. Halstead, C. Gamble and P. Elefanti (eds.), *Zooarchaeology in Greece: Recent Advances*, British School at Athens 10, 239–247. London.

Hawkes, K. 1991. "Showing off: Tests of Another Hypothesis about Men's Foraging Goals." *Ethnology and Sociobiology* 11: 29–54.

Hecker, H. 1975. *The Faunal Analysis of the Primary Food Animals from the Pre-Pottery Neolithic Beidha (Jordan)*. Unpublished PhD Dissertation, Columbia University. New York.

Helmer, D. 1991. "Etude de la faune de la Phase 1a (Natoufien final) de Tell Mureybet (Syrie), fouilles Cauvin." In O. Bar-Yosef and F. R. Valla (eds.), *The Natufian Culture of the Levant*, International Monographs in Prehistory, 359–370. Ann Arbor.

Helmer, D., L. Gourichon and D. Stordeur. 2004. "À l'aube de la domestication animale. Imaginaire et symbolisme animal dans les premières sociétés néolithiques du nord du Proche-Orient." *Anthropozoologica* 39: 143–163.

Helmer, D. 2008. "Revision de la faune de Cafer Höyük (Malatya, Turquie): apports des methodes de l'analyse des melanges et de l'analyse de Kernel a la mise en evidence de la domestication." In E. Vila, L. Gourichon, A. Choyke and H. Buitenhuis (eds.), *Archaeozoology of the Near East VIII*, Travaux de la Maison de l'Orient et de la Méditerranée 49, 169–196. Lyon.

Helmer, D., and L. Gourichon. 2008. "Premières données sur les modalités de subsistance à Tell Aswad (Syrie, PPNB Moyen et Récent, Néolithique Céramique Ancien) – Fouilles 2001–2005." In E. Vila, L. Gourichon, A. Choyke and H. Buitenhuis (eds.), *Archaeozoology of the Near East VIII*, Travaux de la Maison de l'Orient et de la Méditerranée 49, 119–151. Lyon.

Henry, D. O., P. F. Turnbull, A. Emery-Barbier and A. Leroi-Gourhan. 1985. "Archaeological and Faunal Evidence from Natufian and Timnian Sites in Southern Jordan, with Notes on Pollen Evidence." *Bulletin of the American Schools of Oriental Research* 257: 45–64.

Hesse, B. 1978. *Evidence for Husbandry from the Early Neolithic site of Ganj Dareh in Western Iran*. Unpublished PhD Dissertation, Columbia University. New York.

Hesse, B. 1989. "Paleolithic Faunal Remains from Ghar-i-Khar, Western Iran." In P. J. Crabtree, D. V. Campana and K. Ryan (eds.), *Early Animal Domestication and its Cultural Context*, The Museum Applied Science Center for Archaeology, 37–46. Philadelphia.

Hildebrandt, W. R., and K. R. McGuire. 2002. "The Ascendance of Hunting during the California Middle Archaic: An Evolutionary Perspective." *American Antiquity* 67: 231–256.

Hole, F., K. V. Flannery and J. A. Neely. 1969. *Prehistory and Human Ecology on the Deh Luran Plain*. Memoir No. 1. Museum of Anthropology, University of Michigan. Ann Arbor.

Hongo, H., and R. H. Meadow. 2000. "Faunal Remains from Prepottery Neolithic Levels at Çayönü, Southeastern Turkey: A Preliminary Report Focusing on Pigs (*Sus* sp.)." In M. Mashkour, A. M. Choyke, H. Buitenhuis and F. Poplin (eds.), *Archaeozoology of the Near East IVA*, ARC Publications, 121–139. Groningen.

Horwitz, L. K. 1996. "The Impact of Animal Domestication on Species Richness: A Pilot Study from the Neolithic of the Southern Levant." *Archaeozoologia* 8: 53–70.

Horwitz, L. K., and N. Goring-Morris. 2000. "Fauna from the Early Natufian site of Upper Besor 6 in the Central Negev, Israel." *Paléorient* 26: 111–128.

Horwitz, L. K., and H. Hongo. 2008. "Putting the Meat Back on Old Bones: A Reassessment of Middle Paleolithic Fauna from Amud Cave (Israel)." In E. Vila, L. Gourichon, A. M. Choyke and H. Buitenhuis (eds.), *Archaeozoology of the Near East VIII*, Travaux de la Maison de l'Orient et de la Méditerranée 49, 45–64. Lyon.

James, S. 2004. "Hunting, Fishing and Resource Depression in Prehistoric Southwest North America." In C. L. Redman, S. R. James, P. R. Fish and J. D. Rogers (eds.), *The Archaeology of Global Change: The Impact of Humans on Their Environment*, Smithsonian Books, 28–62. Washingon, D.C.

Janetski, J. 1997. "Fremont Hunting and Resource Intensification in the Eastern Great Basin." *Journal of Archaeological Science* 24: 1075–1088.

Janetski, J. C., and A. Baadsgaard. 2005. "Shifts in Epipaleolithic Faunal Exploitation at Wadi Mataha 2, Southern Jordan." In H. Buitenhuis, A. Choyke, L. Martin, L. Bartosiewicz and M. Mashkour (eds.), *Archaeozoology of the Near East VI*, ARC Publication 123, 24–32. Groningen.

Kersten, A. M. P. 1989. "The Epipalaeolithic Ungulate Remains from Ksar Akil: Some Preliminary Results." In I. Hershkovitz (ed.), *People and Culture in Change*, BAR International Series 508, 183–198. Oxford.

Kersten, A. M. P. 1991. "Birds from the Palaeolithic Rock Shelter of Ksar Akil, Lebanon." *Paléorient* 17: 99–116.

Kuhn, S. L., M. C. Stiner, E. Guleç, I. Ozer, H. Yilmaz, I. Baykara, A. Acikkil, P. Goldberg, K. M. Molist, E. Unay and F. Suata-Altaslan. 2009. "The Early Upper Paleolithic Occupations at Üçağızlı Cave (Hatay, Turkey)." *Journal of Human Evolution* 56: 87–113.

Lasota-Moskalewska, A. 1994. "Animal Remains from Nemrik, a Pre-pottery Neolithic Site in Iraq." In S. K. Kozlowski (ed.), *Nemrik 9: Pre-pottery Neolithic Neolithic site in Iraq*, Wydawnictwa Uniwersytetu Warszawskiego, 5–52. Warsaw.

Lasota-Moskalewska, A. 1998. "Mammal Bone Remains from M'lefaat, Irak (Ecavation Seasons 1989 and 1990)." *Cahiers de l'Euphrate* 8: 215–221.

Legge, A. J., and P. A. Rowley-Conwy. 2000. "The Exploitation of Animals." In A. M. T. Moore, G. C. Hillman and A. J. Legge (eds.), *Village on the Euphrates*, 423–471. Oxford University Press. Oxford.

Linares, O. F. 1976. "Garden Hunting in the American Tropics." *Human Ecology* 4: 331–349.

Makarewicz, C. 2009. "Complex Caprine Harvesting Practices and Diversified Hunting Strategies: Integrated Animal Exploitation Systems at Late Pre-Pottery Neolithic B 'Ain Jammam." *Anthropozoologica* 44: 79–102.

Marean, C. W., and S. Y. Kim. 1998. "Mousterian Large-Mammal Remains from Kobeh Cave: Behavioral Implications for Neanderthals and Early Modern Humans." *Current Anthropology* 39: 79–113.

Mashkour, M., V. Radu, A. Mohaseb, N. Hashemi, M. Otte and S. Shidrang. 2009. "The Upper Paleolithic Faunal Remains from Yafteh Cave (Central Zagros), 2005 Campaign – A Preliminary Study." In M. Otte (ed.), *Iran Paleolithic*, BAR International Series 1968, 73–84. Oxford.

McCorriston, J., and F. Hole. 1991. "The Ecology of Seasonal Stress and the Origins of Agriculture in the Near East." *American Anthropologist* 93: 46–69.

Monahan, B. H. 2000. *The Organization of Domestication at Gritille, a Pre-Pottery Neolithic B Site in Southeastern Turkey*. Unpublished PhD Dissertation, Northwestern University. Evanston, Illinois.

Monahan, C. M. 1998. "The Hadza Carcass Transport Debate Revisited and its Archaeological Implications." *Journal of Archaeological Science* 25: 405–424.

Munro, N. D. 2003. "Small Game, the Younger Dryas, and the Transition to Agriculture in the Southern Levant." *Mitteilungen der Gesellschaft fur Urgeschichte* 12: 47–71.

Munro, N. D. 2004a. "Small Game Indicators of Human Foraging Efficiency and Early Herd Management at the Transition to Agriculture in South-West Asia." In J.-P. Brugal and J. Desse (eds.), *Petits animaux et societés humaines. Du complément alimentaire aux ressources utilitaires*, 515–531. Antibes.

Munro, N. D. 2004b. "Zooarchaeological measures of hunting pressure and occupational intensity in the Natufian." *Current Anthropology* 45: S5–33.

Munro, N. D. 2009. "Epipaleolithic subsistence intensification in the southern Levant: The faunal evidence." In J.-J. Hublin and M. P. Richards (eds.), *The Evolution of Hominin Diets: Integrating Approaches to the Study of Palaeolithic Subsistence*, Springer, 141–155. Leipzig.

Noy, T., A. J. Legge and E. J. Higgs. 1973. "Recent excavations at Nahal Oren, Israel." *Proceedings of the Prehistoric Society* 39: 75–99.

Noy, T., J. Schuldenrein and E. Tchernov. 1980. "Gilgal I, a pre-pottery Neolithic A site in the Lower Jordan Valley." *Israeli Exploration Journal* 30: 63–82.

Peasnall, B. L. 2000. *The round house horizon along the Taurus-Zagros arc: A synthesis of recent excavations of late Epipaleolithic and early aceramic sites in southeastern Anatolia and northern Iraq*. Unpublished PhD Dissertation, University of Pennsylvania. Philadelphia.

Perkins, D. 1964. "Prehistoric fauna from Shanidar, Iraq." *Science* 144: 1565–1566.

Peters, J., and K. Schmidt. 2004. "Animals in the symbolic world of Pre-Pottery Neolithic Gobekli Tepe, south-eastern Turkey: A preliminary assessment." *Anthropozoologica* 39: 179–218.

Peters, J., A. von den Driesch and D. Helmer. 2005. "The upper Ephrates-Tigris basin: Cradle of agro-pastoralism?" In J.-D. Vigne, J. Peters and D. Helmer (eds.), *The first steps of animal domestication: New archaeological approaches*, Oxbow, 96–124. Oxford.

Rabinovich, R., and D. Nadel. 2005. "Broken mammal bones: Taphonomy and food sharing at the Ohalo II submerged prehistoric camp." In H. Buitenhuis, A. M. Choyke, L. Martin, L. Bartosiewicz and M. Mashkour (eds.), *Archaeozoology of the Near East VI: Proceedings of the sixth international symposium on the archaeozoology of southwestern Asia and adjacent areas*, ARC Publication 123, 33–49. Groningen.

Reed, C. A., and D. Perkins. 1984. "Prehistoric domestication of animals in southwestern Asia." In G. Nobis (ed.), *Die Anfange des Neolithikums vom Orient bis Nordeuropa. Der Beginn der Haustierhaltung in der "Alten Welt"*, Bohlau Verlag, 1–23. Vienna.

Rosenberg, M. 1999. "Hallan Çemi." In M. Özdoğan and N. Başgelen (eds.), *Neolithic in Turkey*, Arkeoloji ve Sanat Yayınları, 25–33. Istanbul.

Russell, N., and S. Meece. 2005. "Animal representations and animal remains at Çatalhöyük." In I. Hodder (ed.), *Çatalhöyük perspectives: Reports from the 1995–99 seasons*, McDonald Institute for Archaeological Research, 209–230. Cambridge, UK.

Sana Segui, M. 2000. "Animal resource management and the process of animal domestication at Tell Halula (Euphrates Valley-Syria) from 8800 BP to 7800 BP." In M. Mashkour, A. M. Choyke, H. Buitenhuis and F. Poplin (eds.), *Archaeozoology of the Near East IVA: Proceedings of the fourth international symposium on the archaeozoology of southwestern Asia and adjacent areas*, ARC Publication 32, 241–256. Groningen.

Sapir-Hen, L., G. Bar-Oz, H. Khalaily and T. Dayan. 2009. "Gazelle exploitation in the early Neolithic site of Motza, Israel: The last of the gazelle hunters in the southern Levant." *Journal of Archaeological Science* 36: 1538–1546.

Savard, M., M. Nesbitt and M. K. Jones. 2006. "The role of wild grasses in subsistence and sedentism: new evidence from the northern Fertile Crescent." *World Archaeology* 38: 179–196.

Saxon, E. C. 1974. "The mobile herding economy of Kebarah Cave, Mt Carmel: An economic analysis of the faunal remains." *Journal of Archaeological Science* 1: 27–45.

Simmons, A. H., I. Kohler-Rollefson, G. O. Rollefson, R. Mandel and Z. Kafafi. 1988. "Ain Ghazal: A major Neolithic settlement in central Jordan." *Science* 240: 35–39.

Speth, J., and S. L. Scott. 1989. "Horticulture and large-mammal hunting: The role of resource depletion and the constraints of time and labor." In S. Kent (ed.), *Farmers as Hunters: The implications of sedentism*, Cambridge University Press, 71–79. Cambridge, UK.

Stampfli, H. R. 1983. "The fauna of Jarmo with notes on animals bones from Matarrah, the Amuq, and Karim Shahir." In L. S. Braidwood and R. J. Braidwood (eds.), *Prehistoric Archaeology Along the Zagros Flanks*, Oriental Institute Publications 105, 431–483. Chicago.

Starkovich, B. M., and M. C. Stiner. 2009. "Hallan Çemi Tepesi: High-ranked game exploitation alongside intensive seed processing at the Epipaleolithic-Neolithic transition in southeastern Turkey." *Anthropozoologica* 44: 41–62.

Stiner, M. C., N. D. Munro and T. A. Surovell. 2000. "The tortoise and the hare." *Current Anthropology* 41: 39–73.

Stiner, M. C. 2006. *The faunas of Hayonim Cave, Israel: A 200,000 year record of Paleolithic diet, demography, and society.* Peabody Museum Press. Cambridge, Massachusetts.

Stutz, A. J., N. D. Munro and G. Bar-Oz. 2009. "Increasing the resolution of the Broad Spectrum Revolution in the southern Levantine Epipaleolithic (19-12 ka)." *Journal of Human Evolution* 56: 294–306.

Szuter, C. R., and F. E. Bayham. 1989. "Sedentism and prehistoric animal procurement among desert horticulturalists of the North American southwest." In S. Kent (ed.), *Farmers as Hunters: The implications of sedentism*, Cambridge University Press, 80–95. Cambridge, UK.

Tchernov, E., and O. Bar-Yosef. 1982. "Animal Exploitation in the Pre-Pottery Neolithic B Period at Wadi Tbeik, Southern Sinai." *Paléorient* 8: 17–37.

Tchernov, E. 1991. "The impact of sedentism on animal exploitation in the southern Levant." In H. Buitenhuis and A. T. Clason (eds.), *Archaeozoology of the Near East. Proceedings of the first international symposium on the archaeozoology of southwestern Asia and adjacent areas*, Universal Book Services, 10–26. Leiden.

Tchernov, E. 1993. "From sedentism to domestication: A preliminary review for the southern Levant." In A. Clason, S. Payne and H.-P. Uerpmann (eds.), *Skeletons in her Cupboard: Festschrift for Juliet Clutton-Brock*, Oxbow Monographs 34, 189–233. Oxford.

Tchernov, E. 1994. *An Early Neolithic Village in the Jordan Valley Part II: The fauna of Netic Hagdud.* Peabody Museum of Archaeology and Ethnology. Cambridge, Massachusetts.

Tchernov, E. 1997. "Are Late Pleistocene environmental factors, faunal changes and cultural transformations causally connected? The case of the southern Levant." *Paléorient* 23: 209–228.

Turnbull, P. F., and C. A. Reed. 1974. "The fauna from the terminal Pleistocene of Palegawra cave, a Zarzian occupation site in northeastern Iraq." *Fieldiana: Anthropology* 63: 81–146.

von den Driesch, A., and J. Peters. 1999. "Vorläufiger Bericht über die archäozoologischen Untersuchungen am Göbekli Tepe und am Gürcütepe bei Urfa, Türkei." *Istanbuler Mitteilungen* 49: 23–39.

Whitcher, S. E., J. C. Janetski and R. H. Meadow. 2000. "Animal bones from

Wadi Mataha (Petra Basin, Jordan): The initial analysis." In M. Mashkour, A. M. Choyke, H. Buitenhuis and F. Poplin (eds.), *Archaeozoology of the Near East IVA: Proceedings of the fourth international symposium on the archaeozoology of southwestern Asia and adjacent areas*, ARC Publication 32, 39–48. Groningen.

Yeshurun, R., G. Bar-Oz and M. Weinstein-Evron. 2007. "Modern hunting behavior in the early Middle Paleolithic: Faunal remains from Misliya Cave, Mount Carmel, Israel." *Journal of Human Evolution* 53: 656–677.

Zeder, M. A. 2008. "Domestication and early agriculture in the Mediterranean Basin: Origins, diffusion, and impact." *Proceedings of the National Academy of Sciences* 105: 11597–11604.

Living in a Marginal Environment: Climate Instability and Possible Lathyrism in the Syrian Neolithic

Deborah C. Merrett and Christopher Meiklejohn

Abstract

Since the first demographic transition c. 11000 BC, humans have pushed the boundaries of regions thought habitable. In the Middle Euphrates Valley movement of the steppe/desert boundary has, on occasion, transformed community status from sustainable with rain-fed agriculture to marginal. We examine the adaptive strategy of plant-based food-producing economies for survival in fluctuating marginal environments. Evidence for reliance on the neurotoxic grass pea (*Lathyrus sativus*) in the Near East is reviewed and expected human skeletal manifestations of consumption of this famine food are described. The Neolithic transition resulted in reliance on plant-based agriculture, posing risks to human health in regions prone to droughts and floods (e.g. northeast Africa, the Near East and south Asia), risks still high today. We suggest that the triad of climate instability, marginal environment and population size beyond local carrying capacity combine to increase risk of lathyrism when grass peas are indigenous.

Introduction

Since the first demographic transition, c. 11000 BC,[1] humans have pushed boundaries of regions thought habitable. In the Middle Euphrates Valley, steppe/desert boundary shifts have on occasion transformed communities at the edge from being sustainable, using rain-fed agriculture, to being marginal. In some cases population pressure through declining regional carrying capacity necessitated settlement abandonment.

---

[1] We report all dates as calibrated, using Calib 5.02 and the IntCal04 curve.

We examine here the survival strategy of Near Eastern plant-based, food-producing economies when drought extended beyond a survivable one or two years to decades or centuries (Hole 2007). Reliance on a plant-based agriculture resulting from the Neolithic transition can pose risks to human health, specifically a risk that is still high in regions of regular droughts and floods (e.g. northeast Africa, the Near East and south Asia), that of lathyrism, a disease resulting from consumption of the neurotoxic legume *Lathyrus sativus*, the grass pea. At such times the grass pea can become a toxic hazard resulting in paraparesis[2] or weakness of the lower limb. When all other crops fail, from drought or flooding of agricultural land, this hardy small seeded legume survives as the sole human food source (Butler 1999; Purseglove 1968; Tekle-Haimanot et al. 2005). In the 1990s, lathyrism increased substantially in Ethiopia, affecting up to 2.4 per cent of the population (Getahun et al. 2002), a trend continuing currently (Tekle-Haimanot et al. 2005).

Human agency contributes to this process. Creation of the local human niche changes the proportions of indigenous plants, magnifying effects of climate change. Early in the history of crop cultivation, plants such as the grass pea were probable low level contaminants of legume crops such as lentils, inadvertently domesticated along with the larger seeded crops. Use of the grass pea as fodder for newly domesticated ovicaprines may have aided this transformation, giving the Neolithic grass pea a usage similar to that in the modern sub-tropical Old World (Erskine et al. 1994). It is argued that once the grass pea became established, climatic stress in marginal regions propelled it from weed status and animal fodder crop to ubiquitous famine food with the potential to produce lathyrism.

Butler (1999) and Hansen (1999) have called for vigilance during analyses of human skeletal remains to the possible presence of lathyrism. However, to our knowledge, there are no guidelines for its archaeological recognition, although radiographic identification of osseous sequelae is reported in the clinical literature (Streifler and Cohn 1981; Haque et al. 1997; Misra et al. 1993; Paissios and Demopoulos 1962) and in experimental work with rats (Ponseti and Shepard 1954; Selye 1957).

We examine climate instability, its effects on human occupation of the Middle Euphrates Valley, a marginal environment bordering the Arabian Desert, and the response of its inhabitants to drought. Archaeological

---

2 A glossary of medical terms is provided immediately before the references cited.

evidence of grass pea use in the early to middle Holocene is also explored. Through review of the clinical literature and presentation of a case study from Bouqras, Syria (Meiklejohn et al. 1983), occupied from c. 7400 to 6200 BC (Akkermans and Schwartz 2003), we discuss criteria for and difficulties in the recognition of lathyrism in human skeletal remains.

Climate instability and marginal environments in the Near East

Late glacial climate oscillations are known in the Near East (Brooks 2004; COHMAP Members 1988; Staubwasser and Weiss 2006; Weninger et al. 2006) and affected marginal areas especially (Hole 1997, 2007). Although stable ecotones (regions between two ecosystems) usually exhibit high species diversity that makes them attractive for habitation (Dincause 2000: xxiv), problems arise in arid regions. Climate instability easily disrupts plant communities as ecotone boundaries fluctuate. Of interest here is the steppe-desert interface between the 200 and 250 mm precipitation isohyets, marking the arid extreme of rain-fed agriculture.

Even today, climate instability causes fluctuation of the 200 mm isohyet (Hole 1997, 2007). Between 1981 and 1996, the annual precipitation near Bouqras ranged from 50 to 100 mm (fig. 1) making the steppe-desert ecotone to the north, in the Khabur basin of northeastern Syria. Variation in rainfall leads to successful crops in normal years and diminished crop yields in drought years (Hole 1997, 2007). People occupying sites in this ecotone in the early-middle Holocene would likely have experienced similar droughts and crop losses.

Both palaeobotanical (van Zeist and Waterbolk-van Rooijen 1985) and zooarchaeological (Clason 1983) analyses show that the Bouqras people practised mixed farming and heavy reliance on plant cultivation. The combination of farming activities (Clason 1983; van Zeist and Waterbolk-van Rooijen 1985), the site location at the arid eastern extreme of the Middle Euphrates Valley, and palaeoclimatic reconstructions (Staubwasser and Weiss 2006; Weninger et al. 2006) place the ecotone farther south than at present. Unpredictable northward movement of the isohyet would have left Bouqras crops subject to periodic water stress. If droughts were of short duration, the site would remain viable with plant cultivation possible on the nearby Euphrates floodplain. Fluctuations in precipitation at the headwa-

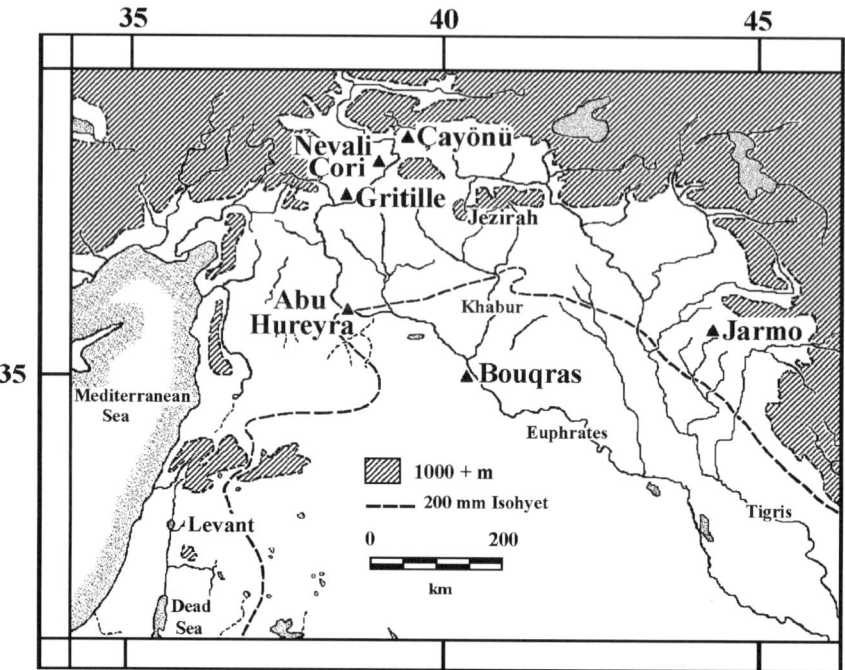

FIGURE 1. *Map of Near East showing sites where grass peas have been recovered: Abu Hureyra, Nevali Çori, Çayönü, Gritille, Jarmo, and location of case site of Bouqras. Modern 200 mm precipitation isohyet is indicated.*

ters of the Euphrates would also have left Bouqras crops subject to periodic flooding.

When hyper-aridity extended into decades, extensive stress on the local carrying capacity would elicit human responses ranging from settlement abandonment to increased cultural complexity, playing into the cycle of the rise and collapse of Near Eastern civilisations (Brooks 2004; Staubwasser and Weiss 2006; Weninger et al. 2006). Indeed, abandonment of Bouqras c. 6200 BC is consistent with regionwide settlement retraction (Akkermans and Schwartz 2003) coinciding with the 6250 BC aridity event (Brooks 2004; González-Sampériz et al. 2009; Staubwasser and Weiss 2006; Weninger et al. 2006). Outbreaks of lathyrism in this environment, at the margin of possible rain-fed plant cultivation, can be both expected and predicted.

Cultural reactions to drought

Cultural responses to long-term drought can vary (Hole 2007), and include retraction of settlements, transhumance, enhanced mobility, and risk dispersion behaviour such as shifts in livestock raised and subsistence strategy (Brooks 2004; Weninger et al. 2006: 416). In shorter droughts, coping strategies can range from storing surpluses to increasing trade and raiding (Hole 2007). Focus on the hardiest local foodstuffs is another possible response to environmental stress, regardless of normal use or function in times of adequate rainfall. For example, at present in marginal environments of northeastern Africa through the Near East to the Indian subcontinent, the famine food of choice is the grass pea (*Lathyrus sativus*), despite its potential for lathyrism (Steel 1884, cited in Miles 2005; Tekle-Haimanote et al. 2005).

Reliance on the grass pea as risk dispersion

Lathyrism is both a modern disease and of historical antiquity. Nineteenth- and early 20th-century European documents, as well as the Hindu text Bhava Prakash (c. 1550), discuss paralysis following ingestion of grass peas, demonstrating wide recognition of toxic effects. The 19th-century Mejah Cripples' Asylum in Allahabad, India, was established primarily for treatment of lathyrism with regional prevalence estimated at four per cent of the population (Steel 1884, cited in Miles 2005). Earlier references to lathyrism include the Greek physician Hippocrates and the Roman Pliny (Dwivedi 1994; Gardner and Sakiewicz 1963; Paissios and Demopoulos 1962; Selye 1957). Exploration of the time depth of grass pea consumption is thus pertinent to analysis of Near Eastern Neolithic sites following the adoption of agriculture.

Although poorly preserved archaeologically, charred grass pea remains occur on many Near Eastern Neolithic sites, with earliest records at Abu Hureyra (c. 9500 BC), c. 200 km upstream from Bouqras (Hillman 2000). Other examples include Nevali Çori and Çayönü (c. 8800–7700 BC) in the Upper Euphrates Valley of Turkey and the slightly later Jarmo (c. 7500 BC) in northeastern Iraq (see Renfrew 1973; Zohary and Hopf 2000), the last contemporary with Bouqras. Low concentrations of *Lathyrus* at such sites could represent the use of whole plants as animal fodder (Ladizinsky 1989), of dung as fuel (Miller 1996) or its presence as a weed in other legume crops (Erskine et al. 1994). However, at Gritille in Turkey, 450 km upstream from and contemporary with Bouqras (c. 6900 BC), 200 grass pea seeds were recovered from storage bins (Voigt 1988), suggesting both cultivation and storage for the subsequent year's planting and/or human consumption (Miller 1991).

Grass pea was not recovered at Bouqras (van Zeist and Waterbolk-van Rooijen 1985). However, the contemporary presence suggests that its use would be known. Absence in the record also probably reflects the tiny legume sample that was recovered. Possible contributing factors include the generally poor archaeological preservation of pulses and the general opinion in the 1980s that cereal domestication was dominant. More recent work has, however, revealed a much higher proportion of legume calories in Neolithic diets than previously believed (Hansen 1999; Hillman 2000; Zohary and Hopf 2000). Hansen (1999) sees legumes as important in the Near Eastern diet since c. 11000 BC. Kislev and Bar-Yosef (1988) assert that legumes were the initial cultivars. Taken together, we can conclude that pulses were underrepresented in the palaeobotanical finds from Bouqras. The presence of grass pea is a reasonable inference, especially given its recovery in more recent excavations in the region (Hillman 2000; Miller 1996; Voigt 1988).

In combination, the presence of grass pea, fluctuating climate and marginal environment, with the potential for rapid reduction of regional carrying capacity, place early-mid Holocene occupants of the steppe-desert ecotone (as at Bouqras) at high risk of lathyrism. Skeletal manifestations reported in modern populations (Paissios and Demopoulos 1962; Streifler and Cohn 1981; Weintroub et al. 1980) suggest that the effects of grass pea use might also be recognised in skeletal remains.

Lathyrism: Skeletal manifestations

Grass pea contains two potent neurotoxins: β-N-oxalyl-L-α,β diaminopropionic acid (β-ODAP) and β-aminopropionitrile (β-APN). β-ODAP targets the central nervous system. Susceptible individuals can show upper pyramidal tract neural damage, resulting in varying degrees of lower limb paraparesis (neurolathyrism) (Barceloux 2008; Khan et al. 1995; Ludolph et al. 1987; Paissios and Demopoulos 1962; Streifler and Cohn 1981). Locomotor difficulties include slight flexion and adduction of hip and knees producing a characteristic scissoring gait, and walking on the balls of the feet with torso angled posteriorly (Ludolph et al. 1987; Misra et al. 1993). As the disease worsens, plantar flexors of the foot contract continually with eventual complete loss of leg function, the crawler stage. Continued rapid contraction of foot muscles exacerbates injury. Individuals resort to use of the arms as the only source of locomotor power (Khan et al. 1995; Ludolph et al. 1987; Paissios and Demopoulos 1962; Streifler and Cohn 1981), developing substantial upper body strength (Ludolph et al. 1987), lower limb emaciation (Miles 2005) and bone atrophy. In contrast, β-APN disrupts collagen synthesis and thus bone growth and repair (osteolathyrism) (Haque et al. 1997), intensifying the development of osseous lesions including exostoses and excavations at entheses, the attachment sites of tendons and ligaments, and delayed or absent epiphyseal fusion.

Skeletal lesions are thought to arise from enthesopathic injury related to disruption of collagen metabolism, compromising enthesis integrity, and/or related to repetitive stress resulting from chronic contraction of the plantar flexors (Dawson et al. 2002; Paissios and Demopoulos 1962; Streifler and Cohn 1981; Weintroub et al. 1980). Injuries would be predicted at the origins and insertions of the muscles of plantar flexion (gastrocnemius, soleus and plantaris) (fig. 2). Abnormal bone morphology would be predicted on the posterior of the distal femur, proximal tibia and fibula, and calcaneus. Locomotor impairment in ambulatory individuals can also result in spinal injury such as spondylolysthesis (separation of neural arch from vertebral body) (Paissios and Demopoulos 1962), and deformation of knee, ankle and toe joints (Haque et al. 1997). In addition, epiphyseal non-fusion in adults has been observed radiographically (Paissios and Demopoulos 1962; Weintroub et al. 1980).

Skeletal effects depend on age of onset of motor impairment, severity and duration of disease symptoms and age of observation[3] relative to age of disease onset (Haque et al. 1997; Paissios and Demopoulos 1962). If onset occurs before skeletal maturity, evidence of neurolathyrism and enthesis avulsion would be expected. If survival extends past adolescence, epiphyseal non-union of iliac crest, ischial tuberosity and vertebral bodies would be expected. If symptoms were mild, especially related to continual muscle contractions, enthesopathies might be minor with limited sequelae observable in some affected individuals. In contrast, if locomotor function

3 Age of observation for skeletal remains is age at death, whereas for the living it is the age of the patient at time of medical intervention and clinical observations.

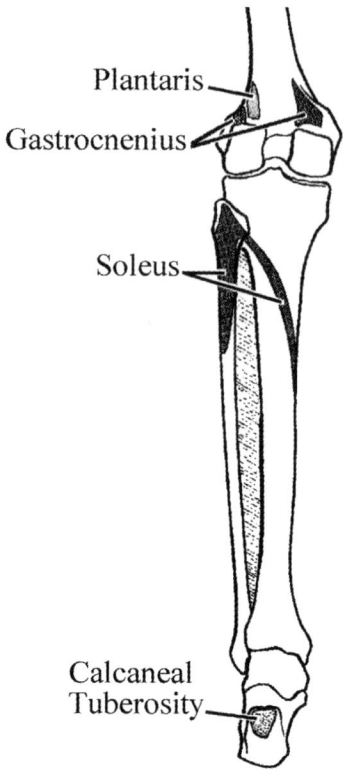

FIGURE 2. Diagram of location entheses of posterior knee (femur, tibia and fibula) and heel (calcaneus) discussed in text (modified from Platzer 2004: 263).

loss is for a substantial time prior to death, lower limb disuse atrophy and increased upper limb robusticity would be expected.

Currently lathyrism does not affect everyone equally, whether in a community or family unit. Epidemiological observations in modern outbreaks show non-adults more affected than adults, males more than females. The proportion of lathyritic individuals in susceptible populations ranges from 0.2 to 4 per cent (Getahun et al. 2002; Haque et al. 1997; Steel 1884, cited in Miles 2005; Tekle-Haimanot et al. 2005) with 9.5 per cent of interviewed households affected (Getahun et al. 2002). In an exceptional wartime (WWII) case, 57 per cent of individuals in a forced labour camp experienced symptoms (Paissios and Demopoulos 1962). Prevalence in past populations might be expected to be within this overall range.

Case study

Our case study of possible skeletal lathyrism is from Bouqras, a moderate sized tell on the west bank of the Middle Euphrates, c. 10 km south of where the Khabur River enters (fig. 1). The site occupies c. 2.75 ha, lying on a Late Pleistocene terrace. Cultural deposits of c. 4.5 m were divided into ten levels spanning the millennium from 7400 to 6200 BC (Akkermans et al. 1981, 1983; Akkermans and Schwartz 2003).

Human skeletal material came from a single structure, House 12, level III, dated to 6950–6900 BC (Meiklejohn et al. 1983, 1992; Merrett and Meiklejohn 2001, 2007) midway through site occupation. Five individuals, two adults, an adolescent and two children were recovered. Other than a single Romano-Byzantine intrusive burial, these were the only human remains recovered (Meiklejohn et al. 1983, 1992; Merrett and Meiklejohn 2007). The skeletal anomalies noted here were only found on the young adolescent, individual 2 (Meiklejohn et al. 1983) (field number TB/1977 in the site records).

Individual 2 consists of a partial skull and infra-cranial skeleton in "generally poor condition" (Meiklejohn et al. 1983: 366), with most bones partially present except the right hand, right patella and fibula, and most of the left foot. Other than the anomalies discussed below, no other skeletal lesions were identified (e.g. *cribra orbitalia*, porotic hyperostosis, enamel hypoplasia, Harris' lines, inflammatory responses, joint deformation). An age range of 12 to 14 years is suggested with the erupted second molars showing light wear facets, while presence of a narrow sciatic notch suggests that the individual was possibly male (Meiklejohn et al. 1983).

All the anomalies are located at entheses, the bone to ligament and bone to tendon interfaces on both clavicles and long bones of the left leg (figs. 3 and 4). Bony excavations, with destruction of cortical bone and scalloped margins, are present on inferior surfaces of both medial clavicles at the insertion of the costoclavicular ligament (fig. 3), creating a rhomboid fossa. Remodelling of the margins without sclerosis, observed radiographically, and thickening of the exposed trabeculae provide evidence of partial healing.

In the lower limb, depressions in the compact bone surface occur at the origins of the soleus muscle, a strong plantar flexor (fig. 4). On both tibia and fibula, the compact bone surface at the area affected is intact with no expo-

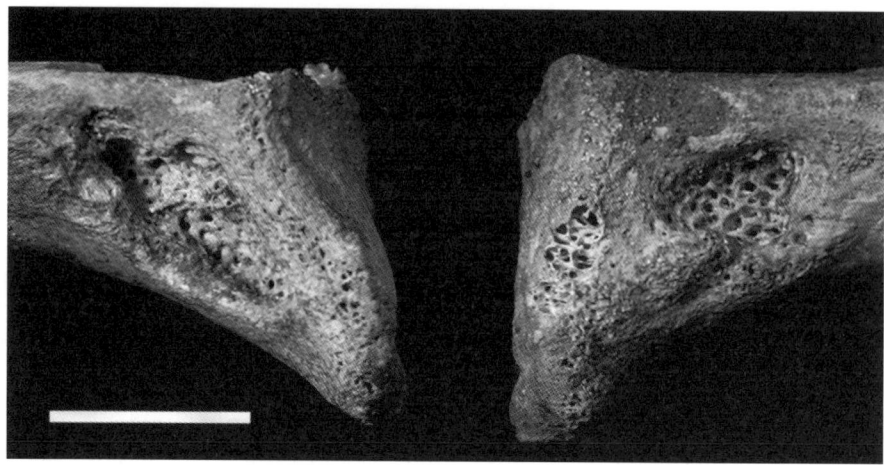

FIGURE 3. *Inferior medial surfaces of clavicles of individual 2. Scale bar = 10 mm.*

sure of underlying trabeculae (cf. clavicular appearance). The depression on the tibia, the soleal line, is slight and smooth-floored. In contrast, the fibular depression is extensively excavated. The fibular depression is larger in the area reflecting the larger size of lateral soleus origin. It is also deeper, creating a U-shaped rather than circular cross-section of the proximal fibular diaphysis. The floor of the fibular cavitation is rough and covered with small bony spicules (exostoses), suggesting that the bone contour at the lateral soleal enthesis resulted from tissue injury and protracted cycles of healing and re-injury (Donnelly et al. 1999; Niepel and Sit'aj 1979). The fragmentary nature of individual 2 means that the origins of gastrocnemius and plantaris (other plantar flexors) on the posterior of the distal femoral metaphysis, and their insertions on the calcaneus, are absent as are the bones of the opposite limb. Cortical excavations at these locations would be expected in individuals with plantar flexor hyperactivity (Resnick and Greenway 1982). The proposed cause of the fibular morphology, an avulsion injury, is seen clinically (Tehranzedah 1987). As set out above, these changes in morphology are consistent with clinical observations in some lathyritic individuals (Paissios and Demopoulos 1962; Streifler and Cohn 1981; Weintroub et al. 1980).

Because of the age of individual 2 (young adolescent), fused epiphyses of

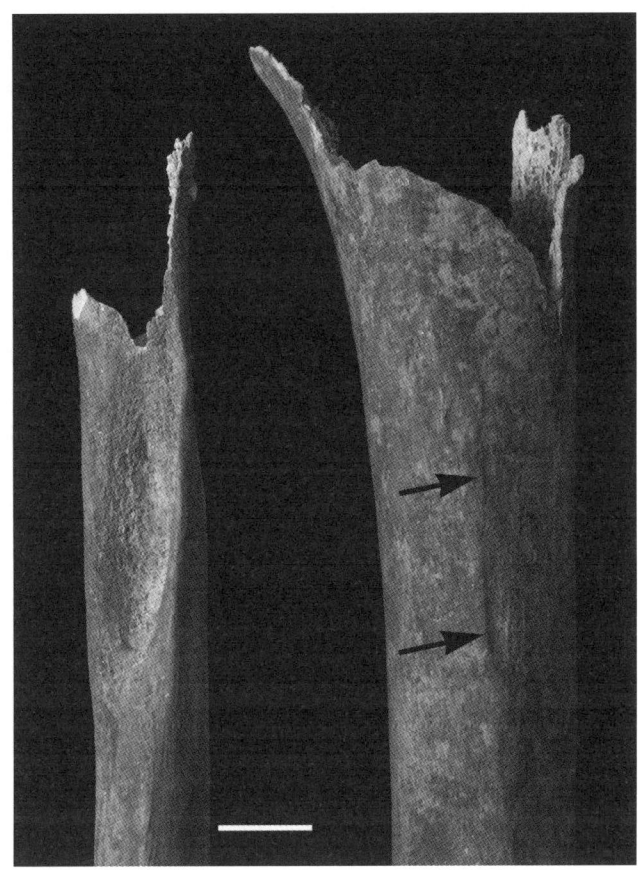

FIGURE 4. *Proximal fibular and tibial entheses (medial indicated by arrows). Scale bar = 10 mm.*

iliac crest, ischial tuberosity and vertebral bodies are not expected; the adult epiphyseal non-union criterion of lathyrism is not applicable.

Identifying lathyrism in dry bone

The skeletal markers predicted for lathyrism are non-specific, whose exact cause may be difficult to ascertain and with multiple possible interpretations. They may also be seen as part of a continuum of normal variation, a simple reflection of their causative mechanism. As a result, interpretation is difficult. The range of appearance of the costoclavicular ligament attachment on the medial clavicle includes both rough and smooth surfaces, from elevated through flat to various depths of a rhomboid fossa (Cave 1961; Jit and Kaur 1986; Mann and Hunt 2005). The fossa has been interpreted as a variant of normal morphology (Mann and Hunt 2005; Stirland 1998), or as evidence of habitual strenuous upper limb activity (Hawkey and Merbs 1995; Peterson 1998). Similar lesions noted in the forensic literature have focused on age and sex correlates rather than musculoskeletal activity (e.g. Rogers et al. 2000).

The costoclavicular ligament connects the first costal cartilage to the inferior surface of the medial clavicle, the site of cortical irregularities in the Bouqras adolescent. It stabilises the sternoclavicular joint and minimises elevation of the medial clavicle when tension forces are transferred from arms to clavicle (e.g. when supporting body weight on the arms) (Cave 1961). Rhomboid fossa frequency varies both today and in the past. Modern north Indian human cadavers had the fossa in 70 per cent of adults (N=789) (Jit and Kaur 1986). In contrast, it was present in only 28.5 per cent of modern British adult cadavers from London (N=153) (Cave 1961). Similarly, it occurred in 18.2 per cent of a modern Tennessee sample (N=344, age range 10–92 years) (Rogers et al. 2000). In the Indian study, the fossa depth in children was minimal compared to that of adults (Jit and Kaur 1986), also minimal compared to the Bouqras adolescent. Thus while presence of the fossa in Bouqras 2 is not unexpected, its morphology is more extensive than normally seen in non-adults.

The popliteal line on the tibia is visible to some extent on all tibiae. In contrast soleal origin on the fibula is often not visible, though its presence is considered normal in some groups (C. Knüsel, pers. comm., 2007). However, avulsion injuries at the fibular origin are also possible (Tehranzedah 1987), providing an alternate cause of the bone appearance. In addition, although a roughened cavitation may result from injury to an enthesis (Donnelly et al. 1999), the extent of bony changes may not be linearly related to the extent of injury or activity (Rogers et al. 1997). This confounds interpretation and caution should be taken in identifying specific activities in skeletal remains (Villotte et al. 2010). Molleson (2007) suggests a baseline range of morphology be established within a population under study for comparative purposes. Unfortunately, the small sample size at Bouqras precludes this.

Only a small portion of the site was excavated, hindered in part by the presence of a modern cemetery in its midst. As well, house floors in the level with the skeletal remains were not excavated (Akkermans et al. 1981), a level lying at the Pre-Pottery (PPN) to Pottery (PN) Neolithic transition, a time when ancestor cults were well developed. At other sites, burials typically occurred below plastered house floors (Akkermans and Schwartz 2003; Coqueugniot 1998; Moore and Molleson 2000). As argued elsewhere, lack of burials in the rest of the site probably means that they were beneath house floors (Merrett and Meiklejohn 2007). Only further excavation can confirm this.

Examination of entheses by functional muscle group (e.g. plantar flexors in this study) connects bone form with biomechanical function of the limb (Molleson 2007). However, degree of bone preservation places limitations on interpretation. Entheses of the posterior knee are located on bone metaphyses where extreme thinness of cortical bone predisposes it to damage in life, particularly in adolescence (Benjamin et al. 2002) and after death. Thus, good bone preservation increases the probability that all relevant entheses in a muscle group are available for study, but not in the case of Bouqras 2.

Both clinical and epidemiological studies show that lathyritic paraparesis can affect bone morphology (Paissios and Demopoulos 1962; Streifler and Cohn 1981; Weintroub et al. 1980). However, due to difficulties in the interpretation of dry bone appearance, support for possible presence of lathyrism in the ancient Near East requires synthesis of independent clinical, ethnographic and archaeological evidence at the local and regional level (Molleson 2007). In this case study, the age (12–14 years) and sex (possibly male) (Meiklejohn et al. 1983) are consistent with epidemiological observations in modern outbreaks of lathyrism, with younger individuals more affected than adults, males more than females (Getahun et al. 2002; Tekle-Haimanot et al. 2005). Furthermore, grass pea is clearly present at sites contemporary with and adjacent to Bouqras (reviewed in Renfrew 1973; Voigt 1988).

Discussion and conclusions

Sustainable life in the steppe-desert ecotone of the Middle Euphrates Valley means bridging years of crop failure and avoiding site relocation. In the modern steppe-desert ecotone of the Khabur basin, decreased crop yields are expected in dry years and an extensive famine every decade (Hole 2007).

During drought and flooding, with restricted food choices, reliance on famine foods may be the only viable adaptive behaviour. In the Middle Euphrates Valley, northward movement of the 200 mm isohyet can transform grass pea from weed and fodder crop to toxic hazard.

Living at the edge results in costs to human health. Both in the past and the present, the grass pea has had maladaptive consequences. In the marginal zone, it predisposes to lathyrism, a risk for a major proportion of the population. Primarily affecting adolescent and young adult males, it thereby targets a young, productive sector of the community (Getahun et al. 2002).

Although diagnosis of lathyrism from dry bone is difficult, we argue that the suite of criteria used on the Bouqras adolescent may aid identification of prehistoric outbreaks of lathyrism and the use of the grass pea as human food. We would welcome examination of other larger skeletal samples in the region for supporting evidence of this disease at this early date, allowing for establishment of population baselines of enthesis morphology (Molleson 2007). A more complete understanding of adaptive mechanisms to climate fluctuations in the steppe-desert ecotone might then be achieved. As drought extended to decadal and millennial duration, the stage was set for adaptive strategies that led within a millennium of Bouqras's occupation to the rise of larger centres between the Euphrates and Tigris rivers. Marginal areas like Bouqras were abandoned.

The Neolithic transition, with reliance on plant-based agriculture, posed substantial risks to human health; one of them lathyrism. With environmental degradation in arid environments people face a "choiceless" choice: eat grass pea and risk the paraparesis of lathyrism, or die of starvation. The triad of climate instability, marginal environment and population size beyond the carrying capacity of the local and regional environment, combine to increase the risk of lathyrism. As alluded to above, lathyrism is still a major scourge in similar environments of the modern Near and Middle East, and is still the subject of major research to find solutions acceptable to modern populations.

Although "there was no single trajectory followed by societies as they adapted to increasing aridity" (Brooks 2004: 28), to live one must eat. If the only food available is grass pea, everyone in the community faces associated health risks in order to survive. Grass pea can be truly transformed from weed and fodder crop to toxic hazard.

Acknowledgements

The authors thank the Institute of Pre- and Protohistory, State University of Amsterdam, under whose auspices the Bouqras excavation was undertaken. This research was funded, in part, by a University of Manitoba Graduate Fellowship and a Social Sciences and Humanities Research Council of Canada Doctoral Fellowship to Deborah Claire Merrett. We also thank Chris Knüsel at the University of Exeter and Julie M. Hansen at Boston University for comments on an earlier draft.

Glossary

Adduction: Limb movement toward the midline of the body.

Avulsion: A type of enthesopathy (see below): in which force exerted by a ligament or tendon tears away or forcibly separates the bony attachment site from the remainder of the bone.

Costo-/costal: Referring to ribs as in the costoclavicular ligament which joins the clavicle to the first rib.

Cortex/cortical bone: The usually dense outer surface or shell of a bone.

Diaphysis: Shaft or body of a long bone (compare epiphysis and metaphysis below).

Enthesis/entheses: Region(s) of bone where a ligament (bone to bone connection) or tendon (bone to muscle) attaches.

Enthesopathy/enthesopathic: Injury to bone at the site of attachment of a ligament or tendon.

Epiphysis/epiphyses/epiphyseal: The separate bony centre at the end(s) of the shaft of a long bone and normally fuse to the shaft or diaphysis during adolescence.

Exostoses: Projections of bone above the normal bone surface. At entheses, it can be an indicator of prior injury with subsequent healing.

Flexion: Decrease in a joint angle (e.g. to bend as at the knee or elbow) (opposite of extension).

Fossa: In anatomical terminology, a depression or hollow on a bone surface.

Metaphysis: Flared end of the diaphysis of a long bone, the area where active growth occurs in childhood and early adolescence.

Paraparesis: Weakness (or partial paralysis) of the lower extremities

Plantar flexors: Muscles that when contracted cause pointing of the foot and toes.

Sclerosis (in bone): Increase in bone density that can sometimes occur during bone healing.

Trabeculae/trabecular bone: Referring to the spongy porous structure of the interior of bone.

Upper pyramidal tract: Nerves of the spine that control movement, primarily of the lower limbs. Damage results in loss of muscle control in the affected limbs.

Bibliography

Akkermans, P. A., J. A. K. Boerma, A. T. Clason, S. G. Hill, E. Lohof, C. Meiklejohn, M. Le Mière, G. M. F. Molgat, J. J. Roodenberg, W. Waterbolk-van Rooyen and W. van Zeist. 1983. "Bouqras Revisited: Preliminary Report on a Project in Eastern Syria." *Proceedings of the Prehistoric Society* 49: 335–372.

Akkermans, P. A., H. Fokkens and H. T. Waterbolk. 1981. "Stratigraphy, architecture and lay-out of Bouqras." In P. Sanlaville and J. Cauvin (eds.), *Préhistoire du Levant,* Centre National de la Recherche Scientifique, 485–501. Paris.

Akkermans, P. M. M. G., and G. M. Schwartz. 2003. *The Archaeology of Syria: From Complex Hunter-Gatherers to Early Urban Societies (c. 16,000–300 BC).* Cambridge University Press. Cambridge.

Barceloux, D. G. 2008. "Grass pea and neurolathyrism (*Lathyrus sativus* L.)." In D. G. Barceloux (ed.), *Medical Toxicology of Natural Substances: Foods, Fungi, Medicinal Herbs, Toxic Plants, and Venomous Animals,* John Wiley & Sons, 62–66. Hoboken, New Jersey.

Benjamin, M., T. Kumai, S. Milz, B. M. Boszczyk, A. A. Boszczyk and J. R. Ralphs. 2002. "The Skeletal Attachment of Tendons – Tendon 'Entheses'." *Comparative Biochemistry and Physiology Part A* 133: 931–945.

Brooks, N. 2004. "Beyond collapse: The role of climatic desiccation in the emergence of complex societies in the middle Holocene." In S. Leroy and P. Costa (eds.), *Environmental Catastrophes in Mauritania, the Desert and the Coast, Abstract Volume and Field Guide.* Mauritania, 4–18 January 2004. First Joint Meeting of ICSU Dark Nature and IGCP 490: 26–30.

Butler, A. 1999. "The ethnobotany of *Lathyrus sativus* L. in the highlands of Ethiopia." In M. van der Veen (ed.), *The Exploitation of Plant Resources in Ancient Africa,* Kluwer Academic, 123–136. New York.

Cave, A. J. E. 1961. "The nature and morphology of the costoclavicular ligament." *Journal of Anatomy* 95: 170–179.

Clason, A. T. 1983. "Faunal remains." In Akkermans et al., "Bouqras Revisited: Preliminary Report on a Project in Eastern Syria." *Proceedings of the Prehistoric Society* 49: 359–362.

COHMAP Members. 1988. "Climatic changes of the last 18,000 years: Observations and model simulations." *Science* 241: 1043–1052.

Coqueugniot, E. 1998. "Dja'de el Mughara (Moyen-Euphrate), un village

Néolithique dans son environnement naturel à la veille de la domestication." In M. Fortin and O. Aurenche (eds.), *Espace naturel, espace habité en Syrie du nord (10ᵉ–2e millénaires av. J.-C.)*, Canadian Society for Mesopotamian Studies 33: 109–114.

Dawson, D. A., A. C. Rionaldi and G. Pöch. 2002. "Biochemical and toxicological evaluation of agent-cofactor reactivity as a mechanism of action for osteolathyrism." *Toxicology* 177: 267–284.

Dincause, D. F. 2000. *Environmental Archaeology: Principles and Practice*. Cambridge University Press. Cambridge.

Donnelly, L. F., G. S. Bisset, C. A. Helms and D. L. Squire. 1999. "Chronic avulsive injuries of childhood." *Skeletal Radiology* 28: 138–144.

Dwivedi, M. P. 1994. "Lathyrism – A historical review." *Nutrition Foundation of India*. Accessed 12 July 2006. http://nutritionfoundationofindia.res.in/pdfs/BulletinArticle/nfi_07_94.pdf

Erskine, W., J. J. Smartt and F. J. Muehlbauer. 1994. "Mimicry of lentil and the domestication of common vetch and grass pea." *Economic Botany* 48(3): 326–332.

Gardner, A. F., and N. Sakiewicz. 1963. "A review of neurolathyrism including the Russian and Polish literature." *Experimental Medicine and Surgery* 21: 164–191.

Getahun, H., F. Lambein, M. Vanhoorne and P. van der Stuyft. 2002. "Pattern and associated factors of the neurolathyrism epidemic in Ethiopia." *Tropical Medicine and International Health* 7: 118–124.

González-Sampériz, P., P. Utrilla, C. Mazo, B. Valero-Garcés, M. C. Sopena, M. Morellón, M. Sebastián, A. Moreno and M. Martínez-Bea. 2009. "Patterns of human occupation during the early Holocene in the Central Ebro Basin (NE Spain) in response to the 8.2 ka climatic event." *Quaternary Research* 71(2): 121–132.

Hansen, J. M. 1999. "Palaeoethnobotany and palaeodiet in the Aegean region: Notes on legume toxicity and related pathologies." In S. J. Vaughan and W. D. E. Coulson (eds.), *Palaeodiet in the Aegean*, Oxbow, 13–27. Oxford.

Haque, A., M. Hossain, F. Lambein and E. A. Bell. 1997. "Evidence of osteolathyrism among patients suffering from neurolathyrism in Bangladesh." *Natural Toxins* 5: 43–46.

Hawkey, D. E., and C. F. Merbs. 1995. "Activity-induced musculoskeletal stress markers (MSM) and subsistence strategy changes among ancient

Hudson Bay Eskimos." *International Journal of Osteoarchaeology* 5: 324–338.

Hillman, G. C. 2000. "The plant food economy of Abu Hureyra 1 and 2. Abu Hureyra 1: The Epipalaeolithic." In A. M. T. Moore, G. C. Hillman and A. J. Legge (eds.), *Village on the Euphrates: From Foraging to Farming at Abu Hureyra*, Oxford University Press, 327–399. Oxford.

Hole, F. 1997. "Evidence for mid-Holocene environmental change in the western Khabur drainage, northeastern Syria." In H. N. Dalfes, G. Kukla and H. Weiss (eds.), *Third Millennium BC Climate Change and Old World Collapse*, Springer Verlag, 39–66. Berlin.

Hole, F. 2007. "Agricultural sustainability in the semi-arid Near East." *Climate of the Past* 3: 193–203.

Jit, I., and H. Kaur. 1986. "Rhomboid fossa in the clavicles of North Indians." *American Journal of Physical Anthropology* 70: 97–103.

Khan, J. K., Y.-H. Kuo, A. Haque and F. Lambein. 1995. "Neurotransmitters in the cerebrospinal fluid and serum of neurolathyrism patients." In H. K. M. Yusuf and F. Lambein (eds.), *Lathyrus sativus and Human Lathyrism: Progress and Prospects*, University of Dhaka, 111–117. Dhaka.

Kislev, M. E., and O. Bar-Yosef. 1988. "The legumes: The earliest domesticated plants in the Near East?" *Current Anthropology* 29(1): 175–179.

Ladizinsky, G. 1989. "Origin and domestication of the southwest Asian grain legumes." In D. R. Harris and G. C. Hillman (eds.), *Foraging and Farming: The Evolution of Plant Exploitation*, Unwin Hyman, 374–389. London.

Ludolph, A. C., J. Hugon, M. P. Dwivedi, H. H. Schaumburg and P. S. Spencer. 1987. "Studies on the aetiology and pathogenesis of motor neuron diseases. 1. Lathyrism: Clinical findings in established cases." *Brain* 110: 149–165.

Mann, R. W., and D. R. Hunt. 2005. *Photographic Regional Atlas of Bone Disease: A Guide to Pathologic and Normal Variation in the Human Skeleton*, 2nd edition. CC Thomas. Springfield, Illinois.

Meiklejohn, C., A. Agelarakis, P. A. Akkermans, P. E. L. Smith and R. Solecki. 1992. "Artificial cranial deformation in the Proto-Neolithic and Neolithic Near East and its possible origin: Evidence from four sites." *Paléorient* 18: 83–98.

Meiklejohn, C., G. M. F. Molgat and S. G. Hill. 1983. "Human Skeletal Material." In Akkermans et al., "Bouqras Revisited: Preliminary Report on a Project in Eastern Syria." *Proceedings of the Prehistoric Society* 49: 365–370.

Merrett, D. C., and C. Meiklejohn. 2001. "Palaeopathology at the Origins of Agriculture in Central Syria." *American Journal of Physical Anthropology* S32: 108.

Merrett, D. C., and C. Meiklejohn. 2007. "Is House 12 at Bouqras a Charnel House?" In M. Faerman, L. Kolska Horwitz, T. Kahana and U. Zilberman (eds.), *Faces from the Past: Skeletal Biology of Human Populations from the Eastern Mediterranean*, British Archaeological Reports International Series 1603, 127–139. Oxford.

Miles, M. 2005. "Resilience of South Asian disabling conditions: A glimpse of lathyrism among comparative histories." *Lathyrus Lathyrism Newsletter* 4: 18–21.

Miller, N. F. 1991. "The Near East." In W. K. van Zeist, K. Wasylikowa and K.-E. Behre (eds.), *Progress in Old World Palaeoethnobotany*, Balkema, 133–160. Rotterdam.

Miller, N. F. 1996. "Seed eaters of the ancient Near East: Human or herbivore?" *Current Anthropology* 37: 521–528.

Misra, U. K., V. P. Sharma and V. P. Singh. 1993. "Clinical aspects of neurolathyrism in Unnao, India." *Paraplegia* 31: 249–254.

Molleson, T. 2007. "A method for the study of activity related skeletal morphologies." *Bioarchaeology of the Near East* 1: 5–33.

Moore, A. M. T., and T. Molleson. 2000. "Disposal of the dead." In A. M. T. Moore, G. C. Hillman and A. J. Legge (eds.), *Village on the Euphrates*, Oxford University Press, 277–299. London and New York.

Niepel, G. A., and Š. Sit'aj. 1979. "Enthesopathy." *Clinics in Rheumatic Diseases* 5(3): 857–872.

Paissios, C. S, and T. Demopoulos. 1962. "Human lathyrism: A clinical and skeletal study." *Clinical Orthopaedics* 23: 236–249.

Peterson, J. 1998. "The Natufian hunting conundrum: Spears, atlatls, or bows? Musculoskeletal and armature evidence." *International Journal of Osteoarchaeology* 8: 378–389.

Platzer, W. 2004. *Color Atlas of Human Anatomy, Vol. 1: Locomotor System*. 5th edition. Thieme. New York.

Ponseti, I. V., and R. S. Shepard. 1954. "Lesions of the skeleton and other mesodermal tissues in rats fed sweet-pea (*Lathyrus odoratus*) seeds." *Journal of Bone and Joint Surgery* 36A: 1031–1058.

Purseglove, J. W. 1968. *Tropical Crops: Dicotyledons 1*. Wiley and Sons. New York.

Renfrew, J. M. 1973. *Palaeoethnobotany: The Prehistoric Food Plants of the Near East and Europe*. Methuen. London.

Resnick, D., and G. Greenway. 1982. "Distal femoral cortical defects, irregularities, and excavations." *Radiology* 143: 345–354.

Rogers, J., L. Shepstone and P. Dieppe. 1997. "Bone formers: Osteophyte and enthesophyte formation are positively associated." *Annals of the Rheumatic Diseases* 56: 85–90.

Rogers, N. L., L. E. Flournoy and W. F. McCormick. 2000. "The rhomboid fossa of the clavicle as a sex and age estimator." *Journal of Forensic Sciences* 45(1): 61–67.

Selye, H. 1957. "Lathyrism." *Revue Canadienne de Biologie* 16: 1–82.

Staubwasser, M., and H. Weiss. 2006. "Holocene climate and cultural evolution in late prehistoric-early historic West Asia." *Quaternary Research* 66: 372–387.

Stirland, A. J. 1998. "Musculoskeletal evidence for activity: Problems of evaluation." *International Journal of Osteoarchaeology* 8: 354–362.

Streifler, M., and D. F. Cohn. 1981. "Chronic central nervous system toxicity of the chickling pea (*Lathyrus sativus*)." *Clinical Toxicology* 18: 1513–1517.

Tehranzadeh, J. 1987. "The spectrum of avulsion and avulsion-like injuries of the musculoskeletal system." *RadioGraphics* 7(5): 945–974.

Tekle-Haimanote, R., A. Feleke and F. Lambein. 2005. "Is lathyrism still endemic in northern Ethiopia: The case of Legambo Woreda (district) in the South Wollo Zone, Amhara National Regional State." *Ethiopian Journal of Health Development* 19(3): 230–236.

van Zeist, W., and W. Waterbolk-van Rooijen. 1985. "The palaeobotany of Tell Bouqras, Eastern Syria." *Paléorient* 11: 131–147.

Villotte, S., D. Castex, V. Couallier, O. Dutour, C. J. Knüsel and D. Henry Gambier. 2010. "Enthesopathies as Occupational Stress Markers: Evidence from the Upper Limb." *American Journal of Physical Anthropology* 142: 224–234.

Voigt, M. M. 1988. "Excavations at Neolithic Gritille." *Anatolica* XV: 216–232.

Weintroub, S., D. F. Cohn, R. Salama, M. Streifler and S. L. Weissman. 1980. "Skeletal findings in human neurolathyrism." *European Neurology* 19: 121–127.

Weninger, B., E. Alram-Stern, E. Bauer, L. Clare, U. Danzeglocke, O. Jöris, C. Kubatzke, G. Rollefson, H. Todorova and T. van Andel. 2006. "Climate forcing due to the 8200 cal yr BP event observed at Early Neolithic sites in the eastern Mediterranean." *Quaternary Research* 66: 401–420.

Zohary, D., and M. Hopf. 2000. *Domestication of Plants in the Old World: The origin and spread of cultivated plants in West Asia, Europe and the Nile Valley*. 3rd edition. Oxford University Press. Oxford.

Perceptions of pasture: The Role of Skill and Networks in Maintaining Stable Pastoral Nomadic Systems in Inner Asia

Joshua Wright and Cheryl Makarewicz

Abstract

Nomadic pastoralism is a complex subsistence system in which herder decisions regarding pasture access and mobility are key to their survival. This paper examines several key components that influence pastoralist movement, including graze availability and the configuration of herder social networks. We discuss how these factors, combined with individual herder perceptions of pasture qualities, contribute to the maintenance of stable pastoral systems in the present and the past. We eschew the 'exhausted pasture binary' as a device for understanding herding choices, focusing instead on skilled herders' detailed knowledge of animals' dietary preferences and graze condition, as well as the factors underlying herder decisions to make either long- and short-range movements. Overall, the technical herding outcomes of mobile herders are defined by knowledge of local landscapes, herding skill and social networks. More often than not, change in pastoral systems is brought about by a series of incremental adaptions, alterations of local tactics, and shifts in networks of communication and relationships that build up over the long term. We present indirect archaeological evidence for social networks comparable to and different from modern pastoralist social networks, and argue that the emergence of nomadic pastoralism was accompanied by the development of long-range social systems that sustain it.

Introduction

"Wandering from place to place pasturing their animals. The animals they raise consist mainly of horses, cows, and sheep, but include such rare beasts as camels, asses, mules, and the wild horses ... They move about in search of water and pasture." Shiji 110[1]

This often cited statement, written 2,100 years ago, is part of a political ethnography of the Xiongnu, the iconic nomads of early Asian history, as collected by officials of the Han Empire. For many scholars, it serves as a touchstone for discussions of Inner Asian pastoral nomads, and draws into focus the essential qualities of mobility and a focus on domestic animal herding central to the pastoral nomadic existence. However, the image of pastoral nomads wandering aimlessly throughout the steppe, as reported by Han Dynasty officials, has also contributed to a profound misunderstanding of herder decision-making in many discussions of pastoralist practices, where Inner Asian pastoralists are often perceived as moving indiscriminately across landscapes with homogenous resource distributions. Here, we re-evaluate outdated impressions of Inner Asian pastoral nomadic mobility through an examination of several key components that influence pastoralist movement, including graze availability and the configuration of herder social networks. We then examine how these factors, combined with individual herder perceptions of pasture qualities, contribute to the maintenance of stable pastoral systems, and use this as a springboard for the development of new models of the social landscape of early nomadic pastoralists.

Archaeologists and human behavoural ecologists frequently construct pastoralist decision-making as a series of binary choices, and place a particular emphasis on those decisions regarding pasture quality (Dyson-Hudson and Smith 1976; Bolling and Göbel 1997). In such models, pasture is simply categorized, implicitly or explicitly, either as "useful pasture" i.e. a graze that can support some quantity of animals for a defined period of time, or 'depleted pasture' i.e. a space with no edible graze for any herd animals. Within this framework, important variables that determine the overall character and quality of pastures, including species composition, density, timing of growth, and patch distribution of pastures, are discounted. Such a broad categorization of graze resource as either "present" or "absent" con-

---

[1] Watson, 1961 155–156

sequently forces herder decision-making into a highly canalized decision-making process that is dictated by a threshold of graze conditions which, when crossed, triggers a specific and limited response action i.e. herders move to a new pasture. This simple model has heavily influenced both modern development approaches to rangeland management and anti-desertification campaigns (Liu et al. 2001; Xie and Li 2008), as well as discussions characterising nomadic pastoralism in Inner Asia (Shishlina 2001; Khazanov 1994) and the Near East (Bar-Yosef and Khazanov 1992; Lees and Bates 1974).

Although such binary constructs can serve as a heuristic device for categorising the most basic responses of nomadic pastoralists when encountering graze patches, such a simple model does not adequately characterise the potentially wide range of herder responses to pasture quality. No less significantly, these models leave much to be desired with respect to herder agency. In general, they do not adequately recognise that herders plan pasture rounds in advance, and may attenuate these plans according to social and economic considerations, in addition to environmental ones. Herders regularly take into consideration a wide range of environmental, subsistence, and social factors, which may operate in concert or in conflict with each other, in order to acquire adequate graze for their animals.

We argue that there is no simple relationship between graze availability and the locus of people's habitation. This thesis is not a new one, although binary constructs such as those described above have dominated archaeologically-oriented models, and follows in the footsteps of generations of scholars who have recognised the complex relationship between pastoral nomads, their livestock and rangeland (Lattimore 1940; Krader 1957; Dyson-Hudson and Dyson-Hudson 1980; Roe et al. 1998; Williams 1997; Anthony 1985; Muller et al. 2007; Allsopp et al. 2007). Previous work exploring the dynamics between these elements have drawn from, in general, analytical frameworks derived from human ecology or cultural anthropology. The human ecological perspective concentrates on the technical mastery of landscape – by which we mean that these scholars see successful nomadic pastoralists as technicians of animals and biotic communities skilfully maintaining an equilibrium (Krader 1957; Ekvall 1968; Roe et al. 1998; Bollig and Göbel 1997). Cultural anthropological approaches point out that technical skill is not enough, and suggest that complex social relationships and particular cultural values are a critical factor in defining nomadic pastoralists (Dyson-Hudson and Dyson-Hudson 1980; Salzman 2002; Dyson-Hudson 1972; Murphy 2012). However, most archaeological analyses align more closely with the human ecological camp, not necessarily out of a sectarian divide, but out of a general uncertainty regarding how herders as individuals and their social networks may be extracted from the archaeological record.

Graze variation and diversity in herder perception

Access to high quality pastures is a critical issue for pastoralists, who rely heavily on natural graze patches in order to maintain their animal flocks. The distribution of graze, however, is not uniform across the landscape, but varies according to the composition of the floral biome and precipitation regimes, and herders must adjust their mobility strategies accordingly in order to ensure their animals receive adequate forage. Furthermore, the overall condition of pasture within a particular pasture locale often varies widely, within a single season as well as over the longer term, and pastoralists must take this in account when calculating their movements.

Natural fluctuations in pasture conditions due to seasonality and broader environmental change are important factors in determining graze availability, but human elements also play a part in shaping pastures as well. By virtue of serving as a graze resource for domestic herd animals, pasture composition and quality is determined by the number and type of animals grazing (different domesticates selectively graze on different types of graze within a single pasture), and the duration and frequency of pasture visits. One critical factor that impacts graze resources is pasture degradation, which we define as a largely anthropogenic process characterised by declining quality and quantity of graze resources that is initiated and promulgated by overgrazing, although climate change can play an important part in this process.

Overgrazing is a regular occurrence across Mongolia, but, unlike other regions of the globe, overgrazing here reflects, in general, the seasonal exhaustion of pastures rather than persistent, heavy grazing pressures that prevent graze regeneration, which ultimately creates new, significantly less productive floral biomes dominated by plants that are non-palatable or less nutritious to grazing animals. The relative absence of extreme overgrazing in Mongolia reflects in part the orientation of the general economy and land tenure rights within the country. Unlike the centrally managed, or quasi-free market, pastoral economies of China and Russia, which emphasise intense over-production of animal products in order to enter most of these products into markets in exchange for hard currency, the Mongolian pastoral economy balances healthy subsistence production with select production of some products, in particular wool and cashmere, intended for the market. In addition, land ownership in Mongolia is largely restricted to urban areas, and rangelands are freely open, although land tenure rights do exist for winter pastures. The combination of self-regulated pastures and responsible responses to the market means that over-grazing, in the traditional sense where pastures have trouble regenerating due to incessant intensive grazing, occurs only very rarely.

With respect to selecting pastures, herders carefully take into account the preferences of different animals, the relative nutritional value of different grasses, chenopods and other plant types, and their worth as fodder (Kakinuma et al. 2008; Fernandez-Gimenez 2000). Skilled herders have a broad knowledge base that serves to help find the most likely places where patches of sufficient graze will be found (Williams 1997). Notably, this knowledge base varies between herders, depending on their previous experiences, new observations and depth of cumulative knowledge, i.e. the degree to which information is passed down from one generation to the next (Fernandez-Gimenez 2000). Consequently, in terms of graze resources, a pasture that is "denuded" or useless to one herder is highly productive to another. This diversity in perception of graze quality is subsequently reflected in herder decisions regarding movement – a pasture ground may be of sufficient quality for some herders to stay while others would move on. No less importantly, the way in which herders perceive graze quality, and the mobility decisions associated with that assessment, impacts the type and effectiveness of additional management decisions down the line and their productive outcome.

When too many herders simultaneously utilise a pasture ground, graze and water supplies are heavily pressured. Most models of herder behaviour in a crowded pastoralist landscape suggest that herders will engage in long-range migration or transform their subsistence strategy to reduce the emphasis on herding (Lattimore 1940; Markov 1973; Shishlina 2001). However, rather than turning immediately to a long-range mobility strategy under circumstances of overgrazing and crowding, individual herders often instead alter their strategies on more local scales over the short term.

It is local strategies that are the outlet valve for all but the most extreme crowding. These local strategies rely on two main tactics: graze assessment and local knowledge, and systems of rangeland rights and access at a local level (Xie and Li 2008). In general, most of these mechanisms are enacted as short range movements of herds and herder households. During the summer months, herders may increase the frequency of highly localised moves with the aim of allowing pastures to recover quickly. Some herders choose strategies that, although providing benefits to their herds over the short term, have more costly effects down the road. For example, trespassing on reserved winter pastures during the summer months simply delays risk and, in this case, amplifies it by increasing the likelihood that there will be no graze available for animals during the winter months. If those winter pastures are tenured by another herding family, this decision may sow potential social discord within the local herding community, and fray the social networks of the offending herder.

Herders may also pursue local mobility strategies, even if they are not optimal ones, under conditions of pastoralist crowding for reasons related to regional animosity and local identity, rather than herd management (see Murphy, in press a). Contrary to the general assumption that herders will move as needed in order to maintain their flocks, herders often will deal with poor local graze conditions when they are less inclined to move for social reasons, e.g. people are unfriendly in a different region. In many cases, this reflects the limits, or extent, of herder social networks. Herders who remain local often do not have the strong social relationship required to move further afield.

Long-range mobility, where herders and their animals relocate over several hundred kilometres in a single move on a semi-regular basis, is one of the iconic traits of the nomadic pastoralist. However, while individuals' experience of major long-range movements is common, it is infrequently pursued. Although many herders will undertake a multi-hundred kilometre movements a few times in their lives, the hassles of undertaking a long-range movement are so great that the decision to partake in a long-range move is not made lightly.

Long-range moves are governed by a foreknowledge of local conditions in other regions, knowledge that is established either through word-of-mouth from other herders who have moved or from personal experiences of long-range movement into more productive areas. Interestingly, it is not the logistical technicalities of moving that impede long-range mobility, but the social landscape – the people you meet along the way – that confronts herders. Whenever possible, herders move to areas where they know people, usually relatives or others with whom, for example, they have reciprocal winter barn arrangements. This not only lessens the social discomfort of meeting and negotiating pasture access with new neighbours, but it also offers a source for the critical local knowledge that sustains herding practice (Fernandez-Gimenez 1999; Murphy in press a).

Although a herding family must negotiate a complex social landscape when undertaking long-distance moves, these moves do happen, albeit at a slow pace, and are the primary factor promoting population turnover in a given region. In general, as a single family group moves to a new locale and their relatives gradually follow them year by year, those herding families that originally preceded the newcomers leave. Over time, the population in a region turns over completely. For example, the Egiin Gol Valley in Bulgan Aimag has a population of around 250 people. Among the oldest generation of a dozen or so people, only two were born there. The rest had arrived over the preceding five or six decades as part of a long-term slow process of long- and middle-range movements by their families. Overall, however, although long-range social connections are critical for the survival of herding families particularly under conditions of environmental crisis, social networks are a limiting factor on nomadic mobility. In all but the most drastic cases, one moves only to areas where one already has a stable connection.

Responses to the extremes of pasture: *Zud* and drought

So far we have illustrated ways in which Mongolian nomadic pastoralists cope with the day-to-day ups and downs of their environment and the active tactics and strategies they take up to maintain herds. As a last point in our discussion of contemporary herders, we examine how occasional, but regular, potentially catastrophic environmental events impact herder decision-making and the pastoral system. While working in the region, we were able to observe how information travelled through local networks and triggered responses among the regional herding community. We have two examples: one is a response to a drought in the northern Gobi in 2005; and the second to the phenomenon of *zud*, or the killing winter storms that are a constant issue for herders in temperate Eurasia (Murphy, in press b; Jacobs 2010).

The summer of 2005 experienced the worst summer in a decade long dry spell that had been affecting the region. Although the 2005 troubles did not arise from any direct action by humans on pasture land or water supplies in Mongolia, but global climate change, the responses of the herders of the north Gobi were particularly informative. The northern Gobi is primarily a steppe-desert situated within an Irano-Turanian phytogeographic zone containing moderately high quality, seasonally available graze comprised of grasses and chenopods. There is variation in the distribution and density of graze across the region, which largely depends on local topographic and precipitation conditions. Baga Gazaryn Chuluu is one such area that is geologically and hydrologically different from the surrounding Gobi, and consequently is a subtly lusher locale in most seasons than the surrounding desert steppe. It is widely known amongst herders in this region that Baga Gazaryn Chulluu predictably produces graze even when regional precipitation levels are deficient. More specifically, herders related that it is widely known that after three rainfalls in the region around Baga Gazaryn Chuluu, ample graze will grow there.

During this particular summer, the north Gobi was quite dry and graze conditions relatively poor, but highly localised rain events did occur, especially around Baga Gazaryn Chuluu. Here, the day after the third rain of the summer, two things happened: the grass began to grow and mature very quickly, and new herding families began to appear. Initially, only advance parties of young men scouting for grass and camp sites arrived, and then, within a few days, whole herds and families arrived in order to set up camp (for more discussion of these type of movements, see Xie and Li 2008; Murphy in press b). Within a week, there were three to four times as many herders encamped around Baga Gazaryn Chuluu than the usual summer population. The results that we saw here, when the third rain came, were like ripples in a pond; nearby herders arrived quickly, those from farther away (some from as far away as 80 km) arrived more slowly.

This use of Baga Gazaryn Chuluu is a regional example of a system of nested patches of pasture, where one herder's drought stricken desert is a knowledgeable herder's uncrowded range, on a larger scale. Although Baga Gazaryn Chuluu became a destination for many herders during a period of severe drought, it is significant that not all herders flocked with their animals to this outcrop of rock and graze. Some herders who chose not to move to Baga Gazaryn Chuluu, while acknowledging that pastures would be good, felt that graze would be rapidly depleted and therefore not worth the move. To these herders, graze in Baga Gazaryn Chuluu was perceived as not being robust enough to withstand the surge of incoming herding families. Other herders who moved to Baga Gazaryn Chuluu targeted different pastures within the areas (i.e. high pastures, valley pastures, pastures at the open steppe – rocky outcrop interface) due to different perceptions of graze molded by previous experiences and knowledge passed down to them from relations who had previously visited this site. Once located in Baga Gazaryn Chuluu, differing levels of local knowledge of graze, routes and water sources stabilised the pastoral system in the newly crowded environment.

Another example of how ecological and social knowledge play a role in herders' responses to crisis is how herders are able to cope with the *zud*, a heavy late winter storm that delivers a type of snow or ice such that animals cannot get access to graze and therefore die quickly. A *zud* is a catastrophic event that, more often than not, can wipe out entire livestock herds with quick and deadly efficiency. Examples of the mass death events associated with a *zud* are commonly encountered on the landscape, with the skeletons of scores of animals heaped where they fell near the herders' winter campsites. This would appear to be a grim outcome, although, typically, a *zud* occurs on a relatively regular basis, every decade or so. Such catastrophic events, while unpredictable in their timing, are predictable in their certainty of occurrence.

Herder knowledge, networks and skill are key to surviving such extreme events, in particular how to shepherd your herds at these dire moments, how to make long moves ahead of losses, to know the safe spots and make the long-term choices necessary to rebuild a herd following the inevitable storms. These are the central factors for surviving climatic extremes, and herders readily acknowledge that skill and knowledge are paramount to successfully seeing through extreme events. Indeed, the unprecedented multiple *zud* winters that occurred at the beginning of this century imparted massive damage to the Mongolian pastoralist economy (see Jacobs 2010). However, one thing that herders consistently noted is that those families who suffered the most were the new herders who had only recently moved out from the cities during times of prosperity, and founded fledgling herds and joined into the pastoral lifestyle on the steppe. Lacking the deep knowledge of herding practice necessary to survive even a single major crisis, it was these new families who lost entire herds and were forced to return to the urban cityscape.

In summary, what do these observations of modern herders demonstrate to us? They show us that technical outcomes are defined by local knowledge, herding skill and social networks. From this, we follow Ingold's ideas of the relationship between skill and the body in an environment, the perception of landscape, and immersion in the task (Ingold 2000). From this perspective, the herder acquires knowledge about their landscape, which includes an understanding of the range of potential activities that a particular landscape can support in terms of environmental, subsistence and social action, and knows how to follow the situations and forces of that landscape towards their own goals. In other words, they are skilled within a landscape. The course they develop is variable because herder activities in this mutual construct include different elements of landscapes, in different situations and at different times. This mutually constructed relationship between the person and the landscape consequently means one herder's exhausted land is another's productive territory; or, two herders might receive the same information in different ways, interpret it differently, but the outcome for both is the same. This knowledge of one's immediate local landscape – knowing where the grass grows when times are hard, for example, and knowledge of long-range movement possibilities – is what sustains nomadic pastoralism systems.

Initially, we argued that the responses of nomadic pastoralists to environmental stresses are effective in creating long-term stability in systems of nomadic pastoralism, and that the "exhausted pasture binary" is not a good way to look at issues of long-term change among nomadic pastoralists. More often than not, change is brought about by a series of incremental adaptions, alterations of local tactics and changes in networks of communication and relationships that build up over the long term.

Archaeological approaches to Rangeland responses

We have identified three factors as key to maintenance of stable pastoral systems: graze availability, herder perceptions of pasture quality and herders' social networks. It is possible to make inferences about graze availability in the past (fig. 2 and 3) and to find proxies for social networks in the archaeological record. It is this second point that we would like to highlight here. In archaeological contexts, social networks are frequently reconstructed based on the spatial distribution of ceramic styles. Here, the sources of those ceramics form nodes in the network, and the relationships between the nodes are inferred from patterns observable in the ceramic assemblages (Mizoguchi 2009; Johnson 1972).

As an illustration of the depth and range of the networks of prehistoric pastoralists, a comparison is offered here between the Bronze Age and early Iron Age (4600–2800 BP), the period associated with the initial adoption and spread of mobile pastoralism, and the Xiongnu (or late Iron Age period; 2400–1700 BP), during which many of the techniques of historic nomadic pastoralism were established (Makarewicz 2011; Watson 1961: 155–156). Figure 4 shows a contrast between the ceramics of the Bronze Age and those of the Xiongnu period. Bronze Age pottery is heterogeneous and rather idiosyncratic. Although the forms are roughly the same over a wide region, the material fabrics and decorations are widely variable across the region. In contrast, during the succeeding Iron Age and Xiongnu periods (2500–1600 BP) in Mongolia and south Siberia, ceramics take on very similar forms, decorations and fabrics, over a large region.

These broad trends are borne out in the two intensively surveyed small areas of Mongolia, the Lower Egiin Gol (Wright et al., forthcoming; Turbat et al. 2003) and Baga Gazaryn Chuluu (Wright et al. 2007; Amartuvshin and Honeychurch 2010: fig. 1) where the same patterns are seen in the large samples of ceramics from many contexts. Similarities in ceramics are widely seen as proxies for the communication and movement of people in communities of practice, social and exchange networks (Mizoguchi 2009, Mills et al. 2013).

This brings us back to our premise that skill, knowledge and networks are the foundation of stable nomadic pastoralist systems. The general regional patterns, and the specific details in each region, that have been examined in detail may indicate broad differences between Bronze /Early Iron Age and Xiongnu social networks. Simply stated, the disparate nature of the Bronze and Early Iron Age cultural material suggests that herders from each cultural period participated in widely different social networks. There appears to be an important technical difference between Bronze Age nomadic

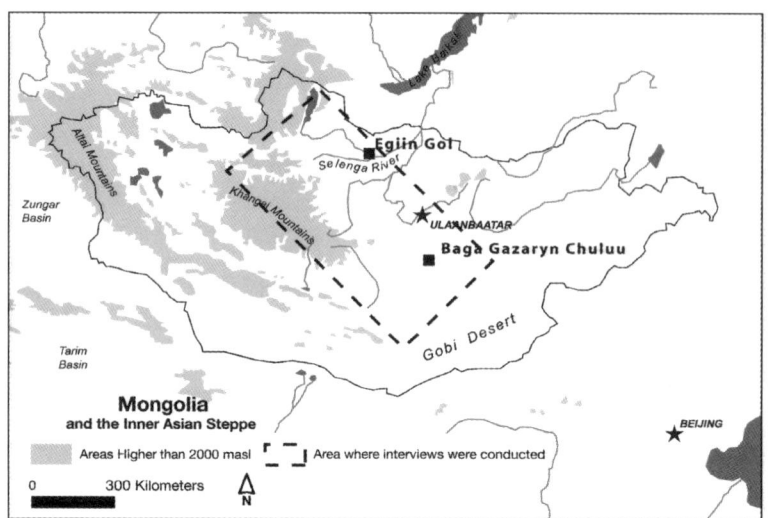

FIGURE 1. Mongolia and the northeast Asian Steppe indicating the two specific areas mentioned in the text.

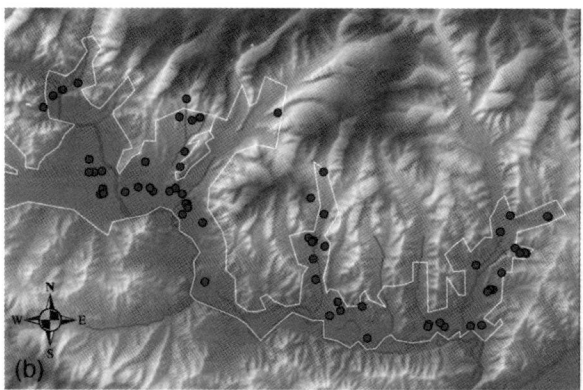

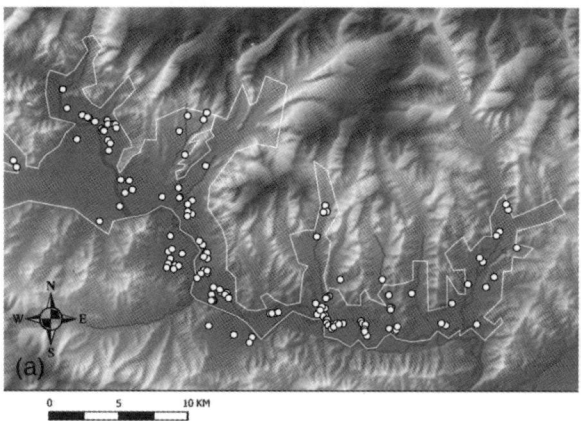

FIGURE 2. The lower Egiin Gol valley, Bulgan Aimag, northern Mongolia. Data recorded by the Egiin Gol Survey showing the parallel distribution of campsites of modern nomadic pastoralists (a) and prehistoric and medieval nomadic pastoralists (b). The similar patterns of settlement suggest that similar economic and social needs have structured landscape over the long term.

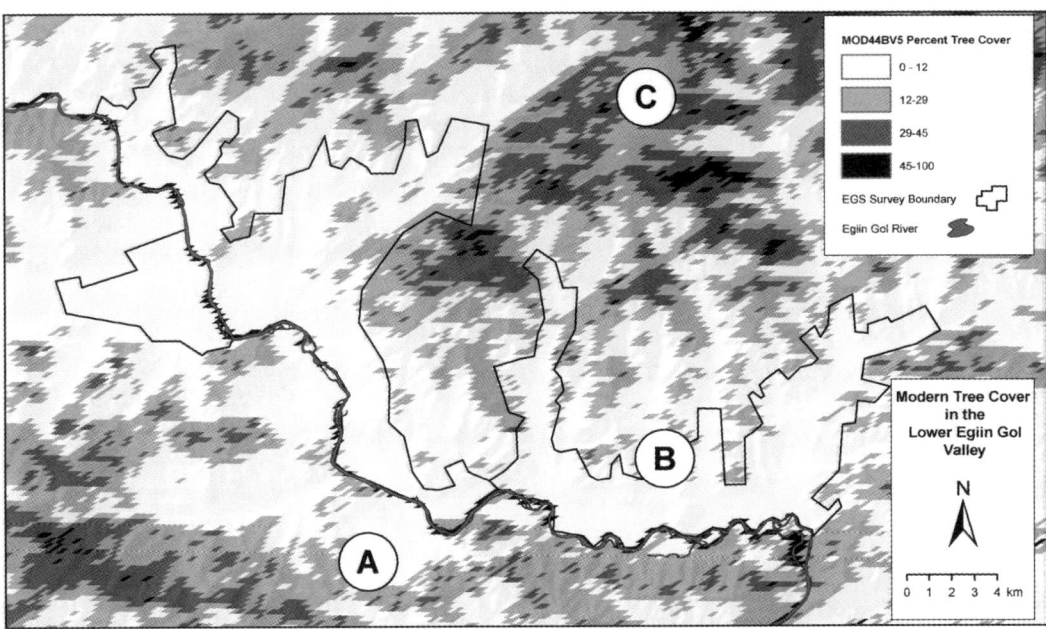

FIGURE 3. *MODIS satellite data on tree cover in the lower Egiin Gol. Lower elevation tree cover is a mixed birch and pine forest. Area A is the southern bank of the Egiin Gol river, where large continuous blocks of tree cover fill the slopes above the river valley. Area B is the more densely inhabited northern bank, where smaller discontinuous blocks of tree cover are found on the slopes above the main river valley. Phytoliths from on- and off-site samples, wood charcoal from archaeological deposits and local geomorphology (see Wright et al., forthcoming) suggest that a similar vegetation regime has been present in the valley since the late Holocene. Area C is a high plateau with dense conifer forests in the high elevations not at all like the river valley.*

pastoralists and Iron Age Nomadic pastoralism, with the former participating in networks characterised by less depth, extensivity and durability compared to the latter. The material culture data suggest that the Bronze Age pastoralists did not enter into the extensive regional social networks that foster the social connections which help make contemporary nomadic pastoralism so robust for some herders. There are certainly syncretic elements in the Bronze Age, but they may not be related to protecting the animals. These early pastoralists may not have been as able to adapt to and recover

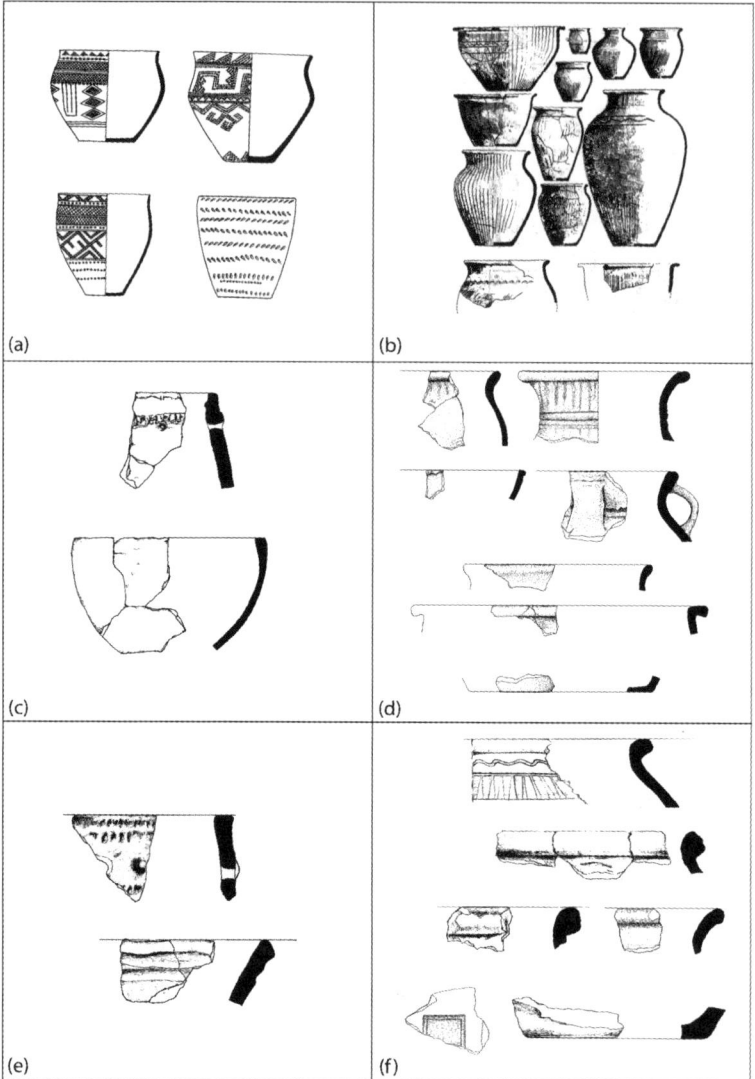

FIGURE 4. *A comparison between the heterogenous ceramics of early nomadic pastoralists of the Central Asian Bronze Age (a, c, e) and the homogeneous ceramics of the established Iron Age pastoralists of the Xiongnu period (b, d, f) from two regions of Mongolia. Bronze Age surface treatments and forms differ from region to region, while Xiongnu vessels are generally similar in forms, rim details and decorations over a wide area. The ceramics shown in c and d are from the Lower Egiin Gol Valley, and those in e and f are from Baga Gazaryn Chuluu. Illustrations a (after Мошкова 1992: fig. 106) and b (after Davydova 1968: fig. 68-2) are provided as reference for whole vessels that fit these general patterns (see also Wright 2011).*

from crises as well as their successors, and are akin to the nomads from the opening quote of this paper; an idea of profound importance to the discussion of the adoptions and spread of nomadic pastoralism.

It is not uncommon, even if only for reasons of the simple efficiency of model building, to assume that nomadic pastoralism as we know it now sprang fully formed from the minds of the first pastoralists many millennia ago. However, early nomadic pastoralists may not have had the long-range social and support networks for their herds, the regional political organisation (Murphy, in press a) or the winter barns that now protect the herders of Inner Asia. Without these key elements, early nomadic pastoralism and the political relationships and risk reduction strategies (Winterhalder et al. 1999; Bollig and Göbel 1997; see also Roe et al. 1998) that underpinned it may well have been very different to those we see in the historical and ethnographic records. In this context, perception of rangeland and local management of graze, short-range nomadic movement (fig. 5) and relationships in a mosaiced social and subsistence landscape (cf. Popova 2009) take on a proportionally greater importance.

Responses to environmental or ecosystem crisis are made up of many short-term choices made by individuals, rather than sweeping major responses dictated by overarching rules of climatic behaviour. We began with the thesis that there is not a simple way to address, or even describe, an exhausted pasture. A whole range of factors affects the ways in which individual pastoralists perceive their available local and regional resources and act on those perceptions. We also argue that the complexity of these responses to immediate short-term local issues is a primary factor in the stability of the system of pastoral nomadism over the long term. Finally, the relationship of risk management in local networks, and the resulting necessities for particular political and social spaces to accommodate those practices, is an important element of the study of the development and emergence of a system of nomadic pastoralism.

Acknowledgements

We would like to celebrate the life and work of our dear friend and classmate Stine Rossel. She was a constant spring of happiness, insightful, cheerful, creative, good beyond every measure and a friend to all who brought out the best in others.

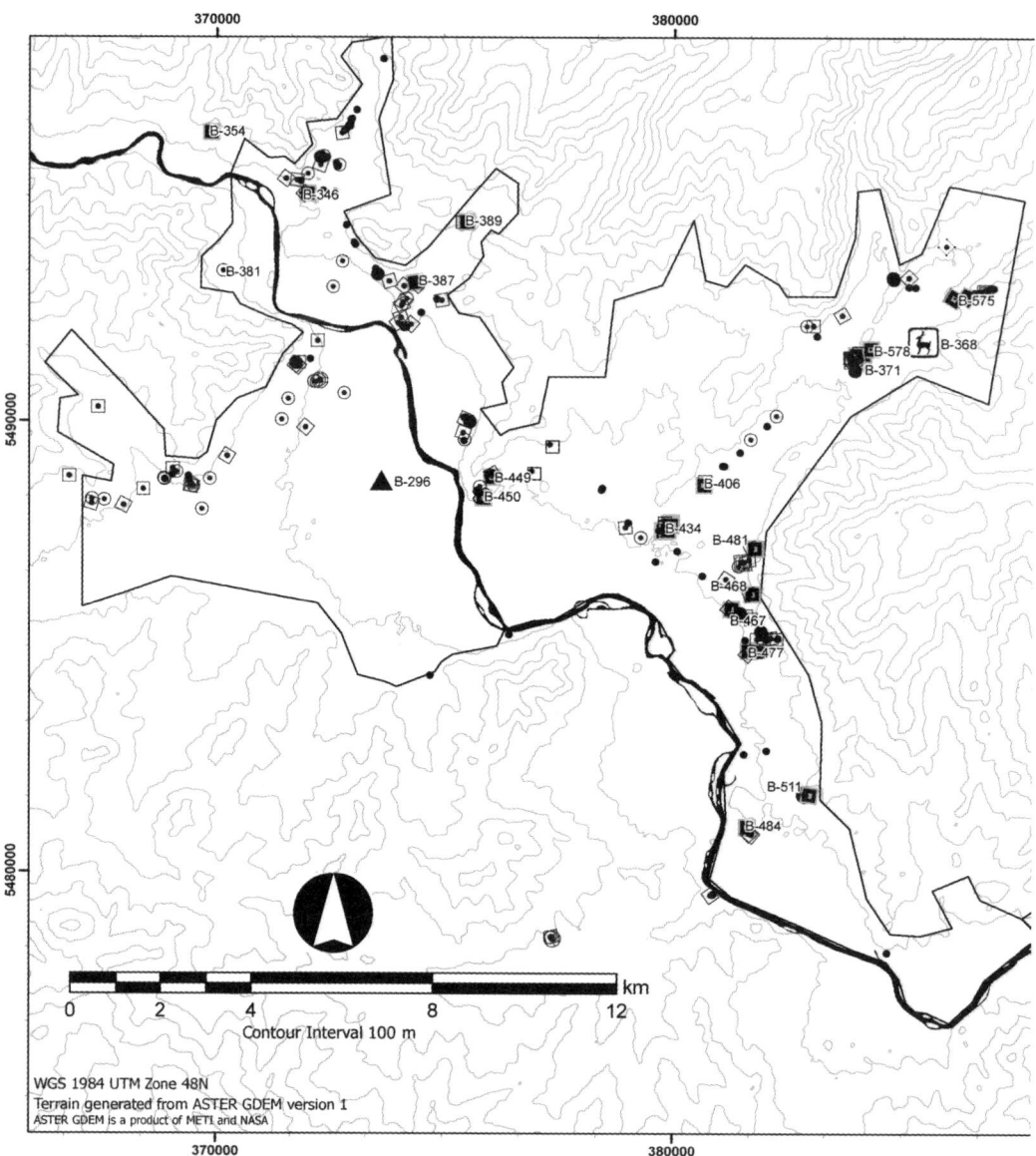

FIGURE 5. *The Bronze Age site distribution from the lower Egiin Gol valley. This map shows a wide selection of stone monument structures including the distinctive slab burial graves of the first Bronze Age pastoralists in the valley. These form a monumental landscape in which Bronze Age nomads lived. Sherd scatters and burials combined show possible self-contained groups of pastoralist yearly move-*

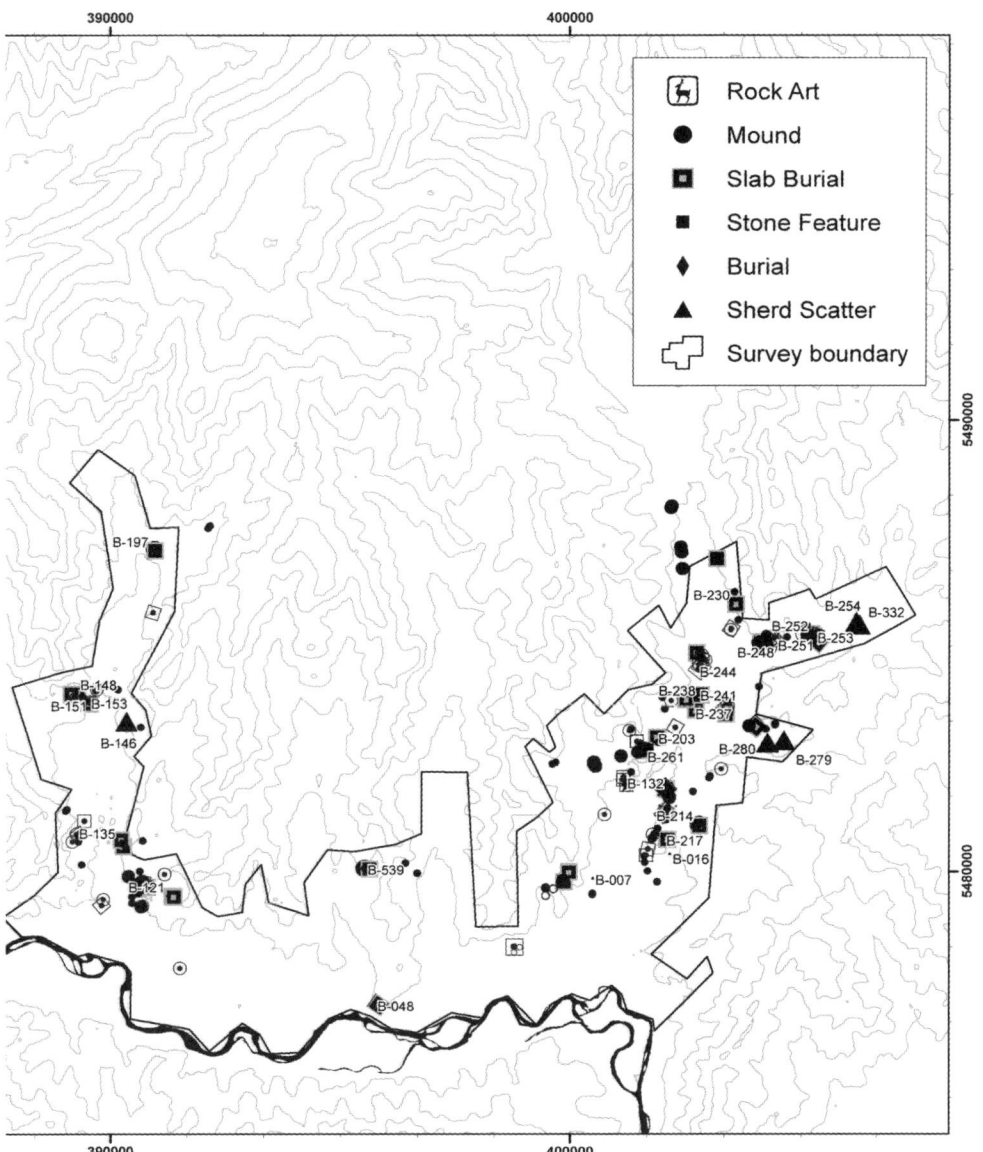

ment rounds in the three major north-south tributary valleys (see also Honeychurch et al. 2009). This territorial system is rooted in habitation areas of the pre-pastoralist population of the valley and was transformed as larger Iron Age political systems affected the lower Egiin Gol (Wright et al., forthcoming; Honeychurch and Amartuvshin 2006).

Bibliography

Allsopp, N., C. Laurent, L. M. C. Debeaudoin and M. I. Samuels. 2007. "Environmental Perceptions and Practices of Livestock Keepers on the Namaqualand Commons Challenge Conventional Rangeland Management." *Journal of Arid Environments* 70: 740–754.

Amartuvshin, C. 2003. "An Examination of Khunnu Period Settlement in the Egiin Gol Valley." *Studia Archaeologica* 21: 59–65.

Amartuvshin, C., and W. Honeychurch (eds.). 2010. *Dundgobi Aimagt Hiisen Arkheologiin Sudalgaa: Baga Gazaryn Chuluu* [Archaeological Research in the Middle Gobi: Baga Gazaryn Chuluu]. Mongolian Academy of Science, Institute of Archaeology. Ulaanbaatar.

Anthony, D. W. 1985. "The Social and Economic Implications of the Domestication of the Horse." PhD dissertation, University of Pennsylvania.

Bar-Yosef, O., and A. Khazanov (eds.). 1992. *Pastoralism in the Levant: Archaeological materials in anthropological perspectives*. Prehistory Press. Madison, Wisconsin.

Bentley, R. A., and H. D. G. Maschner. 2001. "Stylistic Change as a Self-Organized Critical Phenomenon: An Archaeological Study in Complexity." *Journal of Archaeological Method and Theory* 8: 35–66.

Boissevain, J., and J. C. Mitchell (eds.). 1973. *Network Analysis: Studies in Human Interaction*. Mouton. The Hague.

Bollig, M., and B. Göbel. 1997. "Risk, uncertainty and pastoralism: An introduction." *Nomadic Peoples* 1: 5–21.

Davydova, A. V. 1968. "The Ivolga Gorodishche (A monument of the Hsiung-Nu culture in the Trans-Baikal region)." *Acta Archaeologica Academiae Scientiarium Hungaricae* 20: 209–245.

Dyson-Hudson, N. 1972. "The Study of Nomads." In W. Irons and N. Dyson-Hudson (eds.), *Perspectives of Nomadism*, 2–29. E. J. Brill. Leiden.

Dyson-Hudson, R., and N. Dyson-Hudson. 1980. "Nomadic Pastoralism." *Annual Review of Anthropology* 9: 15–61.

Dyson-Hudson, R., and E. A. Smith. 1978. "Human territoriality: An ecological reassessment." *American Anthropologist* 80: 21–41.

Ekvall, R. B. 1968. *Fields on the Hoof, Nexus of Tibetan nomadic pastoralism*. Holt, Rinehart, and Winston. New York.

Fernandez-Gimenez, M. E. 1999. "A Geographical History of Pastoral Land Use in Mongolia." *Geographical Review* 89: 315–342.

Fernandez-Gimenez, M. E. 2000. "The Role of Mongolian Nomadic Pastoralists' Ecological Knowledge in Rangeland Management." *Ecological Applications* 10: 1318–1326.

Frohlich, B., T. Amgalantögs, J. Littleton, D. Hunt, J. Hinton and K. Goler. 2009. "Bronze Age Burial Mounds in the Khövsgöl Aimag, Mongolia." In J. Bemmann, H. Parzinger, E. Pohl and D. Tseveendorzh (eds.), *Current Archaeological Research In Mongolia*, 99–116, Vor- und Frühgeschichtliche Archäologie, Rheinische Friedrich-Wilhelms-Universität Bonn. Bonn.

Gorenflo, L. J., and T. L. Bell. 1991. "Network Analysis and the study of past regional organization." In C. D. Thrombold (ed.), *Ancient Road Networks and Settlement Hierarchies in the New World*, 80–98. Cambridge University Press. Cambridge.

Hodder, I. 1982. *Symbols in action: Ethnoarchaeological studies of material culture*. Cambridge University Press. Cambridge and New York.

Honeychurch, W. 2010. "Pastoral nomadic voices: A Mongolian archaeology for the future." *World Archaeology* 42: 405–417.

Honeychurch, W., and C. Amartuvshin. 2006. "States on Horseback: The Rise of Inner Asian Confederations and Empires." In M. Stark (ed.), *Archaeology of Asia*. Blackwell. Malden, Massachusetts.

Honeychurch, W., J. Wright and C. Amartuvshin. 2009. "Re-Writing Monumental Landscapes as Inner Asian Political Process." In B. Hanks and K. Linduff (eds.), *Social Complexity in Prehistoric Eurasia: Monuments, Metals, and Mobility*, 330–357. Cambridge University Press. New York.

Ingold, T. 2000. *The Perception of the Environment: Essays in livelihood, dwelling and skill*. Routledge. New York.

Jacobs, A. 2010. "Winter Leaves Mongolians a Harvest of Carcasses." *The New York Times*. May 19, 2010.

Jacobson-Tepfer, E., J. E. Meacham and G. Tepfer. 2010. *Archaeology and Landscape in the Mongolian Altai: An Atlas*. ESRI Press. Redlands, California.

Johnson, G. A. 1972. "A Test of the Utility of Central Place Theory in Archaeology." In P. J. Ucko, R. Tringham and G. W. Dimbleby (eds.), *Man, Settlement and Urbanism*. Duckworth. London.

Kakinuma, K., T. Ozaki, S. Takatsuki and J. Chuluun. 2008. "How Pastoralists in Mongolia Perceive Vegetation Changes Caused by Grazing." *Nomadic Peoples* 12: 67–73.

Khazanov, A. M. 1994. *Nomads and the Outside World*. The University of Wisconsin Press. Madiscon.

Krader, L. 1957. "Culture and Environment in Interior Asia." In L. Krader and A. Palerm (eds.), *Studies in Human Ecology*. The Anthropological Society of Washington. Washington, D.C.

Lattimore, O. 1940. *Inner Asian Frontiers of China*. American Geographical Society. New York.

Lees, S. H., and D. G. Bates. 1974. "The Origins of Specialized Nomadic Pastoralism: A systematic model." *American Antiquity* 39: 187–193.

Liu, H., H. Cui and Y. Huang. 2001. "Detecting Holocene movements of the woodland-steppe ecotone in northern China using discriminant analysis." *Journal of Quaternary Science* 16: 237–244.

Markov, G. E. 1973. "Nekotorye problemy vozniknoveniia i rannykh etapov kochnevnichestva v. Azii. [Some problems in the origin and early stages of nomadism in Asia]". *Sovetskaia etnografiia*:101–113.

Mizoguchi, K. 2009. "Nodes and edges: A network approach to hierarchisation and state formation in Japan." *Journal of Anthropological Archaeology* 28: 14–26.

Moskova, M. G. [М. Г. МОШКОВА] (ed.) 1992. *Степная полоса Азиатской части СССР в скифо-сарматское время* [The Steppe Zone of the Asian part of the USSR to the Scythian-Sarmatian time]. Наука. Moscow.

Muller, B., A. Linstadter, K. Frank, M. Bollig and C. Wissel. 2007. "Learning from Local Knowledge: Modeling the Pastoral-nomadic Range Management of the Himba, Namibia." *Ecological Applications* 17: 1857–1875.

Murphy, D. J. in press (a). "Reviving the Xoroo: Risk, uncertainty, and the re-emergence of group mobility in rural Mongolia." *Journal of Anthropological Archaeology*.

Murphy, D. J. in press (b). "Encountering the Franchise State: Dzud, Otor, and Transformations."

Newman, M. E. J. 2003. "The structure and function of complex networks." *SIAM Review* 45: 167–256.

Popova, L. M. S. 2009. "Blurring the Boundaries, Foragers and Pastoralists in the Volga-Urals Region." In B. K. Hanks and K. M. Linduff (eds.), *Social Complexity in Prehistoric Eurasia, Monuments, Metals and Mobility*, 296–320. Cambridge Unversity Press. Cambridge.

Roe, E., L. Huntsinger and K. Labnow. 1998. "High Reliabilty Pastoralism." *Journal of Arid Environments* 39: 39–55.

Salzman, P. C. 2002. "Pastoral Nomads: Some general observations based on research in Iran." *Journal of Anthropological Research* 58: 245–264.

Shelach, G. 2009. *Prehistoric Societies on the Northern Frontiers of China*. Equinox Publishing. London.

Shishlina, N. I. 2001. "The Seasonal Cycle of Grassland Use in the Caspian Sea Steppe during the Bronze Age: A new approach to an old problem." *European Journal of Archaeology* 4: 346–366.

Shishlina, N. I., and F. T. Hiebert. 1998. "The Steppe and the Sown: Interaction between Bronze Age Eurasian Nomads and Agriculturalists." In V. Mair (ed.), *The Bronze Age and Early Iron Age Peoples of Eastern Central Asia*. The Institute for the Study of Man. Washington, D.C.

Sneath, D. 1993. "Social relations, networks and social organizations in post-socialist Mongolia." *Nomadic Peoples* 33: 193–207.

Turbat, T., C. Amartuvshin and U. Erdenbat. 2003. Эгийн Голын Сав Нутаг Дахь Археологийн Дурсгалууд [*Archaeological Monuments of Egiin Gol Valley*], Mongolian Academy of Science, Institute of Archaeology. Ulaanbataar.

Watson, B. (trans.). 1961. *Records of the Grand Historian of China, translated from the Shih Chi of Ssu-ma Ch'ien*. Columbia University Press. New York.

Watson, P. J. 1995. "Archaeology, anthropology, and the culture concept." *American Anthropologist* 97: 683–694.

Wheat, J. B., J. C. Gifford and W. W. Wasley. 1958. "Ceramic Variety, Type Cluster, and Ceramic System in Southwestern Pottery Analysis." *American Antiquity* XXIV: 34–47.

Williams, D. M. 1997. "Patchwork, pastoralists, and perception: Dune sand as a valued resource among herders of inner Mongolia." *Human Ecology* 25: 297–317.

Winterhalder, B., F. Lu and B. Tucker. 1999. "Risk-sensitive adaptive tactics: Models and evidence from subsistence studies in biology and anthropology." *Journal of Archaeological Research* 7: 301–348.

Wright, J. 2011. "Xiongnu Ceramic Chronology and Typology in the Egiin Gol Valley, Mongolia." In U. Brosseder and B. K. Miller (eds.), *Xiongnu Archaeology, Multidisciplinary Perspectives of the First Steppe Empire in Inner Asia*. Vor-und Frühgeschichtliche Archäologie Rheinische Friedrich-Wilhelms-Universität. Bonn.

Wright, J., W. Honeychurch and C. Amartuvshin. 2007. "Initial findings of the Baga Gazaryn Chuluu archaeological survey (2003–2006)." *Antiquity* 81(313): Project Gallery.

Wright, J., W. Honeychurch and C. Amartuvshin (forthcoming). *Continuity and Authority in the Mongolian Steppe, The Egiin Gol Survey 1997–2002.* Yale Anthropological Press. New Haven.

Xie, Y., and W. Li. 2008. "Why do Herders Insist on Otor? Maintaining Mobility in Inner Mongolia." *Nomadic Peoples* 12: 35–52.

Stable Isotope
Analysis in
the Middle East

Understanding the Reasons for Non-Sustainability in Past Agricultural Systems

Simone Riehl

Introduction

With the strengthening of environmental archaeology since the 1970s, past agricultural systems in arid and semi-arid environments have been a focal point for the reconstruction of climate and cultural change in ancient societies, particularly with regard to the collapse of Early Bronze Age civilisations (see the various contributions in Dalfes et al. 1997, and in Marro and Kuzucuoglu 2007, Giosan et al. 2012). The most frequently discussed reasons for settlement abandonment have been overexploitation of resources and political conflict leading to a regionally diversified propensity for crop failure (Wilkinson 1997; Akkermans and Schwartz 2003). Others have linked the collapse to the 4200 BP climatic event (Ristvet and Weiss 2005; Staubwasser and Weiss 2006).

In terms of the problem at hand – the effects of global change – archaeology presents an important tool for understanding how sustainability could be addressed (Tainter 1995; van der Leeuw 1998). Therefore, a favoured topic of environmental archaeology in the Near East today is the influence of environmental dynamics on choices between economic adaptation through technological progress and traditional subsistence strategies. Archaeological debate on the reasons for societal change is sometimes still fed by classic, discipline-immanent misconceptions. As an example, the lack of regional differentiation of climate effects when looking for climate impact based on global climate fluctuations leads some archaeologists to erroneous expectations of evenness in climate effects over wide areas, and in turn to the refusal of climate as a factor in landscape development, because regional differences are recognised in the data sets. Similarly, despite empirical observation giving no justifications for expecting congeneric or simultaneous reactions in different human populations, or different cultural and political unities, to environmental fluctuations, the absence of a supraregionally identical human behaviour is often used as an argument against environmental factors playing a role in societal change. Such biased concepts can prevent a constructive collaboration between traditional and environmental archaeologists.

In addition to presenting some case studies on the interrelationship of environmental/climate change and the development of human societies, this methodological essay will discuss why it is important to disengage from deterministic thinking and to accept the "variability principle" in order to achieve an understanding of the reasons for non-sustainability in past agricultural systems.

The continuous need for fundamental data

In the light of recent discussions on the gloomy scenarios relating to global climate change, history and prehistory have become an important information pool for questions on sustainability today and in the future (van der Leeuw 1998).

Important questions that are addressed in this relation are:

- Which environmental and/or socio-political and/or cultural factors lead either to persistence or change in agricultural systems?
- How did persisting systems adapt their agricultural production to such changes?
- How diverse are procedures of change, i.e. is there a regional or even supraregional unity in adaptation patterns?

To answer such questions, an integration of different lines of evidence is called for. These comprise climatic fluctuations and their effects as well as environmental degradation through human impact and agricultural decision-making based on cultural conventions and economic and political goals.

A large body of data has been accumulated by palaeoclimatological research. The integration of this data into archaeological work, and vice versa, is, however, underdeveloped. Despite the comprehensive data and knowledge about climate fluctuations and the past development of human populations in their cultural and political environment, the missing link between these processes is still the reason for much debate between archaeologists. Bioarchaeology, or more specifically archaeobotany and zooarchaeology, are the main methods for investigating ancient agriculture and have over recent decades generated a large number of works which focus on the dynamics of subsistence and economy, with now, finally, some attempts to integrate both methods (Van Derwarker and Peres 2010; Smith and Miller 2009).

Today, we have access to an abundance of archaeobotanical data from the Near Eastern region to test hypotheses in relation to ecological and economic sustainability. Major advances in ecology and agroecology, for example, allow for the classification of crops and their autecology in relation to their region of cultivation. Data management systems based on the increased amount of data enable the emergence of syntheses and the development of a broader view of archaeobotanical assemblages, allowing for the elaboration of general patterns and for separating out atypical data. Methodological problems in using archaeobotanical databases are, however, numerous and consist of differing data quality caused by differences in sampling strategies and the quantity of analysed remains, the fragmentation of archaeobotanical research in many geographic areas, and a lack of accurate dating. Therefore, the archaeobotanical database is still too small to allow representative syntheses. Further bioarchaeological sampling and analysis is necessary to fill these gaps in the future.

Another critical aspect is that the distribution patterns of archaeobotanical remains may allow for the recognition of changes in agro-production, but it remains unknown so far how different factors interacted, how people were involved in these changes and how their perceptions built the motivation of their decision-making.

The visualisation of archaeological data and the results given by geographic information systems promise the ability in the future to handle the large amount of existing data. A number of geoarchaeological studies, including analysis of satellite imagery that has been used to identify ancient settlement and land use patterns, have been published (Goldhausen and Ricci 2005; Wilkinson 1997, 2003; Wilkinson et al. 2007; Deckers and Riehl 2009). Some of these studies contradict archaeobotanical results by reconstructing a stronger degradation of the landscape than what is indicated in the botanical remains (Deckers and Riehl 2007). In addition, the modelling of the different factors determining yield structures has been developed in relation with geographic information systems applications (e.g. SOTER, Soil Terrain Database), opening up a broad variety of interpretative approaches. The selective nature of such investigations requires the continuous and systematic extension of the geographic locations of research.

The involvement of stable isotope analysis in archaeological studies has also furthered the information basis on ancient nutrition and climate development. Of particular interest to ancient agricultural conditions is the analysis of stable carbon isotopes in archaeobotanical crop remains, which allows for the recognition of water stress signals (Farquhar et al. 1982; Ehleringer et al. 1986; Araus et al. 1997; Riehl et al. 2008). Although there is good correlation between the 4200 BP event and stress signals at the transition from the Early to the Middle Bronze Age, a major methodological problem is the fact that the absence of a stress signal through $\delta^{13}C$ values is in itself not enough to allow for concluding whether higher precipitation or sophisticated irrigation were the reason.

While ancient texts are limited in their explanatory power, they should be analysed in relation to the development of agricultural systems and technology. This has been done in a number of research projects, such as at Tell Schech Hamad (Röllig 2008) or as an integral part of environmental archaeology (Riehl et al. 2012). Some attempts have also been undertaken to interpret ancient texts for their information on ancient irrigation and agriculture (Wiggermann 2000; Lafont 2000; van Koppen 2001). A systematic consideration of agricultural aspects and the ancient environment, however, still waits to be generated. This includes the investigation of human perceptions and goals, particularly those of the elite, as a common background to the manufacture of texts. This is, at the same time, a major aspect in sustainability research, as societal development, to a large extent, is propelled by the decision-making of the elites (Tainter 1995; Diamond 2005).

We can conclude, therefore, that due to the regionally diverse character of past agricultural systems, our understanding of the processes within such systems will only grow with the continuous increase of archaeological and environmental data.

How determinism shaped our interpretations

To make sense of data, we can only use our brain, the thinking of which has been formed by specific concepts in our family, academic and cultural environments. We need, therefore, to be aware of the perceptions and preconceptions that we apply to interpreting data. This is particularly important when looking for causal relations of different processes. Since Aristotle (384–322 BC), the relation of cause and effect has been considered to be unidirectional in nature, where the effect-event is the direct or transferred consequence of a spatially attached efficient cause, assuming that there are laws regulating this relationship. This is the principle of determinism, which appears in a number of variations, depending on which specific all-explaining cause is considered to be true by the researcher with a deterministic view.

Environmental determinism, which goes back to the Greek geographer Strabo (c. 63 BC–24 AD), gained increasing prominence in the geosciences during the late 19th and early 20th centuries. With its short-sighted application between 1920 and 1940, it was abused to justify racism which led to an abandonment of the environmental determinism hypothesis for many decades. This is one of the reasons today for many archaeologists' negative reactions to any suggestion of considering environmental influences on the development of human society. Usually, denying environmental determinism means supporting cultural determinism, which assumes that humans are responsible for the development of society solely through the power of thought, socialisation and cultural environment.

Anthropologists, in the last decades, have favoured cultural determinism over environmental determinism. One of the most common opinions was expressed by Hubbard in an overview on the development of agriculture in Europe and the Near East, "[T]he patterns of crop exploitation seem interpretable in terms of natural ecology only in the very earliest periods" (Hubbard 1980: 51), implying that, by the Bronze Age, agricultural decision-making was mainly a matter of free will, of cultural interests or economic and political goals, with little impact from climate change. Another famous argument of cultural determinism is the interpretation of the absence of simultaneous abandonment of settlements in the Near East as proof against the impact of environmental on settlement collapse (Meijer 2007), based on the underlying but false assumption that climate and/or environmental change would result in simultaneous and identical consequences over a wide area.

Like environmental determinism, however, cultural determinism also has the potential to be abused to support ideas of colonialism and racism. Furthermore, applications of environmental and cultural determinism in archaeology are often based on the faulty perception that there is a beginning and an end to the transformation of human societies and, as discussed above, that responses to change must be linear, in order to prove a causal relation.

A further example of a biased understanding of environmental influence on human societies is the flawed assumption that only linear responses to external forces indicate causality between these forces and observed effects, as claimed by Coombs and Barber (2005; see also Butzer 1997). Their argument is that the processes of cultural decline exhibit patterns characteristic of complexity cascading within self-organised systems, and that, therefore, the nonlinear nature of the systems' responses cannot be causally related to the processes themselves. This is not in agreement with the principle that complex systems rarely produce linear or replicable responses. Environmental change, whether naturally, politically or culturally induced, may create the necessity for change in human society; however, it does not determine the precise trajectory or the timescale for that change. Neither are differences in human decision-making foreseeable among various groups due to differing perceptions of the environment or differences in innovation capacity or in political goals. Both, environmental and cultural determinism, are too limited to fully explain causality in complex systems.

When dealing with causation in human systems, we need instead to expect an infinite, multi-causal structure with a broad range of variability in natural and process-related effects and human actions (fig. 1), an aspect which is consistent with the compatibilistic world-view of cultural ecology, i.e. the acknowledgement of differences in perception of the environment by different human groups, which leads to different attitudes toward environmental change and resource management.

The explanation of the development of archaeobotanical crop assemblages dated to a sequence between 3000–1800 BC provides a case study of a deterministic endeavour in archaeology (Riehl 2009a). In this study, crop plant assemblages from the Early and Middle Bronze Ages in the Near East have been compared, showing the reduction of drought-susceptible

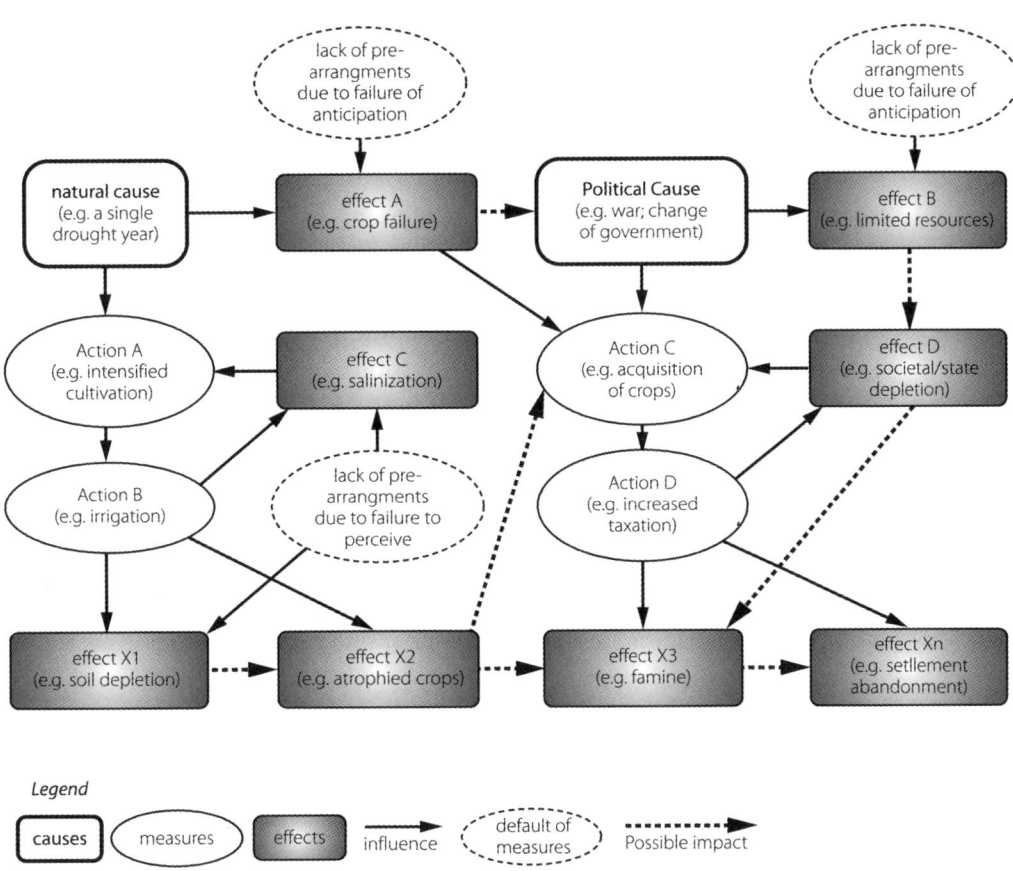

FIGURE 1. *Infinite structure of a cause-and-effect model for societal collapse; selection from a pool of interchangeable causes, effects and actions (figure from Riehl 2009b).*

crops during the Middle Bronze Age. This is particularly visible in flax seeds. Beside the fact that all archaeological data is fragmentary and limits a realistic reflection of original past conditions, a major problem of flax seeds in the archaeobotanical record is that nothing is known of the original purpose of the find alone, as it may have been consumed as a seed, used for oil production, or the plant fibres used in linen production. Ancient texts and iconography, however, do indicate that linen was an important product, particularly for religious purposes, during the Bronze Age (Waetzoldt 1983).

Following the argument of environmental determinism, it may be concluded that the disappearance of flax seeds during the Middle Bronze Age could be related to increasing aridity due to the frequently cited global 4200 BP event (Staubwasser and Weiss 2006; deMenocal 2001; Gasse 2000; Booth et al. 2005; An et al. 2005, etc.), because fibre flax requires more than 700 mm water during the growing season (El-Nakhlawy and El-Fawal 1989); a level of water that would not have been available in the area of consideration. Even when irrigation would have been practised, increasing aridity would probably have resulted in a considerable decrease of yield. Wool may have been a profitable alternative, thus explaining a shift in textile preferences by a change in the availability of the resources caused by environmental change.

The culturally determined argument could consider a higher preference for wool to be responsible for a shift from linen to wool textiles. Preference for wool may have been related to its qualities for dying, i.e. it can take brighter colours as compared to linen (cf. Roth et al. 1992). However, linen textiles were rare and generally considered luxuries reserved for an elite (Waetzoldt 1983), indicating their high value despite or due to their restricted availability. This, however, does not give the reasons for decreasing cultivation, and it is obvious that the only way to answer this question is by including other sources of information.

In the history of flax production, climate change and economy seem to be strongly intertwined. In most cases, economic decisions are made with respect to cost-benefit calculations, and when the costs for production increase, alternatives become more interesting. In the case of flax, costs for production would rise due to increasing aridity or the deterioration of the soil. These factors are determinable by isotopic studies and geoarchaeology.

Stable carbon isotope measurements provide a signal of drought stress experienced by plants from arid and semi-arid environments (e.g. Araus et al. 1997; Ferrio et al. 2005). Roughly 70 $\delta^{13}C$ values of barley grains from seven Near Eastern Bronze Age sites do indeed suggest a decreasing availability of soil moisture during the Middle Bronze Age and would support the hypothesis of the increasing economic inefficiency of flax production (Riehl et al. 2008). However, it remains unknown how the ancient society perceived this development, an aspect that may be evaluated by direct or indirect information in ancient texts. Generally, however, such specific questions are difficult to answer from the textual evidence and are thus rarely addressed.

In reconsidering the development of archaeological science over the last decades, it becomes clear that the only way to avoid biased conclusions through deterministic arguments is through multi- and interdisciplinarity. Fortunately, interdisciplinary research in archaeology is well developed and we have the necessary tools in hand already. The main challenge is to integrate information of different methodological origin without letting our preconceptions of an all-explaining reason drown the multi-causal nature of the evidence.

The acceptance of systemic variability

When accepting the multi-causal nature of deterministic causation, as well as differences in human perception leading to different actions, we are automatically confronted with variable effects. But these are not the only aspects of variability, however. The landscape and environmental parameters of the Near East are extremely diverse, creating a high degree of regional differentiation and which probably also determined living conditions in the past.

The eastern Mediterranean coast in the west, the Anatolian highlands in the north, and the desert plains in the southeast characterise the diverse climate of Syria and the eastern part of southern Turkey. In summer, the subtropical trade wind zone causes constant weather conditions, while in the winter, cyclonic activity is responsible for strong regional differences between the coastal area and central Syria, resulting in distinctive climate regions. The Levant is characterised by a Mediterranean climate, although to the south mean annual precipitation drastically decreases resulting in geographic climate classifications of Mediterranean humidity in the Bay of Iskenderun, Mediterranean semi-humid at the Syrian coast, Mediterranean semi-arid in coastal Lebanon and Israel north of Tel Aviv, and arid in the area south of Tel Aviv (Wagner 2001). When focusing on global climate events during the Holocene, regional differences in climate development can be found comparing the isotopic records from Lake Van and Soreq Cave after 4200 BP (Wick et al. 2003; Litt et al. 2009; Bar-Matthews et al. 1997). Staubwasser and Weiss (2006) also see developmental differences in comparing the aforementioned records in the northwest with results from Oman (Fleitmann et al. 2003), where signals for the 4200 BP event could not be detected. Another frequently discussed geographically differing climate development is visible in the results from Iran (Stevens et al. 2001, 2006).

Precipitation is the most important climate factor, because of its regional and seasonal diversity and inter-annual fluctuations. The mean actual precipitation in a series of drought years may be 100 to 200 mm below the long-term mean annual precipitation (i.e., the 250 mm isohyet in such drought years is equivalent to the 400 mm isohyet of the long-term mean), causing a continuous sequence of crop failures in areas with long-term mean annual precipitation below 400 mm (Wirth 1971). The occurrence of droughts for the duration of three years or longer appears in a multi-year to decadal cyclicity (Touchan et al. 1999, 2003), exposing settlements in regions below the critical 400 mm isohyet to permanent risk of crop failure. Most of the Early and Middle Bronze Age sites in Syria lie in an area of 200 to 400 mm of modern mean annual precipitation, i.e. in an area that today is very sensitive to any decrease in precipitation (fig. 2).

Investigations of archaeological charcoal in the area (van Zeist and Bakker-Heeres 1985; Engel 1993; Willcox 1999, 2002; Deckers and Riehl 2004; Pessin 2004) demonstrate that the shrub land, grassland steppe and desert dominating the Near Eastern area today is a relatively young development. This is confirmed by information on the distribution of dense woodland in ancient western Syria (Wirth 1971) and by the few palynological works available (Bottema and van Zeist 1980; Rossignol-Strick 1999). This data also correlates with other palaeoclimatological results (Bar-Matthews et al. 1997, 1998; Goodfriend 1999; Fontugne et al. 1999; Roberts et al. 2001; Hazan et al. 2005; Pustovoytov et al. 2007).

Considering Holocene environmental development, interpretation of palaeoclimate data becomes more complex after 5000 BC, because of an increasing human impact on the environment, a factor that, at least as far as the vegetation cover is concerned, is difficult to distinguish from climatic increase in aridity and which may in fact be interrelated (Roberts et al. 2004, Roberts et al. 2011).

Furthermore, cultural and environmental influences cannot be considered separately. Agricultural sustainability depends on natural decadal to centennial climatic variability, as well as on annual to decadal social, economic and technological changes that impact the land (cf. Hole 2007), thus introducing another considerable factor of variation in the interrelationship of the environment and human society.

Diamond (2005) lists four categories of factors that contribute to failures of group decision-making and which may have led to the collapse of ancient civilisations:

(1) The failure to anticipate a problem before it arrived. This may be because of the lack of prior experience with similar problems, because of a similar experience which was not adequately transmitted as it occurred too long ago, or because of false analogy.
(2) The failure to perceive. This may happen when the problem is invisible in its initial stage (e.g. incipient salinisation) or imperceptible through a gradual shift of "normalcy" toward worse conditions, such as long-term climate trends that are concealed within noisy fluctuations ("creeping normalcy").
(3) Rational bad behaviour arising from clashes of interest between people. This can include inconsiderate behaviour and the self-enrichment of the elite or leaders of societies, sometimes in relation with resource bottlenecks.
(4) Irrational behaviour by people acting on disastrous values, often found in extremist religion, which detain a community from acting on perceived environmental problems.

This shows how environmental, socio-political and cultural problems are interwoven and illuminates the complex, multi-faceted character of human decision-making.

To summarise, the presetting for reconstructions of past Near Eastern environmental development are determined by at least four variable parameters. These are:

- Geomorphological diversity of coastal and inland areas; and related to this an east-west gradient of climate conditions, a north-south gradient of precipitation, and differing river systems.
- Inter-annual fluctuation of precipitation.
- Regionally variable human impacts generally difficult to distinguish from climate impacts in pollen diagrams.
- Variable human decision-making due to differing socio-cultural background and political goals.

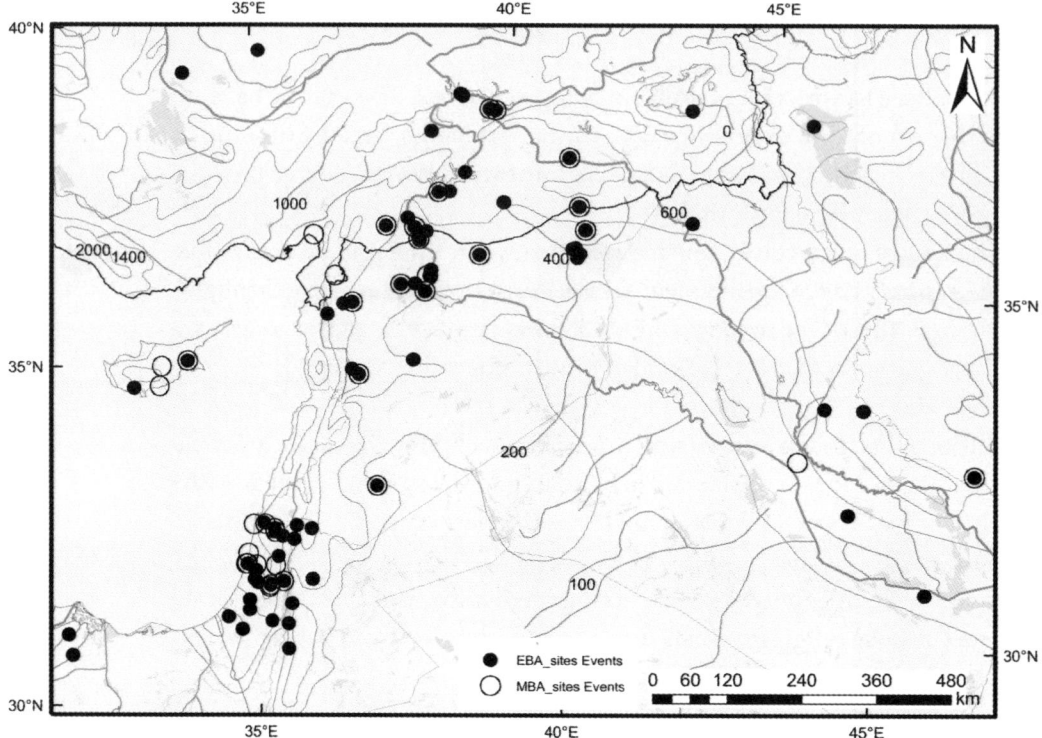

FIGURE 2. *Area under consideration with Early and Middle Bronze Age sites in their relation to modern isohyets.*

Finally, chronological imprecision, which occurs in archaeological sites as well as in the palaeoclimatic archive, complicates the comparison of data from different geographical areas.

Some of the above aspects have been discussed within an archaeobotanical study on the variability in human adaptation to changing environmental conditions in Upper Mesopotamia during the Early and the Middle Bronze Age (Riehl and Bryson 2007). This is a case study focused mainly on the two different river systems of the Khabur and Euphrates and on the area of the north-south gradient of precipitation of roughly 100 km, with the archaeological sites of Tell Mozan (200–250 mm of winter precipitation; Riehl 2010), Tell Brak, Tell Kerma, Tell Atij (100–150 mm of winter pre-

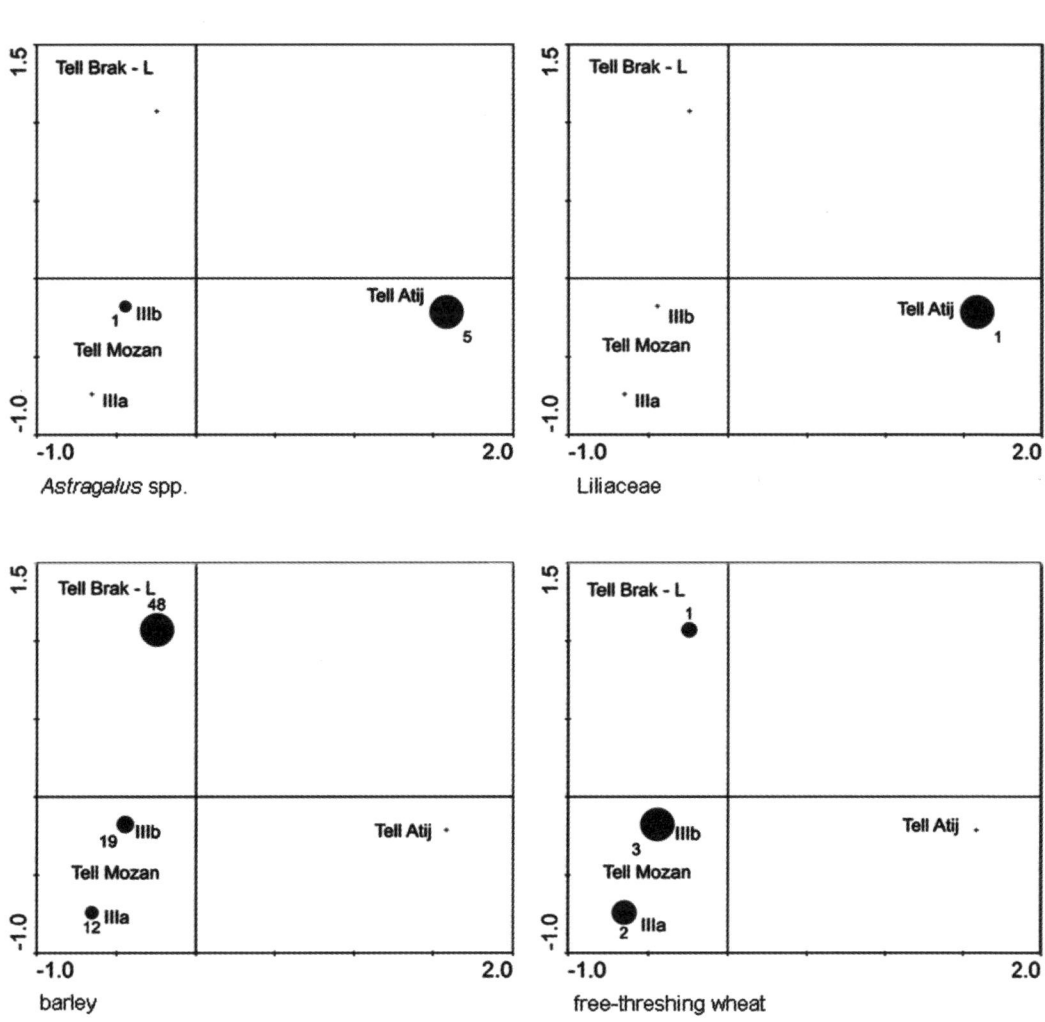

FIGURE 3. *Attribute plots of some selected taxa to the correspondence analysis of the Early Bronze Age (c. 2600–2300 BC) assemblages of Tell Mozan, Tell Brak and Tell Atij.*

cipitation) and Tell al-Raqa'i. The study showed that the archaeobotanical assemblages cluster according to the different river systems and to the precipitation ranges they are situated in. The latter aspect is particularly well visible in the attribute plots of taxa extracted from canonical correspondence analysis, with high proportions in wild taxa of degraded vegetation units at southern Tell Atij, and comparatively high proportions of more demanding, free-threshing wheat at northern Tell Mozan (fig. 3).

Although we do not know, as yet, the ancient precipitation patterns and whether or not the basic outline of the isohyets changed, we can assume a similar gradient of precipitation in the past as today, at least for some areas such as the Khabur. The abovementioned study also showed that the different river systems of the Euphrates and the Khabur were related to differences in economic strategies, with broader crop assemblages in the Khabur sites and an extreme focus on barley at the Euphrates sites. This opens up a broad range of possible human actions that create change.

Conclusions

Even slight changes in environmental conditions may result in multiple transformations in economic organisation, and the interrelatedness of these is difficult, if not impossible, to disentangle without an integrative and interdisciplinary methodology. The examples outlined above also make clear that we need to cease searching for single causes in the development of ancient societies, since linear cause-and-effect explanations are artificial philosophical constructs of single disciplines and are insufficient for detecting interactions between diverse human societies and a variable environment. Increasing the number of local studies and the acceptance of diversity even within small boundary units may help to overcome these problems in the future.

Acknowledgements

The German Research Council (DFG) enabled this methodological consideration by funding numerous archaeobotanical projects in the Near East (RI 1193/6).

Bibliography

Akkermans, P. M. M. G., and G. M. Schwartz. 2003. *The archaeology of Syria: From complex hunter-gatherers to early urban societies (ca. 16000–300 BC).* Cambridge University Press. Cambridge.

An, C.-B., L. Tang, L. Barton and F.-H. Chen. 2005. "Climate change and cultural response around 4000 cal yr BP in the western part of the Chinese Loess Plateau." *Quaternary Research* 63: 347–352.

Araus, J. L., A. Febrero, R. Buxo, M. D. Camalich, D. Martin, F. Molina, M. O. Rodriguez-Ariza and I. Romagosa. 1997. "Changes in carbon isotope discrimination in grain cereals from different regions of the western Mediterranean basin during the past seven millennia: Palaeoenvironmental evidence of a differential change in aridity during the late Holocene." *Global Change Biology* 3: 107–118.

Bar-Matthews, M., A. Ayalon and A. Kaufman. 1997. "Late Quaternary paleoclimate in the Eastern Mediterranean region from stable isotope analysis of speleothems at Soreq Cave, Israel." *Quaternary Research* 47: 155–168.

Bar-Matthews, M., A. Ayalon and A. Kaufman. 1998. "Middle to late Holocene (6500 yr. period) paleoclimate in the Eastern Mediterranean region from stable isotopic composition of speleothems from Soreq Cave, Israel." In A. S. Issar and N. Brown (eds.), *Water, environment and society in times of climatic change*, Kluwer Academic Press, 203–214. Dordrecht, Boston and London.

Booth, R. K., S. T. Jackson, S. L. Forman, J. E. Kutzbach, E. A. I. Bettis, J. Kreig and D. K. Wright. 2005. "A severe centennial-scale drought in mid-continental North America 4200 years ago and apparent global linkages." *The Holocene* 15: 321–328.

Bottema, S., and W. van Zeist. 1980. "Palynological evidence for the climatic history of the Near East 50000–6000 BP." *Colloques Internationaux du C.N.R.S.* 598: 111–132.

Butzer, K. W. 1997. "Sociopolitical discontinuity in the Near East c. 2200 BC: Scenarios from Palestine and Egypt." In H. N. Dalfes, G. Kukla and H. Weiss (eds.), *Third Millennium BC Climate Change and Old World Collapse*, Springer Verlag, 245–296. Berlin and Heidelberg.

Coombes, P., and K. Barber. 2005. "Environmental determinism in Holocene research: Causality or coincidence?" *Area* 37: 303–311.

Dalfes, H. N., G. Kukla and H. Weiss. 1997. *Third Millennium BC Climate Change and Old World Collapse*. NATO ASI Series. Springer Verlag. Berlin and Heidelberg.

Deckers, K., and S. Riehl. 2004. "The development of economy and environment from Bronze Age to the Early Iron Age in Northern Syria and the Levant: A case-study from the Upper Khabur region." *Antiquity* 78 (302) Project Gallery. www.antiquity.ac.uk/projgall/deckers302/

Deckers, K., and S. Riehl. 2007. "Fluvial environmental contexts for archaeological sites in the Upper Khabur basin (northeastern Syria)." *Quaternary Research* 67: 337–348.

Deckers, K., and S. Riehl. 2009. "Tells in the Upper Khabur Basin of northeastern Syria: Their hydrological and agrarian contexts." *Paléorient* 34: 173–189.

deMenocal, P. B. 2001. "Cultural responses to climate change during the late Holocene." *Science* 292: 667–672.

Diamond, J. 2005. *Collapse: How Societies Choose to Fail or Succeed*. Penguin. New York.

Ehleringer, J. R., C. B. Field, Z.-f. Lin and C.-y. Kuo. 1986. "Leaf carbon isotope and mineral composition in subtropical plants along an irridiance cline." *Oecologia* 70: 520–526.

El-Nakhlawy, F. S., and M. A. El-Fawal. 1989. "Tolerance of five oil crops to salinity and temperature stresses during germination." *Acta Agronomica Hungarica* 38: 59–65.

Engel, T. 1993. "Archaeobotanical analysis of timber and firewood used in third millennium houses at Tall Bderi / Northeast-Syria." In K. R. Veenhof (ed.), *Houses and Households in Ancient Mesopotamia*, Netherlands Institute of the Near East, 105–115. Leiden.

Farquhar, G. D., M. H. O'Leary and J. A. Berry. 1982. "On the relationship between carbon isotope discrimination and the intercellular carbon dioxide concentration in leaves." *Australian Journal of Plant Physiology* 9: 121–137.

Ferrio, J. P., J. L. Araus, R. Buxó, J. Voltas and J. Bort. 2005. "Water management practices and climate in ancient agriculture: Inferences from the stable isotope composition of archaeobotanical remains." *Vegetation History and Archaeobotany* 14: 510–517.

Fleitmann, D., S. J. Burns, M. Mudelsee, U. Neff, J. Kramers, A. Mangini and

A. Matter. 2003. "Holocene forcing of the Indian monsoon recorded in a stalagmite from southern Oman." *Science* 300: 1737–1739.

Fontugne, M., C. Kuzucuoglu, M. Karabiyikoglu, C. Hatte and J.-F. Pastre. 1999. "From Pleniglacial to Holocene: A ^{14}C chronostratigraphy of environmental changes in the Konya Plain, Turkey." *Quaternary Science Reviews* 18: 573–591.

Gasse, F. 2000. "Hydrological changes in the African tropics since the last glacial maximum." *Quaternary Science Reviews* 19: 189–211.

Giosan, L., D. Fuller, K. Nicoll, R.K. Flad, P. Clift. 2012. *Climates, Landscapes, and Civilizations.* Vol. 198. American Geophysical Union monograph series, Washington DC.

Goldhausen, M., and A. Ricci. 2005. "Political centralisation in the Syrian Jezira during the 3rd millennium: A case study in settlement hierarchy." *Altorientalische Forschungen* 32: 132–157.

Goodfriend, G. A. 1999. "Terrestrial stable isotope records of Late Quaternary paleoclimates in the eastern Mediterranean region." *Quaternary Science Reviews* 18: 501–513.

Hazan, N., M. Stein, A. Agnon, S. Marco, D. Nadel, J. F. W. Negendank, M. J. Schwab and D. Neev. 2005. "The late Quaternary limnological history of Lake Kinneret (Sea of Galilee), Israel." *Quaternary Research* 63: 60–77.

Hole, F. 2007. "Agricultural sustainability in the semi-arid Near East." *Climate of the Past* 3: 193–203.

Hubbard, R. N.L. B. 1980. "Development of agriculture in Europe and the Near East: Evidence from quantitative studies." *Economic Botany* 34: 51–67.

Lafont, B. 2000. "Irrigation agriculture in Mari." In R. M. Jas (ed.), *Rainfall and Agriculture in Northern Mesopotamia,* Nederlands Instituut voor het Nabije Oosten, 129–146. Leiden.

Litt, T., S. Krastel, M. Sturm, R. Kipfer, S. Örcen, G. Heumann, S. O. Franz, U. B. Ülgen and F. Niessen. 2009. "PALEOVAN, International Continental Scientific Drilling Program (ICDP): Site survey results and perspectives." *Quaternary Science Reviews* 28: 1555–1567.

Marro, C., and C. Kuzucuoglu. 2007. *Upper-Mesopotamia: Did the 2100 BC crisis take place?* Varia Anatolica and de Boccard. Istanbul and Paris.

Meijer, D. 2007. "Crisis = collapse? Collapse of what?" In C. Kuzucuoglu and C. Marro (eds.), *Sociétés humaines et changement climatique à la fin du troisième millénaire: une crise a-t-elle eu lieu en Haute-Mésopotamie?*, Varia Anatolica and de Boccard, 40–43. Istanbul and Paris.

Pessin, H. 2004. "Stratégies d'approvisionnement et utilisation du bois dans le Moyen Euphrate et la Damascène. Approche anthracologique comparative de sites historiques et préhistoriques." Unpublished PhD thesis. University of Paris I.

Pustovoytov, K., K. Schmidt and H. Taubald. 2007. "Evidence for Holocene environmental changes in the northern Fertile Crescent provided by pedogenic carbonate coatings." *Quaternary Research* 67: 315–327.

Riehl, S. 2009a. "Archaeobotanical evidence for the interrelationship of agricultural decision-making and climate change in the ancient Near East." *Quaternary International* 197: 93–114.

Riehl, S. 2009b. "A cross-disciplinary investigation of cause-and-effect for the dependence of agro-production on climate change in the ancient Near East." In R. de Beauclair, S. Münzel and H. Napierala (eds.), *Knochen pflastern ihren Weg. Festschrift for Hans-Peter and Margarete Uerpmann*, Verlag Marie Leidorf, 217–226. Rahden, Westphalia.

Riehl, S. 2010. "Plant production in a changing environment: The archaeobotanical remains from Tell Mozan." In K. Deckers, M. Doll, P. Pfälzner and S. Riehl (eds.), *Ausgrabungen 1998–2001 in der Zentralen Oberstadt von Tall Mozan / Urkeš: The Development of the Environment, Subsistence and Settlement of the City of Urkeš and its Region*, Harrassowitz, 173–320. Wiesbaden.

Riehl, S., and R. A. Bryson. 2007. "Variability in human adaptation to changing environmental conditions in Upper Mesopotamia during the Early to Middle Bronze Age transition." In C. Marro and C. Kuzucuoglu (eds.), *Sociétés humaines et changement climatique à la fin du troisième millénaire: une crise a-t-elle eu lieu en Haute-Mésopotamie?*, Varia Anatolica and de Boccard, 523–548. Istanbul and Paris.

Riehl, S., R. A. Bryson and K. Pustovoytov. 2008. "Changing growing conditions for crops during the Near Eastern Bronze Age (3000–1200 BC): The stable carbon isotope evidence." *Journal of Archaeological Science* 35: 1011–1022.

Riehl, S., K. Pustovoytov, A. Dornauer and W. Sallaberger. 2012. "Mid-late Holocene agricultural system transformations in the northern Fertile Crescent: A review of the archaeobotanical, geoarchaeological, and philological evidence." In L. Giosan, D. Fuller, K. Nicoll, R. K. Flad and P. Clift (eds.), *Climates, Landscapes, and Civilizations*, vol. 198, 115–136. Washington DC: American Geophysical Union monograph series.

Ristvet, L., and H. Weiss. 2005. "The Habur Region in the late third and early second millennium BC." In W. Orthmann (ed.), *The History and Archaeology of Syria*. Saarbrücken Verlag. Saarbrücken

Roberts, N., J. Reed, M. J. Leng, C. Kuzucuoglu, M. Fontugne, J. Bertaux, H. Woldring, S. Bottema, S. Black and E. Hunt. 2001. "The tempo of Holocene climatic change in the Eastern Mediterranean region: New high-resolution crater-lake sediment data from central Turkey." *The Holocene* 11: 721–736.

Roberts, N., T. Stevenson, B. Davis, R. Cheddadi, S. Brewster and A. Rosen. 2004. "Holocene climate, environment and cultural change in the Circum-Mediterranean region." In R. W. Battarbee (ed.), *Past Climate Variability Through Europe and Africa*. Springer, 343–362. Dordrecht.

Roberts, N., D. Brayshaw, C. Kuzucuoglu, R. Perez, L. Sadori. 2011. "The mid-Holocene climatic transition in the Mediterranean: Causes and consequences." *The Holocene* 21: 3–13.

Röllig, W. 2008. *Land- und Viehwirtschaft am Unteren Kābūr in Mittelassyrischer Zeit, Berichte der Ausgrabung Tall Šēikh Hamad / Dūr-Katlimmu* 9. Harassowitz. Wiesbaden.

Rossignol-Strick, M. 1999. "The Holocene climatic optimum and pollen records of sapropel 1 in the eastern Mediterranean, 9000–6000 BP." *Quaternary Science Reviews* 18: 515–530.

Roth, L., K. Kormann and H. Schweppe. 1992. *Färbepflanzen, Pflanzenfarben: Färbemethoden, Analytik, türkische Teppiche und ihre Motive*. Ecomed Fachverlag. Landsberg am Lech.

Smith, A., and N. F. Miller. 2009. "Integrating plant and animal data: Delving deeper into subsistence: Introduction to the special section." *Current Anthropology* 50: 883–884.

Staubwasser, M., and H. Weiss. 2006. "Holocene climate and cultural evolution in late prehistoric-early historic West Asia." *Quaternary Research* 66: 372–387.

Stevens, L. R., E. Ito, A. Schwalb and H. E. Wright. 2006. "Timing of atmospheric precipitation in the Zagros mountains inferred from a multi-proxy record from Lake Mirabad, Iran." *Quaternary Research* 66: 494–500.

Stevens, L. R., Wright, H. E. and A. Ito. 2001. "Changes in seasonality of climate during the Lateglacial and Holocene at Lake Zeribar, Iran." *The Holocene* 11: 747–756.

Tainter, J. A. 1995. "Sustainability of complex societies." *Futures* 27: 397–407.

Touchan, R., G. M. Garfin, D. Meko, G. Funkhauser, N. Erkan, M. K. Hughes and B. S. Wallin. 2003. "Preliminary reconstructions of spring precipitation in southern Turkey from tree-ring width." *International Journal of Climatology* 23: 157–171.

Touchan, R., D. Meko and M. K. Hughes. 1999. "A 396-year reconstruction of precipitation in southern Jordan." *Journal of the American Water Resources Association* 35: 49–59.

van der Leeuw, S. 1998. "The Archaeomedes Project: Understanding the natural and anthropogenic causes of land degradation and desertification in the Mediterranean." Office for Official Publications of the European Union. Luxembourg.

van Koppen, F. 2001. "The organisation of institutional agriculture in Mari." *Journal of the Economic and Social History of the Orient* 44: 451–504.

van Zeist, W., and J. A. H. Bakker-Heeres. 1985. "Archaeobotanical studies in the Levant 4. Bronze Age sites on the north Syrian Euphrates." *Palaeohistoria* 27: 247–316.

VanDerwarker, A. M., and T. M. Peres. 2010. *Integrating Zooarchaeology and Paleoethnobotany: A Consideration of Issues, Methods, and Cases.* Springer Verlag. New York.

Waetzoldt, H. 1983. "Leinen." In M. P. Streck (ed.), *Reallexikon der Assyriologie*, De Gruyter, 583–594. Berlin.

Wagner, H.-G. 2001. *Mittelmeerraum.* Wissenschaftliche Buchgesellschaft. Darmstadt.

Wick, L., G. Lemcke and M. Sturm. 2003. "Evidence of Lateglacial and Holocene climatic change and human impact in eastern Anatolia: High-resolution pollen, charcoal, isotopic and geochemical records from the laminated sediments of Lake Van, Turkey." *The Holocene* 13: 665–675.

Wiggermann, F. A. M. 2000. "Agriculture in the northern Balikh valley: The case of Middle Assyrian Tell Sabi Abyad." In R. M. Jas (ed.), *Rainfall and Agriculture in Northern Mesopotamia.* Nederlands Instituut voor het Nabije Oosten. 171–232. Leiden.

Wilkinson, T. J. 1997. "Environmental fluctuations, agricultural production and collapse: A view from Bronze Age upper Mesopotamia." In H.

N. Dalfes, G. Kukla and H. Weiss (eds.), *Third Millennium BC Climate Change and Old World Collapse*, Springer Verlag, 67–106. Berlin and Heidelberg.

Wilkinson, T. J. 2003. *Archaeological Landscapes of the Near East*. University of Arizona Press. Tucson.

Wilkinson, T. J., H. J. Christiansen, J. Ur, M. Widell and M. Altaweel. 2007. "Urbanization in a dynamic environment: Modelling Bronze Age communities in Upper Mesopotamia." *American Anthropologist* 109: 52–68.

Willcox, G. 1999. "Charcoal analysis and Holocene vegetation history in southern Syria." *Quaternary Science Reviews* 18: 711–716.

Willcox, G. 2002. "Evidence for ancient forest cover and deforestation from charcoal analysis of ten archaeological sites on the Euphrates." In S. Thiébault (ed.), *Charcoal Analysis. Methodological Approaches, Palaeoecological Results and Wood Uses*, British Archaeological Reports International Series 1063, 141–145. Oxford.

Wirth, E. 1971. *Syrien: Eine geographische Landeskunde*. Wissenschaftliche Buchgesellschaft. Darmstadt.

AMS ¹⁴C-dated Plants as a Tool for Investigating Palaeoclimate: New Data for Analysing Social Complexity in Ebla and Qatna (Northwestern Syria) in the Light of 3rd Millennium BC Climate Change

Girolamo Fiorentino and Valentina Caracuta

Abstract

The societal "collapse" of the urban-based system, which is part of the rise-flourish-and-fall cycle of the protohistoric societies of the Near East, is investigated here via an integrated approach based on the identification of short-term climate changes and variations in settlement pattern.

Thirty-eight plant samples collected from the Ebla and Qatna archaeological sites were analysed by AMS to simultaneously determine the amount of water received by the plants (as expressed by their $\delta^{13}C$ values) and their radiocarbon date.

Assuming that the plants received no additional water input from irrigation, the variation in $\delta^{13}C$ values provides information on rainfall.

By determining rainfall fluctuations over a period of about 1500 years (4500–3100 BP/ 3200–1400 BC) we were able to establish the palaeoclimatic trends.

The fine resolution of the ¹⁴C analysis showed a pattern of short-term climate change which, when compared with the historical record and our knowledge of socio-economic aspects in the region, raises questions about the links between environmental variations affecting food production and the resilience of complex societies.

Introduction

It is now widely recognised that ancient climate fluctuations are strongly associated with changes in the relationship between humans and their environment, making the integration of human history and climate variability an important task in archaeological research. This assumption is especially valid for societies situated in fragile ecosystems like those in the sub-arid regions of the Near East, where the ancient settlements' capacity for survival was dependent on climate factors such as water availability, the primary resource for food production (Nieuwenhuis et al. 2006).

However, the assumption that environmental stress was the main cause of ancient administrative collapse is far too simplistic. Climate change might only slow down or intensify the effects of human pressure on the landscape. In considering human responses to environmental shifts, historical and political ecologists such as Crumley (1994), Greenberg and Park (1994) and Peet and Watts (1994) have stressed the role played by conflicts between the ruling classes and other segments of society, conflicts about economic strategies and political choices.

The protohistoric societies of the Near East were hierarchically based, and in order to convince primary producers to generate a surplus that could be used to maintain non-primary producers (Redman et al. 2004: 159), they depended on the widespread acceptance of an ideology. Inappropriate strategic choices made by the ruling classes regarding production could lead to social and economic dysfunction and ultimately to crisis (Dincauze 2006: 77). A crisis might then precipitate sudden and massive change, with rapid transformation of both society and culture (Flannery 1972; Renfrew 1979). Traces of such changes are visible in the archaeological record, but the causes and processes must be inferred and interpreted.

Our aim in this study, therefore, is to examine the interactive relationship between ancient Near Eastern societies and short-term changes in their environments, using a new paleoclimate proxy. Social organisation is also considered, together with technology and political and economic factors as inferred from archaeological evidence and epigraphic sources, as well as climate fluctuations and the way these were perceived by human beings.

The study context

The Near East region has a highly diversified environmental pattern in which various climatic regimes (mountain, maritime, steppe, desert) exist very close to each other. The Syrian territories reflect the division proposed for the Near East region, with the coastal Mediterranean belt thinning away eastwards into semi-arid and arid steppes (Pabot 1957; Zohary 1973). The two sites under consideration, Tell Mardikh-Ebla in the Idleb district and Tell Mishrifeh-Qatna in the Hama region, lie at the boundary between the Mediterranean coastal belt and the steppe region, on the eastern side of the Anti-Lebanon mountains that prevent humidity from the west from moving eastwards (fig. 1).

Situated in a sub-arid region, but bordering a more humid and fertile ecosystem on which they depended, the two sites were very sensitive to short-term climatic variations. A few years of drought would cause the steppes to extend westwards, reducing the amount of arable land and severely affecting productivity (Fiorentino and Caracuta 2007; Fiorentino et al. 2008).

Climate fluctuation: Timescale and human response

In environmental studies, the concept of scale plays a fundamental role in the appreciation of the spatial extent, chronological range and magnitude of environmental shifts. Butzer (1982: 24) attempts to classify environmental change as follows: i) short term, which involves variability over years or decades, including seasonal fluctuations in the availability of water and animal biomass; ii) medium term, which involves fundamental changes in hydrology, productivity and all categories of biomass over the course of centuries; and iii) long term, which entails biome shifts occurring over the course of several centuries but persisting for millennia.

In studies of Holocene climate and societies, the concern is primarily with medium- and short-term variability, since it is changes on this scale that directly influenced Bronze Age societies (Rosen 2007: 5). Several years of drought were likely to be more stressful for ancient communities, where the average generation is about thirty years, than modern societies, where the average generation is about sixty to seventy years (Hassan 1978).

The human response to perceived changes is not different from that of other animals: all seek minor adjustments in behaviour or surroundings

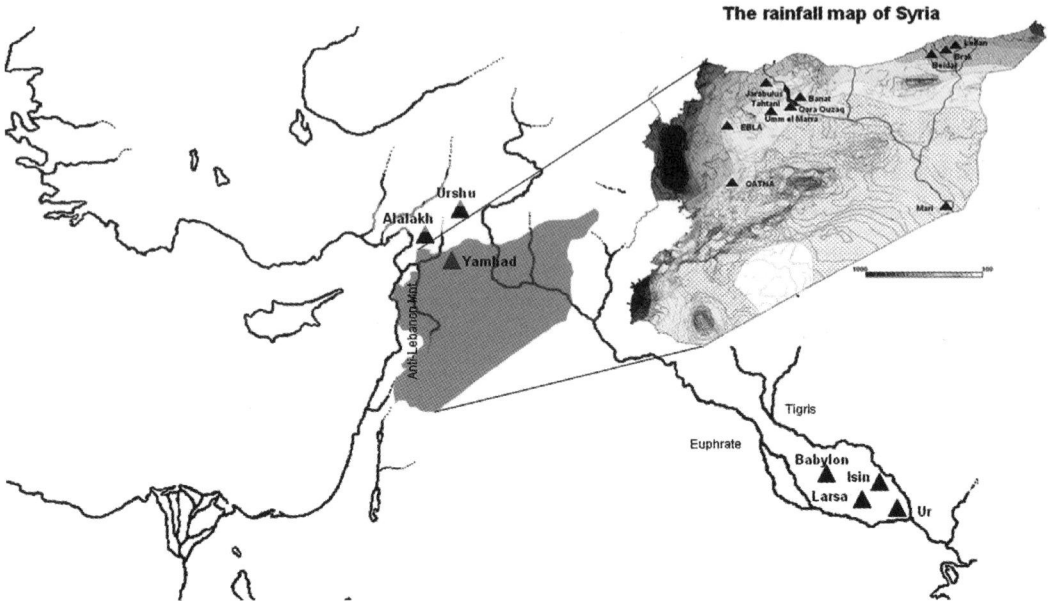

FIGURE 1. *Map of Syria with the location of the sites under study.*

that will restore the relative comfort of a former state. What is unique to humans is the range of strategic choices available to them (Dincauze 2006: 75). The strategies chosen vary according to community resilience, which is the capacity to adapt qualitatively as well as quantitatively to profound, unexpected and continuous perturbations (McAdams 1978).

Human responses also vary according to the nature of the perceived change (good or bad), the amplitude of the change, the flexibility of established lifestyles, population structure and density, the amount of diversity in the environment and the technology available to the group. Decisions made by pre-literate groups are typically conservative and usually the decision-maker first draws on a store of knowledge that is largely derived from personal or communal experience. This compendium of shared knowledge has been termed the social memory of a society and can be crucial in determining the resilience of that society to external pressure (McIntosh et al. 2000: 24).

Unlike human beings, the response of vegetal cover to climate change involves a substantial time lag, usually corresponding to long-term environmental change. Nevertheless, there are some important exceptions concerning the response of vegetation at ecotones, the boundaries between ecosystems (Gosz 1992; Holland et al. 1991). Especially in semi-arid ecotones, which are considered the most sensitive according to the Intergovernmental Panel on Climate Change (1996), vegetation cover shifts extensively and rapidly as a result of mortality in response to a severe drought (Allen and Breshears 1998). Human societies in sub-arid ecotones, where natural resources vary rapidly over time, might thus be expected to exhibit a marked ability to manage risk.

Short-term climate changes and variation in the socio-economic system

In examining the relationship between environmental change and human behaviour from a climatic perspective, it should be kept in mind that ancient societies cannot be regarded as monolithic entities that are impacted uniformly by intense climatic events. On the contrary, communities are segmented in innumerable ways and consist of subgroups and individual players whose goals and motivations may differ, thus leading to differing responses to environmental factors (Rosen 2007: 4). Short-term changes that require an immediate response stress individuals or particular social subgroups by creating shortages or inadequacies of resources or leadership, or by causing temperature or moisture to exceed their comfort or safety range.

Under favourable conditions, some strategies – such as intensifying production from fields already under cultivation, bringing new fields into cultivation, centralising control of food production and increasing imports of goods from satellite settlements – may lead to increased overall food production and a surplus that could be used to support non-food-producing urban dwellers (Redman 2004: 159). Nevertheless, if these strategies are carried too far, or if they are applied for too long, they may seriously degrade the environment and reduce the sustainability threshold of the system, especially during periods of drought.

Concerning the Near East, climate fluctuations from the 4th to the 2nd millennia BC appear to be correlated with episodes of political disintegration, especially those associated with the ends of two major archaeological periods, the Early Bronze Age (EBA) and the Middle Bronze Age (MBA) (Kuzucuoğlu et al. 2007). Signs of "crisis" in near eastern "urban-based" institutions were: i) partial or complete abandonment of settlements, along with the loss or depletion of their centralising functions; ii) breakdown of central authority, as seen in the disappearance of administrative apparatus such as palaces, symbols and other elements of state power; iii) contraction of regional economic systems, attested by a decline in evidence of a tributary economy, indicating the predominance of self-sufficient productive systems; and iv) the eclipse of the ruling elite's ideology (Schwartz 2007: 47; 1993).

Based on this assumption, we looked for extensively studied proto-historical sites which developed in sub-arid ecotones over a time period with recognisable climate fluctuations. Having identified two such sites, we tested the reliability of a new palaeoclimate tool to analyse human responses to abrupt environmental shifts.

Ancient climate fluctuations and human response: Choosing the right tool to identify them

Since there is no instrumental record available for ancient history, scholars have turned to other kinds of evidence. Archaeological data, epigraphic documents and, above all, natural proxies have all been used as paleoclimate tools. A wide range of natural phenomena are climate-dependent and often become sealed in stratified deposits which thus contain built-in proxy measures of past climate (Bradley 2003). Several natural proxies can be used, but none of them are linked to archaeological sites and they do not provide direct information on human response to environmental stress. A possible exception is fossil pollen grains, which provide information on vegetation cover over time when these are well preserved in peat bogs, buried soils and lake sediments. In arid lands, such as the Syrian territories, pollen analysis reveals both environmental shifts and human activities (Valsecchi 2007; Yasuda et al. 2000; Horowitz 1992; Nikleswki and van Zeist 1970; Kaniewski et al. 2010). However, the time resolution of the environmental changes preserved in the pollen record is fairly low, as the response of vegetation cover to climate change generally involves a substantial time lag. Furthermore, despite the fact that vegetation cover responds more rapidly in ecotones, pollen analysis cannot record short-term drought events which affect human activities but do not entail changes in vegetal cover on a regional scale.

The question thus arises of how to investigate climate fluctuations which affect human subsistence but do not entail significant changes in biota. Previous studies indicate that archaeological plant remains, which record chemical-physical environmental properties in the form of stable carbon isotope ratios ($\delta^{13}C$), are highly suitable for this research since they are natural products that have been transformed by human activity (Vernet et al. 1996; Ferrio et al. 2003, 2006, 2007; Riehl et al. 2007; Fiorentino and Caracuta 2007; Fiorentino et al. 2008; Voltas et al. 2008). Since in arid ecosystems these ratios depend on water received (Ehleringer 1989; Ferrio et al. 2005), plant isotope signatures are directly related to environmental conditions. Thus, plant carbon isotope ratios provide information on climate changes, even when these were not severe enough to determine a shift in vegetal cover.

The $\delta^{13}C$ of AMS ^{14}C-dated archaeological plant remains

To determine the $\delta^{13}C$ of archaeological plant remains, we used Accelerator Mass Spectrometry (AMS) instead of the more common and less costly IRMS, because it is the only technique which can simultaneously measure radiocarbon age (Fiorentino et al. 2008, 2009). The average accuracy of AMS $\delta^{13}C$ measurements was ± 0.5‰, only slightly lower than that of IRMS, which is around ± 0.3‰.

Thirty-eight archaeological samples from Ebla and Qatna, covering a time period of fifteen hundred years from the middle of the 4th millennium to the end of the 3rd millennium BC, were analysed (for further details on the carbon isotopes methodology and its reliability as a palaeoclimate tool, see Fiorentino et al. 2008; Roberts et al. 2011). Correspondences were found between the climatic variations revealed by AMS and the general trend highlighted by other palaeoclimate records available for this region (for further detail, see Cremaschi 2007a; Riehl et al. 2007; Wilkinson 2004; Issar 2003; Cullen et al. 2000; deMenocal et al. 2000; Frumkin et al. 1999; Bar-Matthews et al. 1997; Courty 1994; Weiss et al. 1993; Horowitz 1992).

The $\delta^{13}C$ values resulting from the radiocarbon testing show a pattern of variation with three periods of reduced rainfall (4300–4200 BP / 3100–2850 BC; 3800–3700 BP / 2200–2050 and 3500–3300 BP / 1800–1650 BC) separated by two rainy periods (4100–3900 BP / 2550–2350 BC and 3300–3200

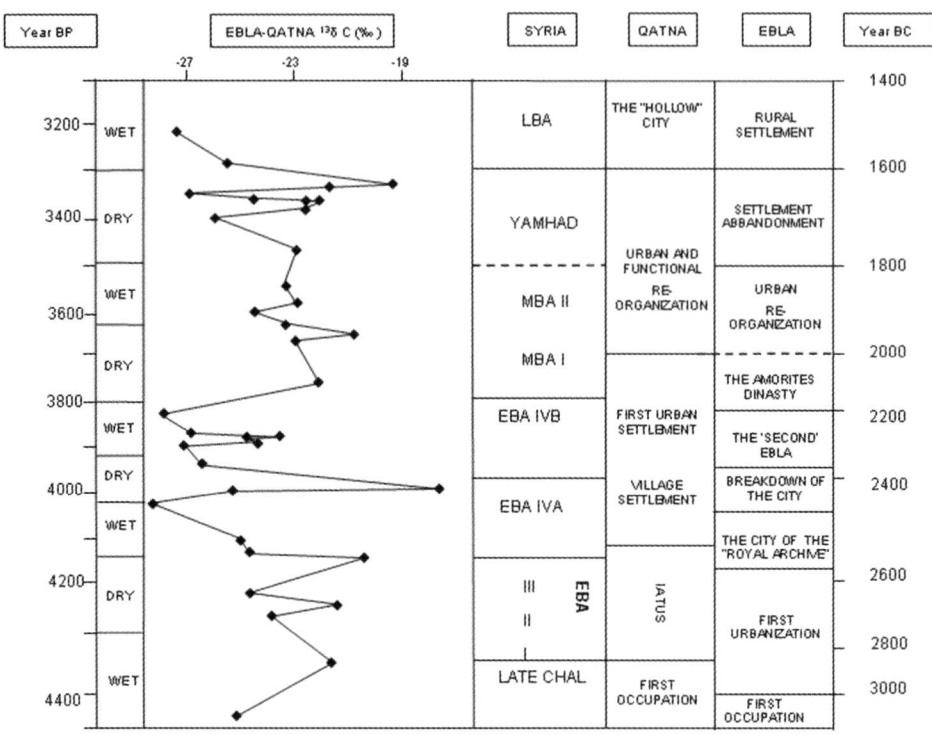

FIGURE 2. Comparison of palaeoclimate trends (as highlighted by carbon isotope analysis) with major changes occurring in Ebla and Qatna and the Near Eastern historical framework (redrawn after Morandi Bonacossi 2007; Matthiae 1995; Rouault and Masetti-Rouault 1993; Rouault and Wäfler 2000; Klengel 1992).

BP / 1600–1500 BC). In addition, an isolated drought event occurred at 4000 BP / 2400 BC, which corresponds with a rapid decrease of soil moisture recorded by the speleothem of Soreq Cave (Bar-Matthews et al. 1997). Between 3650 and 3550 BP (2050–1800 BC), there was a moderate "rainy period" which may be considered wet only when compared to the centuries that preceded and followed it (see fig. 2).

Considering that water availability rather than other factors is the key climatic determinant for plant life in semi-arid areas, and that the more we learn about it the greater its significance for human beings seems to be, we compared the climate pattern inferred from the $\delta^{13}C$ analysis with the major changes which are known to have occurred in Ebla and Qatna from the 4th to the 2nd millennia BC.

Ebla and Qatna: Urban dynamics and environmental changes

As far as Ebla and Qatna are concerned, the analysis of the archaeological record shows they were part of an urban-based system, since they were complex administrative centres which exercised control over a wide range of satellite sites (Morandi Bonacossi 2007; Matthiae 1995). The overall pattern that emerges from the archaeological evidence and epigraphic sources is complex, being geographically as well as socially differentiated. However, the intent was clearly to maximise the flow of food and other resources to the urbanised areas. Ebla and Qatna have been shown to be dependent upon outsiders for the supply of commodities not available locally, thus requiring sophisticated economic management. The Ebla archives reveal that most of the primary products brought to the palace derived from state-owned fields (Archi 1999; Milano 1987, 1990) or from taxation (Milano 1981). The state also controlled the trading system, which ensured inward flows of lapis lazuli and other raw materials that were not available locally (Pinnock 2004).

Reinforcing the impression of imposed centralisation and hierarchical authority is the remarkable quantity of information concerning the redistributive system, by means of which palace workers (i.e. soldiers, administrators, wool workers and ladies of court) received payment in the form of grain and/or flour (Milano 1990). When social and environmental conditions were favourable, the success of this system could be seen in massive production and distribution networks for specific commodities, population growth (especially urban) and the proliferation of monumental public buildings and bureaucratic procedures (McAdams 1978).

Climate is subject to fluctuations which can affect agricultural productivity, resulting in damage to small land owners and problems for the entire system. If we consider the history of Ebla and Qatna, the effects of the short-term environmental shifts inferred from the AMS $\delta^{13}C$ analysis are readable in changes in the socio-economic organisation and are even reflected in the layout of the cities.

The first important urban settlement in Ebla – known as the "Royal archive" city due to the discovery of a large amount of epigraphic texts in the Early Bronze Age (IVA) Palace G – became a regional empire (Matthiae 1995, 2006) during the period of relatively abundant rainfall (2550–2350 BC), as identified by the carbon isotopes "curve" of Ebla-Qatna (Roberts et al. 2011). The combined analysis of archaeological data and epigraphic sources reveals that during this phase (EBA IVA), Qatna and Ebla consolidated their political structure and extended their control on a regional scale (see fig. 3). The success of their gathering and redistribution systems was evident especially in Ebla, where valuable wood (fir and cedar) from the Anti-Lebanon Mountains was used for making luxury products despite the cost of transport (Fiorentino and Caracuta 2013).

The well-structured system in Ebla was thrown into crisis shortly after 2400 BC when the city of Mari destroyed it and burnt down the Royal Palace (Matthiae 1989). Why did Ebla, whose power extended over most of northwestern Syria and whose influence reached as far as Egypt, yield so easily to its enemies? Could the drought that took place in this period (2400 BC) have played a role in accelerating its defeat? Could the ruling class have lost the support of the masses at the crucial moment because it was unable to resolve the famine?

Answers to these questions may come from comparison with other moments in Ebla's history. After the destruction of Palace G and the disruption that followed, the archaeological evidence points to a new phase, the Early Bronze Age (EBA) IVB, the second Ebla period or "Ebla after the Fall", characterised by a certain degree of urban rebuilding. Evidence of new building activity is visible especially in the lower town where the "Archaic" Palace was erected. On a local scale, the EBA IVB saw the spread of settlements into areas which were more marginal in terms of agriculture, but which were more favourable to grazing (i.e. the steppe) (Mazzoni and Felli 2007). The Ebla-Qatna carbon isotope analysis reveals that this period, corresponding to increased rainfall, ended around 2200 BC when a long drought struck the whole of the Near East. In terms of political structure, the Syrian city states reacted in various ways. In the Jabbul plain, the city of Umm el-Marra was partially abandoned and its area of influence was significantly reduced, with a decline of 20 per cent in the area of sedentary occupation (Schwartz and Miller 2007). The same effect was seen in the Tishrin region, where sites such as Banat, Qara Quzaq and Jerablus Tahtani (Peltenburg 2007) were abandoned, as well as in the Khabur valley, where Brak and Leilan underwent a break in continuity (Wilkinson 2004;

Weiss 2000). The city state of Mari survived, however, showing no signs of disruption in terms of urban layout despite the shift in power from the Šakkanakku dynasty to the Amorites (Butterlin 2007). Ebla then experienced a moment of disruption around 2000 BC, as testified by the presence of burnt debris containing EBA IVB pottery in the core of one of the MBA urban ramparts.

On a regional scale, the period during which the 3rd Ur dynasty was overthrown and the Amorite dynasties (Isin, Larsa and Babylon I) rose to power was characterised by decreases and/or seasonal changes in rainfall (see fig. 2). The renaissance which followed was characterised by a change in the cultural pattern, which saw the Early Bronze Age models abandoned in favour of a new system known as the Middle Bronze Age. During these centuries Ebla was reorganised, with the creation of several palaces, temples and domestic quarters for administrative employees. At this time, the city also controlled a large portion of territory. Even in the absence of textual evidence that might shed light on the social structure, the archaeological records clearly point to a system in which the city drew inwards all the resources from the countryside. The magnificence of the bureaucratic machinery can be seen in the group of richly furnished tombs found under Palace FF, built for local noblemen and a princess (Matthiae 1995).

Shortly before 1800 BC the political situation in the region was still fragmented, with several cities struggling for power. The Fertile Crescent had become the theatre of action for the Babylonian king Hamurrabi I, who in the last years of his reign was able to extend his influence over the great kingdom of Yamhad/Aleppo which governed northwestern Syria. The isotopical palaeoclimate reconstruction points to fluctuations in rainfall against a general backdrop of drought. The most severe period of drought occurred around 1600 BC, when there were repeated military expeditions and invasions by the Hittites from Anatolia. The environmental and historical conjuncture proved to be fatal for Ebla and also affected several other important cities in the region, such as Urshu and Alalakh (Matthiae 2006; Weiss et al. 1993; Courty et al. 1994; deMenocal et al. 2000). This drought was likely so severe that Ebla never regained its status, becoming a rural settlement which was wiped out by the Hittite military campaigns of the Late Bronze Age.

To summarise, Ebla's socio-political history appears to be closely linked to environmental shifts occurring in the region from the 4th to the 2nd millennia BC (Fiorentino et al. 2008). Traces of this interrelationship are clearly visible in the city's layout, but how are archaeologists to manage when the effects of great climate change do not have immediately recognisable results?

The Qatna site is exemplary in this regard, because no evident traces of abandonment or a break in the continuity of the archaeological record are recognisable. The first evidence of well-established occupation dates back to the EBA III, when a village settlement was built at the site of Tell Mishrifeh on a flat Pleistocene terrace located at the confluence of a minor and a major wadi (Cremaschi et al. 2003). The transformation of the simple settlement of the EBA III at the onset of the EBA IV was marked by the construction of a large multi-roomed granary, similar to a contemporary public grain store found in the upper city at Tell Beydar (Goddeeris et al. 1997). Qatna upper town became specialised in the intensive processing and mid- to long-term storage of agricultural surpluses, a function which it maintained until the end of the Early Bronze Age IVB.

The emergence of a social elite in Qatna is also confirmed by the discovery, on the western slope of the upper town, of a collective tomb dated to the EBA IVA, which contained 40 individuals and relative funerary objects. Also reinforcing the idea of the emergence of a genuinely urban settlement in the EBA IV acting as a regional centre is the presence of several archaeological sites distributed around Qatna at regular intervals, especially along the two wadis to the east and west of the main site (Morandi Bonacossi 2007).

A new radical change in Qatna's urban layout occurred at the transition between the EBA and the MBA. Qatna experienced a radical change at this time, although the evidence for this change is not as clear cut as in Ebla. Nevertheless, it is likely that the urban and functional reorganisation reflects changes in administrative patterns as well as in the balance of power. The flexibility of Qatna in the face of adverse climate change also emerges in the subsequent phase, when it seems to have emerged unharmed from the drought of the 3rd millennium BC (see fig. 2). The only recorded alteration is the urban revival of Qatna at the end of the MBA and the beginning of the Late Bronze Age. By this time, the city had already lost its position as the capital of a vast regional kingdom that it had held during the early MBA, becoming a so-called "hollow" city. This term refers to a new concept of city which emerged at the beginning of the MBA, devoid of common residents but hosting several administrative buildings (Morandi Bonacossi 2007).

Excluding the urban reorganisation that took place at the beginning of each wetter period, no other relevant traces of climate change are reflected in the city's layout (see fig. 2). Geoarchaeological analysis (Cremaschi 2007b; Cremaschi et al. 2003) points to various state-led initiatives aimed at endowing the city with water storage facilities, suggesting that the survival of Qatna's administrative system was down to the skill shown in managing locally available seasonal and perennial water sources. Nevertheless, traces of the drought crisis (2200–2000 BC) which affected the whole of the Near East are also visible in the Qatna landscape. Analysis of pollen from the nearby lake shows a drastic change in the landscape, in which the (juniper dominated) forested environment disappeared and the landscape became more open, coinciding with a drop in the level of the lake which became a wetland (Cremaschi 2007a; Valsecchi 2007).

Conclusion

Small- and medium-scale environmental changes are of critical importance for understanding the growth, development and mutual relations of ancient societies in the Near East. This is because even a small change in an important variable, such as rainfall, can rapidly alter the availability of natural resources, with effects on local human communities. This is especially true for the complex societies that developed during the middle and late Holocene in sub-arid ecotones where a few years of drought can cause the steppe to extend westwards, thus reducing the amount of arable land.

Assuming that agricultural potential is the key factor in determining the carrying capacity of a system, the identification of rainfall patterns may be regarded as fundamental to revealing the influence of short-term climate fluctuations on complex societies. The use of carbon isotopes in archaeological plant remains is a useful tool for this kind of analysis, since it is an in situ proxy which correlates the palaeoclimatic data inferred from $\delta^{13}C$ values with the chronological pattern as obtained by ^{14}C analysis.

Historical and cultural changes, embodied in the archaeological evidences of Ebla and Qatna, were found related to the variations in the rainfall regime in different ways. Increase in moisture during the Early Bronze Age IVA and the Middle Bronze Age favoured the spread of Ebla and Qatna control on a regional scale. On the other hand, water scarcity during the EBA IVB mined their political stability, leading to the re-organisation of the settlement pattern and changes in their cultural traditions.

Despite drought striking both cities, they reacted in different ways. Ebla, because of its proximity to the arid steppe, was less resilient than Qatna, where the state intervened to reduce the risk of famine by building water storage facilities.

Acknowledgments

The authors are grateful to Carole Crumley at the Stockholm Resilience Centre for her insightful comments. Special thanks also to Paolo Matthiae for allowing the study on Ebla material and to Daniele Morandi Bonacossi for providing the data on Qatna.

Bibliography

Allen, G. D., and D. D. Breshears. 1998. "Drought-induced shift of a forest-woodland ecotone: Rapid landscape response to climate variation." *Proceedings of the National Academy of Sciences (PNAS)* 95: 14839–14842.

Archi, A. 1999. "Cereals at Ebla." *Archív Orientální* 67: 503–518.

Bar-Matthews, M., A. Ayalon and A. Kaufman. 1997. "Late Quaternary Paleoclimate in the Eastern Mediterranean Region from Stable Isotope Analysis of Speleothems at Soreq Cave, Israel." *Quaternary Research* 47(2): 155–168.

Bradley, R. S. 2003. *Climate of the Last Millennium*. Climate System Research Center, University of Massachusetts. Amherst.

Butterlin, P. 2007. "Mari, les Šakkanakkû et la crise de la fin du troisième millénaire." In C. Kuzucuoğlu and C. Marro (eds.), *Sociétés humaines et changement climatique à la fin du troisième millénaire: une crise a-t-elle eu lieu en haute Mésopotamie?* Actes du colloque de Lyon, 5–8 décembre 2005, Varia Anatolica XIX, De Bocard, 227–246. Paris.

Butzer, K. W., 1982. *Archaeology as human ecology: Method and theory for a contextual approach*. Cambridge University Press. Cambridge.

Courty, M. A. 1994. "Le cadre paléogéographique des occupations humaines dans le basin du Haut-Khabur (Syrie du nord-est): Premiers résultats." *Paléorient* 20: 21–59.

Cremaschi, M. 2007a. "Qatna's lake: A geoarchaeological study of the Bronze Age capital." In D. Morandi Bonacossi (ed.), *Urban and Natural Landscapes of an Ancient Syrian Capital: Settlement and Environment at Tell Mishrifeh/Qatna and in Central-Western Syria*, Studi archeologici su Qatna 1, Forum, 93–104. Udine.

Cremaschi, M. 2007b. "The environment of ancient Qatna: Contribution from natural science and landscape archaeology." In D. Morandi Bonacossi (ed.), *Urban and Natural Landscapes of an Ancient Syrian Capital: Settlement and Environment at Tell Mishrifeh/Qatna and in Central-Western Syria*, Studi archeologici su Qatna 1, Forum, 331–336. Udine.

Cremaschi, M., L. Trombino, A. Sala and V. Valsecchi. 2003. "Underfitted streams and Holocene palaeoenvironment in the Region of Tell Mishrifeh-Qatna (Central Syria)." In D. Morandi Bonacossi et al. (eds.), "Tell Mishrifeh-Qatna 1999–2002. A preliminary report of the Italian component of the joint Syrian-Italian-German project." *Akkadica* 124(1): 71–77.

Crumley, C. L. (ed.). 1994. *Historical Ecology: Cultural knowledge and changing landscapes.* School of American Research. Santa Fe.

Cullen, H. M., P. B. deMenocal, S. Hemming, G. Hemming, F. H. Brown, T. Guilderson and F. Sirocko. 2000. "Climate change and the collapse of the Akkadian empire: Evidence from the deep sea." *Geology* 28(4): 379–382.

deMenocal, P., J. Ortiz, T. Guilderson and M. Sarnthein. 2000. "Coherent high- and low-latitude climate variability during the Holocene Warm Period." *Science* 288: 2198–2202.

Dincauze, D. 2006. *Environmental archaeology: Principles and practice.* 5th edition. Cambridge University Press. Cambridge.

Ehleringer, J. R. 1989. "Carbon Isotope Ratios and Physiological Processes in Aridland Plants." In P. W. Rundel, J. R. Ehleringer and K. A. Nagy (eds.), *Stable Isotopes in Ecological Research*, Ecological Studies 68, Springer-Verlag, 41–54. New York.

Ferrio, J. P., N. Alonso, B. López, J. L. Araus and J. Voltas. 2006. "Carbon isotope composition of fossil charcoal reveals aridity changes in NW Mediterranean Basin." *Global Change Biology* 12: 1253–1266.

Ferrio, J. P., J. L. Araus, R. Buxó, J. Voltas and J. Bort. 2005. "Water management practices and climate in ancient agriculture: Inference from the stable isotope composition of Archaeobotanical remains." *Vegetation, History and Archaeobotany* 14: 510–517.

Ferrio, J. P., J. Voltas, N. Alonso and J. L. Araus. 2007. "Reconstruction of climate and crop conditions in the past based on the carbon isotope signature of archaeobotanical remains." In T. E. Dawson and R. T. W. Siegwolf (eds.), *Stable Isotopes as Indicator of Ecological Change*, Elsevier Academic Press, 319–332. Amsterdam.

Ferrio, J. P., J. Voltas and J. L. Araus. 2003. "Use of carbon isotope composition in monitoring environmental changes." *Management of Environmental Quality: An International Journal* 14(1): 82–98.

Fiorentino, G., and V. Caracuta. 2007. "Palaeoclimate signals inferred from carbon stable isotope analysis of Qatna/Tell Mishrifeh archaeological plant remains." In D. Morandi Bonacossi (ed.), *Urban and Natural Landscapes of an Ancient Syrian Capital: Settlement and Environment at Tell Mishrifeh/Qatna and in Central-Western Syria*, Studi archeologici su Qatna 1, Forum, 153–160. Udine.

Fiorentino, G. and V. Caracuta. 2010. "The use of plants in a ritual well at Ebla (Tell Mardikh) – North-Western Syria." In P. Matthiae, F. Pinnock, L. Nigro and N. Marchetti (eds.), *Proceedings of the 6th International Conference of the Archaeology of the Ancient Near East* Vol.1, Harrassowitz Verlag, 307–320. Wiesbaden.

Fiorentino G., and V. Caracuta. 2013. "Use of wood and environment in Bronze Age Ebla (NW Syria): Results of the anthracological analyses." In F. Damblon et al. (eds.), *Proceedings of the IV International Congress of Anthracology*, Brussels, 8th–13th September 2008, BAR International Series 2486, 93–102.

Fiorentino, G., V. Caracuta, L. Calcagnile, M. D'Elia, P. Matthiae, F. Mavelli, and G. Quarta. 2008. "Third millennium B.C. climate change in Syria highlighted by carbon stable isotope analysis of ^{14}C-AMS dated plant remains from Ebla-Tell Mardikh." In D. Grocke and U. Wortmann (eds.), *Investigating climates, environments and biology using stable isotopes, Palaeogeography, Palaeoclimatology, Palaeoecology* (special issue) 266 (1–2): 51–58.

Fiorentino G., V. Caracuta, G. Volpe, M. Turchiano, G. Quarta, M. D'Elia and L. Calcagnile. 2009. "The First millennium AD climate fluctuations in the Tavoliere Plain (Apulia – Italy): New data from the ^{14}C AMS-dated plant remains from the archaeological site of Faragola." *Nuclear Instrument and Methods in Physics Research. Section B* 268(7–8): 1084–1087.

Flannery, K. V. 1972. "The cultural evolution of civilizations." *Annual Review of Ecology and Systematics* 3: 399–426.

Frumkin, A., I. Carmi, A. Gopher, D. C. Ford, H. P. Schwarcz and T. Tsuk. 1999. "A Holocene millennial-scale climatic cycle from a speleothem in Nahal Qanah Cave, Israel." *The Holocene* 9(6): 677–682.

Goddeeris, A., M. Lahlouh and M. E. Sténuit. 1997. "An Early Dynastic official building and Seleucid-Parthian level (field E)." In M. Lebeau and A. Suleiman (eds.), *Tell Beydar, three seasons of excavations (1992–1994), A preliminary report*, Brepols, 105–115. Turnhout.

Gosz, J. R. 1992. "Gradient analysis of ecological change in time and space: Implications for forest management." *Ecological Applications* 2(3): 248–261.

Greenberg, J. B., and T. K. Park. 1994. "Political ecology." *Journal of Political Ecology* 1: 1–12.

Hassan, F. 1978. *Demographic Archaeology*. Academic Press. New York.

Holland, M. M., P. G. Risser and R. J. Naiman (eds.). 1991. *Ecotones: The Role of Landscape Boundaries in the Management and Restoration of Changing Environments*. Chapman and Hall. New York.

Horowitz, A. 1992. *Palynology of Arid Lands*. Elsevier. Amsterdam.

Intergovernmental Panel on Climate Change. 1996. *Climate change 1995, impacts, adaptation and mitigation of climate change: Scientific-technical analyses*. Cambridge University Press, 171–189. Cambridge.

Issar, A. 2003. *Climate changes during the holocene and their impact on hydrological systems*. International hydrology series. Cambridge University Press. Cambridge.

Kaniewski D., E. Paulissen, E. Van Campo, H. Weiss, T. Otto, J. Bretschneider and K. Van Lerberghe. 2010. "Late Second-Early First Millennium BC Abrupt Climate Changes in Coastal Syria and their Possible Significance for the History of the Eastern Mediterranean." *Quaternary Reasearch* 74: 207–215.

Klengel, H. 1992. *Syria 3000 to 300 BC*. Akademie Verlag. Berlin.

Kuzucuoğlu, C., K. Deckers, L. Herveux, J. McCorriston, H. Pessin, S. Riehl, B. Geyer and E. Vila. 2007. "Characteristics and changes in archaeology-related environmental data during the third millennium B.C. in upper Mesopotamia: Collective comments to the data discussed during the symposium." In C. Kuzucuoğlu and C. Marro (eds.), *Sociétés humaines et changement climatique à la fin du troisième millénaire: une crise a-t-elle eu lieu en haute Mésopotamie?* Actes du colloque de Lyon, 5–8 décembre 2005, Varia Anatolica XIX, 573–580, De Bocard. Paris.

Matthiae, P. 1989. "The Destruction of Ebla Royal Palace: Interconnections between Syria, Mesopotamia and Egypt in the Late EB IVA." In P. Åström (ed.), *High, Middle, or Low?* Acts of an International Symposium on Absolute Chronology held at the University of Gothenburg 20th–22nd August 1987, 163–169. Gothenburg.

Matthiae, P. 1995. *Ebla, un impero ritrovato. Dai primi scavi alle ultime scoperte*. 2nd edition. Einaudi. Turin.

Matthiae, P. 2006. "Archaeology of a Destruction: The End of MB II Ebla in the Light of Myth and History." In E. Czerny (ed.), *Timelines: Studies in Honour of Manfred Bietak*, II, 39–51. Leuven.

Mazzoni, S., and C. Felli. 2007. "Bridging the third/second millennium divide: The Ebla and Afis evidence." In C. Kuzucuoğlu and C. Marro (eds.), *Sociétés humaines et changement climatique à la fin du troisième mil-*

lénaire: une crise a-t-elle eu lieu en haute Mésopotamie? Actes du colloque de Lyon, 5–8 décembre 2005, Varia Anatolica XIX, 205–224, De Bocard. Paris.

McAdams, R. 1978. "Strategies of Maximization, Stability, and Resilience in Mesopotamian Society, Settlement, and Agriculture." *Proceedings of the American Philosophical Society* 122(5): 329–335.

McIntosh, R. J., J. A. Tainter and S. K. McIntosh. 2000. "Climate, History and Human Action." In R. J. McIntosh, J. A. Tainter and S. K. McIntosh (eds.), *The way the wind blows: Climate, history and human action*, Columbia University Press, 1–42. New York.

Milano, L. 1981. "Alimentazione e regime alimentari nella Siria preclassica." *Dialoghi di Archeologia* 3: 85–121.

Milano, L. 1987. "Barley for ratio and barley for sowing (ARET II 51 and related matters)." *Acta Sumerologica* 9: 177–201.

Milano, L. 1990. *Testi amministrativi: assegnazione di prodotti alimentari. Archivio L.2712-parte I*. Archivi Reali di Ebla Testi (ARET) IX.

Morandi Bonacossi, D. 2007. "Qatna and its hinterland during the Bronze and Iron Ages: A preliminary reconstruction of urbanism and settlement in the Mishrifeh Region." In D. Morandi Bonacossi (ed.), *Urban and Natural Landscapes of an Ancient Syrian Capital: Settlement and Environment at Tell Mishrifeh/Qatna and in Central-Western Syria*, Studi archeologici su Qatna 1, Forum, 65–92. Udine.

Nieuwenhuis, G. J. A., A. J. W. de Wit, D. W. G. van Kraalingen, C. A. van Diepen and H. L. Boogaard. 2006. "Monitoring crop growth conditions using the global water satisfaction index and remote sensing." *ISPRS Commission VII mid-term symposium 2006 "Remote sensing: From pixels to processes"*. Enschede : ITC, 8–11 May 2006. www.isprs.org/proceedings/XXXVI/part7/PDF/066.pdf.

Niklewski, J., and W. van Zeist. 1970. "A late Quaternary pollen diagram from northwest Syria." *Acta Botanica Neerlandica* 19: 737–754.

Pabot, H. 1957. *Rapport au gouvernement de Syrie sur l'écologie végétale et ses applications*. Rapport no. 663, Food and Agriculture Organisation of the United Nations, Rome.

Peet, R., and M. Watts. 1994. "A political ecology for the 1990's." *Economic Geography* 69(3): 238–242.

Peltenburg, E. 2007. "Diverse settlement pattern changes in the middle Euphrates valley in the later third millennium B.C.: The contribution of Jerablus Tahtani." In C. Kuzucuoğlu and C. Marro (eds.), *Sociétés humaines et changement climatique à la fin du troisième millénaire: une crise a-t-elle eu lieu en haute Mésopotamie?*, Actes du colloque de Lyon, 5–8 décembre 2005, Varia Anatolica XIX, 226–247, De Bocard. Paris.

Pinnock F. 2004. *Lineamenti di Archeologia e Storia dell'Arte del Vicino Oriente Antico, ca. 3500–330 A.C.* Università di Parma: Istituto di Storia dell'Arte. Parma.

Redman, C. L., S. R. James, P. R. Fish and J. D. Rogers (eds.). 2004. *The Archaeology of Global Change: The Impact of Humans on Their Environment.* Smithsonian Institution Press. Washington, D.C.

Redman, C. L. 2004. "Environmental Degradation and Early Mesopotamian Civilization." In C. L. Redman, S. R. James, P. R. Fish and J. D. Rogers (eds.), *The Archaeology of Global Change: The Impact of Humans on Their Environment*, Smithsonian Institute Press, 158–164. Washington, D.C.

Renfrew, A. C. 1979. "System collapse as a social transformation: Catastrophe and anastrophe in early state societies." In A. C. Renfrew and K. Cooke (eds.), *Transformations: Mathematical approaches to culture change*, Academic Press, 481–506. New York.

Riehl, S., R. Bryson and K. Pustovoytov. 2007. "Changing growing conditions for crops during the Near Eastern Bronze Age (3000–1200 BC): The stable carbon isotope evidence." *Journal of Archaeological Science* 20: 1–12.

Roberts, N., and H. Wright. 1993. "Vegetational, lake-level, and climatic history of the Near East and Southwest Asia." In E. Kutzbach, H. E. Wright and T. Webb et al. (eds.), *Global climates since the last glacial maximum*, University of Minnesota Press, 194–220. Minneapolis.

Roberts, C. N., W. Eastwood, C. Kuzucuoğlu, G. Fiorentino and V. Caracuta. 2011. "Climatic, vegetation and cultural change in the eastern Mediterranean during the mid-Holocene environmental transition." *The Holocene* 21(1): 142–167.

Rosen, A. Miller. 2007. *Civilizing climate: Social responses to climate change in the ancient Near East.* Altamira Press. Lanham, Maryland.

Rouault, O., and M. G. Masetti-Rouault. 1993. *L'Eufrate e il tempo. Le civiltà del medio Eufrate e della Gezira siriana.* Electa. Milan.

Rouault, O., and M. Wäfler. 2000. *La Djéziré et l'Euphrate Syriens de la protohistoire à la fin du IIe millénaire AV J.-C. Tendances dans l'interprétation historique des données nouvelles.* Subartu VII. Brepols. Turnhout.

Schwartz, G. 1993. "Il periodo Uruk: rapporti 'internazionali' nel IV millennio a.C. e lo sviluppo (e l'abbandono?) della civiltà urbana." In O. Rouault and M. G. Masetti-Rouault (eds.), *L'Eufrate e il tempo. Le civiltà del medio Eufrate e della Gezira siriana*, Electa, 34–39. Milan.

Schwartz, G. M. 2007. "Taking the long view on collapse: A Syrian perspective." In C. Kuzucuoğlu and C. Marro (eds.), *Sociétés humaines et changement climatique à la fin du troisième millénaire: une crise a-t-elle eu lieu en haute Mésopotamie?* Actes du colloque de Lyon, 5–8 décembre 2005, Varia Anatolica XIX, 45–68, De Bocard. Paris.

Schwartz, G. M., and N. F. Miller. 2007. "The 'crisis' of the late third millennium B.C.: Ecofactual and artifactual evidence from Umm el-Marra and the Jabbul Plain." In C. Kuzucuoğlu and C. Marro (eds.), *Sociétés humaines et changement climatique à la fin du troisième millénaire: une crise a-t-elle eu lieu en haute Mésopotamie?* Actes du colloque de Lyon, 5–8 décembre 2005, Varia Anatolica XIX, 179–204, De Bocard. Paris.

Valsecchi, V. 2007. "Vegetation and environmental changes during the Middle-Late Holocene at Tell Mishrifeh/Qatna: Climate versus land-use." In D. Morandi Bonacossi (ed.), *Urban and Natural Landscapes of an Ancient Syrian Capital: Settlement and Environment at Tell Mishrifeh/Qatna and in Central-Western Syria*, Studi archeologici su Qatna 1, Forum, 93–104. Udine.

Vernet, J. L., C. Pachiaudi, F. Bazile, A. Durand, L. Fabre, C. Heinz, M. E. Solari and S. Thiebaut. 1996. "Le $\delta^{13}C$ de charbons de bois préhistoriques et historiques méditerranéens, de 35 000 BP à l'actuel. Premiers résultats." *Compte Rendu à l'Académie de Sciences* 323(2a): 319–324.

Voltas, J., J. P. Ferrio, N. Alonso and J. L. Araus. 2008. "Stable carbon isotopes in archaeobotanical remains and palaeoclimate." *Contribution to Science* 4(1): 21–31.

Weiss, H. 2000. "Causality and chance: Late third millennium collapse in Southwest Asia." In O. Rouault and M. Wäfler (eds.), *La Djéziré et l'Euphrate Syriens de la protohistoire à la fin du IIe millénaire AV J.-C. Tendances dans l'interprétation historique des données nouvelles*, Subartu VII, Brepols, 207–217. Turnhout.

Weiss, H., M. A. Courty, W. Wetterstrom, F. Guichard, L. Senior, R. Meadow and A. Curnow. 1993. "The genesis and collapse of third millennium North-Mesopotamian civilization." *Science* 261(5124): 995–1004.

White, I. D. 1995. *Climatic change and human society*. Arnold. London.

Wilkinson, T. J. 2004. *On the margin of the Euphrates: Settlement and land use at Tell es-Sweyhat and in the Upper Lake Assad Area, Syria*. The University of Chicago Oriental Institute publication. Volume 124. Chicago.

Yasuda, Y., H. Kitagawa and T. Nakagawa. 2000. "The earliest record of major anthropogenic deforestation in the Ghab Valley, northwest Syria: A palynological study." *Quaternary International* 73/74: 127–136.

Zohary, M. 1973. *Geobotanical Foundations of Middle East*. 2 volumes. Gustav Fischer Verlag. Stuttgart.

Provenance Studies of Ancient Textiles: A New Method Based on The Strontium Isotope System

Karin Margarita Frei

Abstract

The following study is a result of a multidisciplinary approach which combines theoretical and analytical tools from earth sciences with archaeological material culture, and focuses on ancient textiles. The aim of this study has been to develop a novel chemical protocol that potentially allows for provenance studies of ancient textiles. Generally, provenance studies in archaeology has witnessed an increasing popularity during the past 20 years, particularly with respect to tracing migration patterns of humans and animals in prehistory (Price et al. 1994: Price et al. 2010; Montgomery 2010; Evans et al. 2006; Grupe et al. 1997; Price et al. 2011).

Strontium

In archaeology, Strontium (Sr) isotopes serve as geochemical signatures which can, for example, be used to source a prehistoric skeleton to a particular geological area, and thereby to a geographical locality. Strontium isotopic signatures are conveyed from eroding geological material through soils and food chains into the human/animal skeleton, where strontium substitutes for calcium within the mineral lattices of skeletal tissues. The strontium isotopic tracer system relies on the use of two of the four naturally occurring isotopes, namely ^{87}Sr and ^{86}Sr, and particularly on the variations of their ratio, $^{87}Sr/^{86}Sr$. This ratio is related to the natural abundance of these two isotopes and is often referred to thereby in the literature as numbers in the order of ~0.7 (~7% ^{87}Sr/ ~10 % ^{86}Sr). The variations of the $^{87}Sr/^{86}Sr$ ratio are mostly dependent on the age and type of geological material (i.e. rock, soil or mineral). Hence, the knowledge of the local bedrock geology can be used as a first order estimate of the $^{87}Sr/^{86}Sr$ range in a particular area (Capo et al. 1998; Beard and Johnson 2000). However, recent studies have shown that knowledge of the local bedrock geology is not always sufficient to characterise the actual bio-available strontium isotopic composition of an area. Hence, it has been proposed that small animals, plants or waters can be good proxies of the true bio-available strontium (Price et al. 2002; Evans et al. 2009; Voerkelius et al. 2010; Frei and Frei 2011).

Strontium in Hair/Wool

Although strontium isotope analysis is a powerful tool with the potential to provide valuable information regarding the provenance of humans and animals in prehistory (Montgomery 2010), until now very few studies have applied this method to ancient textiles (Benson et al. 2006; von Carnap-Bornheim et al. 2007). One of the obvious reasons is probably the difficulty associated with the analysis of strontium isotopes in hair/wool, as strontium concentrations in these materials are very low, in the order of only a few or less parts per million (between 0.05 and 15 ppm). In contrast, teeth or bone tissues have much higher strontium concentrations of c. 50–1000 ppm. Figure 1 attempts to illustrate a simplified version of the strontium concentration and isotopic composition pathway from the geological strata to human and animal tissues. The strontium concentrations and isotopic ratios used in this diagrammatic sketch have been gathered from various publications (Bentley 2006; Morita et al. 1986; Attar et al. 1990; Christian et al. 1997; Rosborg et al. 2003) as well as my own data (Frei et al. 2009a, 2009b; Frei 2010).

Textiles

Today, textile research is a fast growing field within archaeology and is proving to be an important source of information on ancient societies. The importance of textiles, such as garments, blankets, sails, floor mats or other items, has historically been somewhat underestimated within the archaeological world (Barber 1991; Gleba 2008), despite the fact that the study of textiles probably goes as far back as the first Egyptian antiquarians. Moreover, fibre arts were already known in the Eurasian continent as early as the Upper Paleolithic (Good 2001), and in Denmark, as represented by the Iron

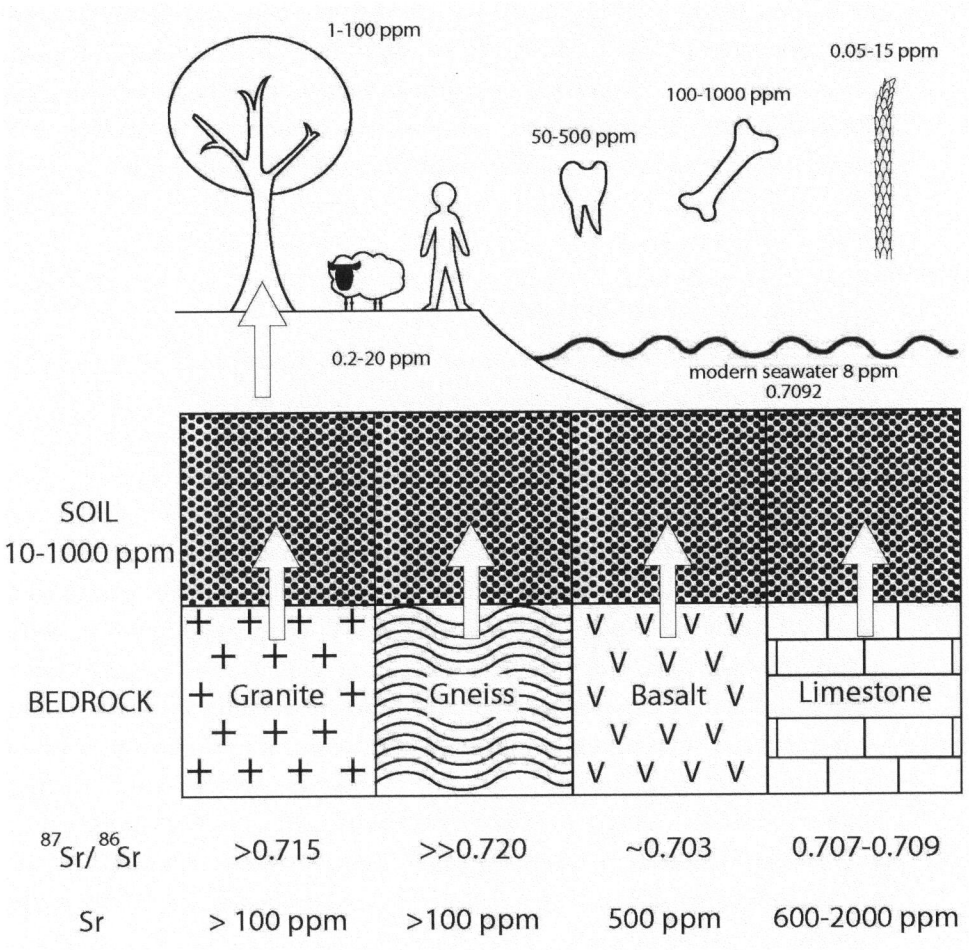

FIGURE 1. *Diagram illustrating the path of strontium from the geological strata to the human/animal hair. Data gathered from Bentley (2006), Attar (1990), Morita et al. (1986), Christian et al. (1997), Rosborg et al. (2003), and Frei et al. (2009).*

Age Huldremose case study presented below, several important textiles date back to the Mesolithic (Bender Jørgensen 1987) and Neolithic periods (Becker 1948). Consequently, the importance of developing a tracing method that would allow for pinpointing the area of origin of a textile's raw material seems obvious.

Prehistoric textiles are mainly made of two types of natural fibres: animal and plant fibres. Animal fibres are composed of proteins which are normally more resistant to decay than plant fibres that are made of cellulose. For that reason, this study focuses chiefly on wool textiles from the Pre-Roman Iron Age in Denmark (500 BC–AD1) as they constitute one of the largest and best preserved collections worldwide (Mannering et al. 2010).

New tracing methodology for ancient textiles

As mentioned above, the aim of the study outlined here is to present the method that I have recently developed for provenance studies of ancient textiles (Frei et al. 2009a; Frei 2010). This method focuses on the importance of a deep chemical pre-cleaning prior to chromatographic separation of strontium, thus allowing for a contamination-free recovery of strontium from the raw material of an ancient textile. The composition of this strontium fraction can potentially pinpoint the area of origin of that material. Attar et al. (1990) showed the importance of removing the lipid fraction of hair as this part of the hair is the most sensitive with respect to contamination from the environment. In the case of ancient textiles, contamination takes place during diagenetic processes as a result of long-term burial and associated exposure to soil particles and percolating waters at the burial site. Frei et al. (2009a) propose a pre-cleaning method involving, among other substances, the use of hydrofluoric acid (HF) which dissolves strontium containing silicate dust particles and removes the lipid fraction of hair fibres. Similarly, hydrochloric acid (HCl) can remove carbonate-rich (and therefore strontium-rich) particles that can potentially mask the primary nutritional strontium isotopic composition of the wool. For the exact chemical protocols, we refer the reader to the work of Frei et al. (2009a, 2009b, 2010) and Frei (2010).

Additional contamination risks, when working with ancient textiles, are dyes. Natural organic dyes were used early in prehistory, e.g. in Mesopotamia, the eastern Mediterranean and Egypt (Cardon 2007). As organic dyestuffs can potentially have provenances that are different from those of the wool fibre threads themselves, and hence mask the original strontium isotopic signature of a wool textile, it is of absolute importance to remove any dyestuff (and mordant if present) prior to strontium isotope analysis. To facilitate this, we have developed an extra step in the chemical pre-cleaning protocol that also allows dyestuff to be removed by a deep oxidative treatment (Frei et al. 2010). Hence, these two new methodologies allow for potentially providing information on the raw material of ancient textiles, even if they have been dyed.

A case study: Huldremose

The Huldremose peat bog site is located in the northeastern part of Jutland, Denmark. Textiles at this site were unearthed on two occasions, the first in 1879 (Huldremose I) in combination with a woman's bog body (Lab. No. K-1396; Sellevold et al. 1984), and the second in 1895 (Huldremose II) by peat diggers (Mannering et al. 2010). The body (Huldremose I find) which is currently on display at the National Museum of Denmark was lying on its back with the head oriented towards the west and her feet to the east. Her legs were drawn up and the right arm was severed from the body, her left arm was bent across the chest and tied to the torso (Brøndsted 1963; Glob 1965). Both finds have recently been ^{14}C re-dated: Huldremose I to 350–341 BC; and Huldremose II to 350–330 BC (Mannering et al. 2010).

During a re-investigation of the Huldremose woman which is displayed at the National Museum of Denmark in the permanent exhibition of Danish Prehistory I had the unique opportunity to study some of the textiles which were recovered with the bog mummy. The Huldremose woman (Huldremose I) was wearing several pieces of clothing, consisting of a chequered skirt, a chequered scarf and two skin capes. In our study, we analysed a thread sample from the chequered scarf, a sample from the mummy's own skin, and finally three spun plant fibre threads discovered during studies on the bog body and on palynological soil samples (fig. 2). This study revealed that the Huldremose woman must have been wearing a garment underneath the other two wool textiles and skin capes. Three of these newly discovered spun plant fibre threads have also been analysed.

The Huldremose II garment was a single find consisting of a large peculiarly shaped tubular textile. In our study, we analysed eleven thread

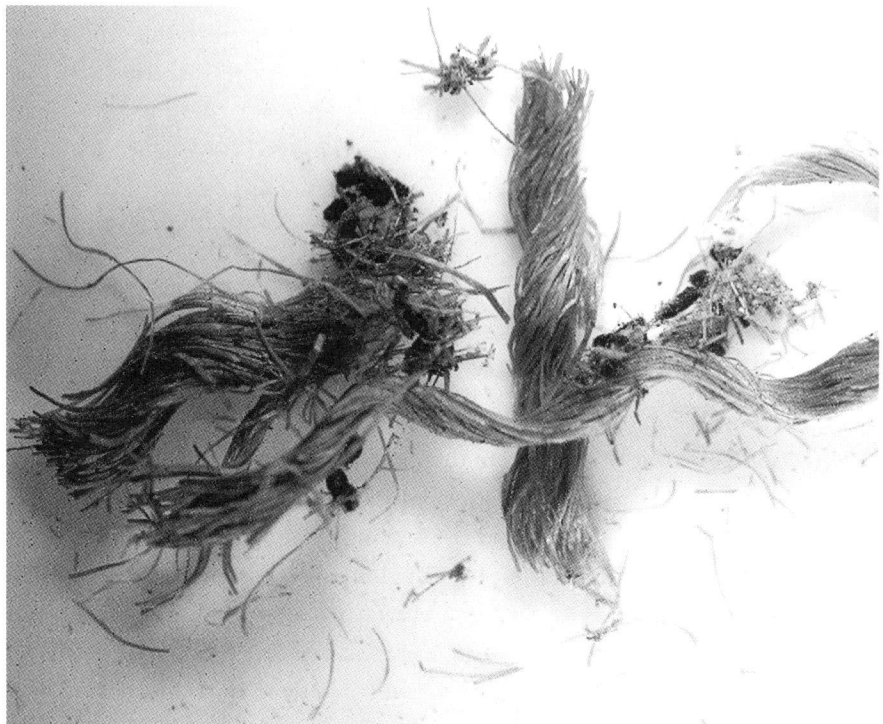

FIGURE 2. *Plant fibre threads found while investigating the Huldremose I find. Length of threads ~ 0.5–2 cm. Photo by Karin Margarita Frei.*

samples from this specific garment, chosen randomly from all edges of the garment (Frei et al. 2009b).

Results

Figure 4 depicts the strontium isotopic results of the Huldremose I find, and figure 5 depicts the strontium isotope results of the Huldremose II find (Frei et al. 2009b). Denmark is quite an ideal target area for isotopic tracing studies as it presents a relatively homogeneous geology (dominated by Cretaceous-Tertiary limestone platforms and glaciogenic cover sediments)

FIGURE 3. *Garments found in connection with the Huldremose woman find (Huldremose I). Photo by Roberto Fortuna (National Museum of Denmark).*

in comparison to the neighbouring countries of Sweden and Norway (which are dominated by old shield areas and fold belts involving lithologies prevalent in such old cratonic areas). To define the local strontium isotope range of Denmark of $^{87}Sr/^{86}Sr \sim 0.708$-$0.711$, several sources of information as well as baseline studies have been used (Frei and Frei 2011; Jørgensen et al. 1999; Frei et al. 2009a; Price et al. 2007).

The strontium isotope analysis of the wool thread sample from the scarf (C 3474) of the Huldremose I find (fig. 4) shows that the sheep from which the wool yarn was spun was most likely feeding on grass from pastures in Denmark. Conversely, the plant fibres are from a place with a basement

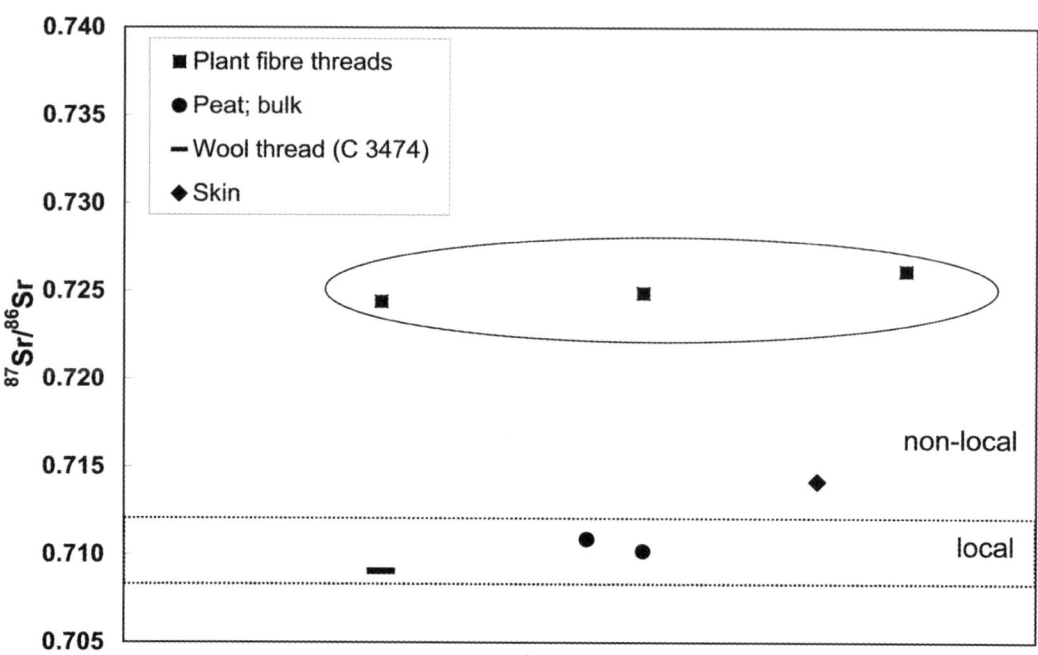

FIGURE 4. $^{87}Sr/^{86}Sr$ ratios of bulk peat, and residues of skin, plant and wool fibres from the Huldremose I find. Plant fibres define a group (encircled by an ellipse) that is characterised by non-local $^{87}Sr/^{86}Sr$ ratios. The skin sample also lies in the non-local $^{87}Sr/^{86}Sr$ range. Peat and wool yarn fibres from the scarf of the Huldremose I find (sample C 3474) lie within the range of $^{87}Sr/^{86}Sr$ ratios that are considered local.

geology much older in age than that found in Denmark (fig. 4). Similarly, values of the skin of the mummy reveal that Huldremose woman had probably been travelling in places outside of Denmark prior to her death. However, as the method for analysing strontium isotopes in human skin has not yet been fully developed, more studies are needed to verify the latter result.

Figure 5 depicts the strontium isotope results of the Huldremose II find (Frei et al. 2009b). Here eleven threads of one single garment were analysed, in order to investigate the strontium isotope heterogeneity within one single large textile piece. As shown by the diagram, some wool threads seem to be local (from within Denmark, excluding Bornholm) and some

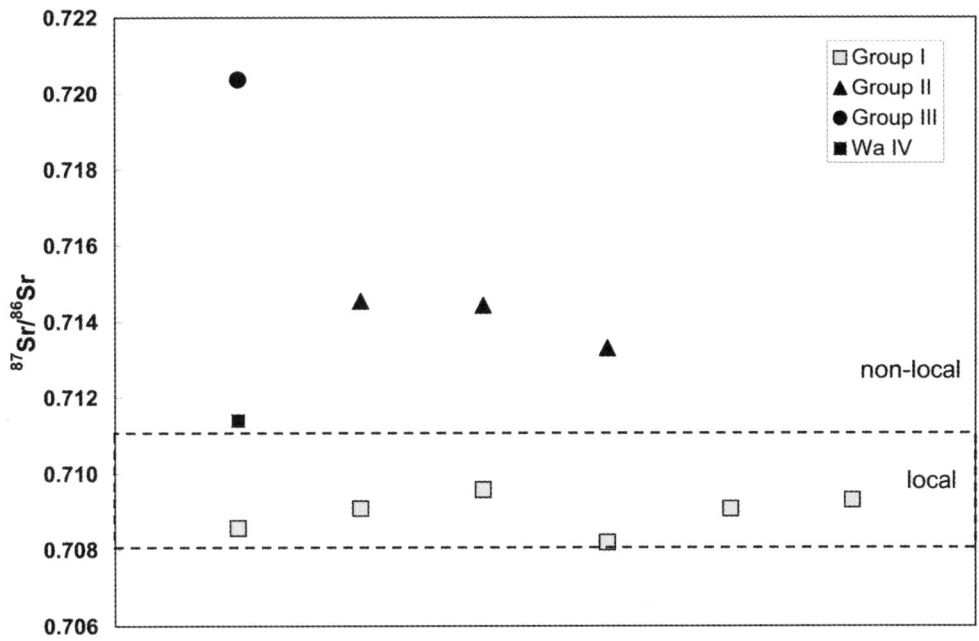

FIGURE 5. $^{87}Sr/^{86}Sr$ ratios of wool residues samples from a single large archaeological garment (Huldremose II find). Three provenance groups can be discerned: Group I wool threads indicate a local source of strontium, with $^{87}Sr/^{86}Sr$ ratios compatible with Danish soil extracts (Frei et al. 2009). Groups II and III comprise wool threads with elevated $^{87}Sr/^{86}Sr$ ratios implying a non-local provenance.

are non-local (from outside Denmark). Consequently, the results unravel an unexpected dimension of the garment, namely that the textile was made of equally fine and homogeneously spun quality wool (personal communcation with Irene Skals, Conservator at the National Museum of Denmark) originating from different areas. Two of these areas potentially lie outside of Denmark (excluding Bornholm). The geographic location of these two non-local areas cannot be pinpointed at this point with certainty, as more base research is needed in the field of delineating bio-available strontium isotopic signatures of different areas within Europe. Nonetheless, we can say that the radiogenic strontium isotope values measured in some of the non-local thread samples are compatible with an origin from old Precambrian terrains, such as those omnipresent in Sweden and Norway as part of the Fennoscandian shield.

However, there are limitations that need to be accounted for when interpreting the data. For example, the strontium isotope tracing system necessitates the knowledge of the local bio-available strontium isotopic range which is still not yet explored in many areas. Furthermore, it is important to establish a good communication with the archaeologist, the textiles experts and archaeometrists to avoid over-interpreting the data.

Conclusion

In conclusion, the strontium isotope results of the Huldremose I and II finds disclose new information and shed new light on Pre-Roman Iron Age textile production in Denmark as well as on the possible trading of textiles and/or their raw material. In addition, the discovery of spun plant fibre thread pieces with a non-local provenance necessitates a re-investigation and renewed study of the bog bodies which have often been interpreted as being offered naked, as is the case with, for example, the Tollund man (on display at the Silkeborg Museum), since there is an emerging new hypothesis that these bodies were wrapped/dressed with plant fibre textiles which, due to their poor preservation, could have simply perished over time. Finally, the Huldremose case study demonstrates the potential of the recently developed tracing method within the field of ancient textiles as a novel tool in archaeological research, as well as presenting new unexpected challenges.

Acknowledgments

I would like to thank the Danish Center of Isotope Geology in the Geoscience and Natural Resources Institute, University of Copenhagen, the National Museum of Denmark, as well as Robert Frei, Ulla Mannering, Irene Skals, Margarita Gleba, Marie Louise Nosch, Eva Andersson Strand, Marianne Bloch Hansen, Roberto Fortuna, Pia Bennike, Poul Otto Nielsen and T. Douglas Price. I acknowledge the financial support of the Danish National Research Foundation grant DNRF-64. An anonymous reviewer is also thanked for constructive comments which helped improve the manuscript.

Bibliography

Attar, K. M., M. A. Abdelaal and P. Debayle. 1990. "Distribution of trace-elements in the lipid and nonlipid matter of hair." *Clinical Chemistry* 36: 477–480.

Barber, E. J. W. 1991. *Prehistoric Textiles*. Princeton University Press. Princeton, New Jersey

Beard, B. L., and C. M. Johnson. 2000. "Strontium isotope composition of skeletal material can determine the birth place and geographic mobility of humans and animals." *Journal of Forensic Sciences* 45: 1049–1061.

Becker, C. J. 1948. *Tørvegravning i ældre jernalder*. Nationalmuseets Arbejdsmark. National Museum of Denmark. Copenhagen.

Bender Jørgensen, L. 1987. "Stone Age textiles in North Europe." In P. Walton and J. P. Wild (eds.), *Textiles in Northern Archaeology*. Archetype Publications. London.

Benson, L. V., E. M. Hattori, H. E. Taylor, S. R. Poulson and E. A. Jolie. 2006. "Isotope sourcing of prehistoric willow and tule textiles recovered from western Great Basin rock shelters and caves – proof of concept." *Journal of Archaeological Science* 33: 1588–1599.

Bentley, R. A. 2006. "Strontium isotopes from the earth to the archaeological skeleton: A review." *Journal of Archaeological Method and Theory* 13: 135–187.

Brøndsted, J. 1963. *Nordische Vorzeit*. K. Walchholtz. Neumünster.

Capo, R. C., B. W. Stewart and O. A. Chadwick. 1998. "Strontium isotopes as tracers of ecosystem processes: Theory and methods." *Geoderma* 82: 197–225.

Cardon, D. 2007. *Natural Dyes; Sources, tradition, technology and science*. Archetype Publications. London.

Christian, J. M., G. Pielack, D. E. Freer and M. J. Lee. 1997. "Reference ranges for 34 trace elements in hair by inductively coupled plasma – Mass spectrometry (ICP-MS)." *Clinical Chemistry* 43 (49th Annual Meeting of the American Association for Clinical Chemistry, Atlanta July 20–24, 1997) 6, pt. 2, p. 277.

Evans, J. A., C. A. Chenery, and A. P. Fitzpatrick. 2006. "Bronze age childhood migration of individuals near Stonehenge, revealed by strontium and oxygen isotope tooth enamel analysis." *Archaeometry* 48: 309–321.

Evans, J. A., J. Montgomery and G. Wildman. 2009. "Isotope domain mapping of Sr-87/Sr-86 biosphere variation on the Isle of Skye, Scotland." *Journal of the Geological Society* 166: 617–631.

Frei, K. M. 2010. *Provenance of Pre-Roman Iron Age Textiles: Methods Development and Applications.* PhD thesis, University of Copenhagen.

Frei, K. M., and R. Frei. 2011. "The geographic distribution of strontium isotopes in Danish surface waters: A base for provenance studies in archaeology, hydrology and agriculture." *Applied Geochemistry* 26: 326–340.

Frei, K. M., R. Frei, U. Mannering, M. Gleba, M. L. Nosch and H. Lyngstrøm, H. 2009a. "Provenance of ancient textiles: Pilot study evaluating the strontium isotope system in wool." *Archaeometry* 51: 252–276.

Frei, K. M., I. Skals, M. Gleba and H. Lyngstrøm. 2009b. "The Huldremose Iron Age textiles, Denmark: An attempt to define their provenance applying the Strontium isotope system." *Journal of Archaeological Science* 36: 1965–1971.

Frei, K. M., I. Vanden Berghe, R. Frei, U. Mannering and H. Lyngstrøm. 2010. "Removal of natural organic dyes from wool: Implications for ancient textile provenance studies." *Journal of Archaeological Science* 37: 2135–2145.

Gleba, M. 2008. *Textile Production in Pre-Roman Italy.* Oxbow Books. Oxford.

Glob, P. V. 1965. "*Mosefolket. Jernalderens Mennesker bevaret i 2000 aar.*" Gyldendal. Copenhagen.

Good, I. 2001. "Archaeological textiles: a review of current research." *Annual Review of Anthropology* 30: 209–226.

Grupe, G., T. D. Price, P. Schroter, F. Sollner, C. M. Johnson and B. L. Beard. 1997. "Mobility of Bell Beaker people revealed by strontium isotope ratios of tooth and bone: A study of southern Bavarian skeletal remains." *Applied Geochemistry* 12: 517–525.

Jørgensen, N. O., J. Morthorst and P. M. Holm. 1999. "Strontium-isotope studies of 'brown water' (organic-rich groundwater) from Denmark." *Hydrogeology Journal* 7: 533–539.

Mannering, U. 2009. "Dragten i tidlig jernalder." *Tollundsmandens verden.* Wormianum og Silkeborg Kulturhistoriske Museum. Silkeborg.

Mannering, U., G. Possnert, J. Heinemeier and M. Gleba. 2010. "Dating Danish textiles and skins from bog finds by means of C-14 AMS." *Journal of Archaeological Science* 37: 261–268.

Montgomery, J. 2010. "Passports from the past: Investigating human dispersals using strontium isotope analysis of tooth enamel." *Annals of Human Biology* 37: 325–346.

Morita, H., S. Shimomura, A. Kimura and M. Morita. 1986. "Interrelationships between the concentration of magnesium, calcium, and strontium in the hair of Japanese schoolchildren." *Science of the Total Environment* 54: 95–105.

Price, T. D., S. H. Ambrose, P. Bennike, J. Heinemeier, N. Noe-Nygaard, E. B. Petersen, P. V. Petersen and M. P. Richards. 2007. "New information on the stone age graves at Dragsholm, Denmark." *Acta Archaeologica* 78: 193–219.

Price, T. D., J. H. Burton and R. A. Bentley. 2002. "The characterization of biologically available strontium isotope ratios for the study of prehistoric migration." *Archaeometry* 44: 117–135.

Price, T. D., J. H. Burton, R. J. Sharer, J. E. Buikstra, L. E. Wright, L. P. Traxler and K. A. Miller. 2010. "Kings and commoners at Copan: Isotopic evidence for origins and movement in the Classic Maya period." *Journal of Anthropological Archaeology* 29: 15–32.

Price, T. D., K. M. Frei, A. S. Dobat, N. Lynnerup and P. Bennike. 2011. "Who was in Harold Bluetooth's army? Strontium isotope investigation of the cementery at the Viking Age fortress at Trelleborg, Denmark." *Antiquity* 85: 476–489.

Price, T. D., G. Grupe and P. Schrotter. 1994. "Reconstruction of migration patterns in the Bell Beaker Period by stable strontium isotope analysis." *Applied Geochemistry* 9: 413–417.

Rosborg, I., B. Nihlgard and L. Gerhardsson. 2003. "Hair element concentrations in females in one acid and one alkaline area in southern Sweden." *Ambio* 32: 440–446.

Sellevold, B. J., U. L. Hansen and J. B. Jørgensen. 1984. *Iron Age Man in Denmark*. Det Kongelige Nordiske Oldskriftselskab. Copenhagen.

Voerkelius, S., D. L. Gesine, S. Rummel, C. R. Quétel, G. Heiss, M. Baxter, C. Brach-Papa, P. Deters-Itzelsberger, S. Hoelzl, J. Hoogewerff, E. Ponzevera, M. van Bocxstaele and H. Ueckermann. 2010. "Strontium isotopic signatures of natural mineral waters, the reference to a simple geological map and its potential for authentication of food." *Food Chemistry* 118: 933–940.

von Carnap-Bornheim, C., M.-L. Nosch, G. Grupe, A.-M. Mekota and M. M. Schweissing. 2007. "Stable strontium isotopic ratios from archaeological organic remains from the Thorsberg peat bog." *Rapid Communications in Mass Spectrometry* 21: 1541–1545.

Contributors

Peter M.M.G. Akkermans
Professor of Near Eastern Archaeology
Chair of Education and Vice-Dean
Faculty of Archaeology – Leiden
 University
P.O. Box 9514, 2300 RA Leiden
The Netherlands

Benjamin Arbuckle
Assistant Professor
The Department of Anthropology
301 Alumni Building
Chapel Hill, NC 27599-3115
USA

Pernille Bangsgaard
Natural History Museum of Denmark
Øster Voldgade 5–7
1350 Copenhagen K
Denmark

Miroslav Bárta
Professor of Egyptology
Czech Institute of Egyptology
Faculty of Arts, Charles University
 in Prague
Nam. J. Palacha 2
110 00 Prague 1
Czech Republic

Peter F. Biehl
Professor and Chair, Department of
 Anthropology
Director, Institute for European and
 Mediterranean Archaeology
State University of New York at Buffalo
380 MFAC- Ellicott Complex
Buffalo, NY 14261-0005,
USA

Tom Boiy
KU Leuven University
Faculteit Letteren
Blijde Inkomststraat 21
3000 Leuven
Belgium

Joachim Bretschneider
Professor for Near Eastern Archaeology
KU Leuven University
Faculteit Letteren
Blijde Inkomststraat 21
3000 Leuven
Belgium

Valentina Caracuta
Laboratory of Archaeobotany and
 Palaeoecology
Department of Cultural Heritage
 University of Salento
Via D. Birago, 64
73100 Lecce
Italy

Elise Van Campo
EcoLab : Laboratoire d'Ecologie
 Fonctionnelle et Environnement
UMR 5245 CNRS UPS INPT
Université Paul Sabatier-Toulouse 3
Bâtiment 4R1
118 Route de Narbonne
31062 Toulouse cedex 9
France

Claudio Casati
Bornholms Museum
Sct. Mortensgade 29
3700 Rønne
Denmark

Louis Chaix
Emeritus Professor in Archaeology and
 Population in Africa Laboratory
Université de Genève
Department of Genetics and Evolution
118 (Acacias I, Anthropology Unit)
12 rue Gustave-Revilliod
1211 Geneva 4
Switzerland

Rachael J. Dann
Associate Professor, Egyptian and
 Sudanese Archaeology
Department of Cross-Cultural and
 Regional Studies
Karen Blixens Vej 4
2300 Copenhagen S
Denmark

Maurits Ertsen
Water Resources
Civil Engineering and Geosciences
Delft University of Technology
PO Box 5048
2600GA Delft
The Netherlands

Girolamo Fiorentino
Associate Professor in New Advances
 in Archaeological Research
Laboratory of Archaeobotany and
 Palaeoecology
Department of Cultural Heritage
 University of Salento
Via D. Birago, 64
73100 Lecce
Italy

Karin Margarita Frei
Senior Researcher
National Museum of Denmark
Environmental Archaeology and
 Material Science
Ny Vestergade 11
1471 Copenhagen K
Denmark

Matthieu Honegger
Professeur et chaire d'archéologie
 préhistorique
Institut d'Archéologie
Université de Neuchâtel
Laténium
Espace Paul Vouga
2068 Hauterive
Switzerland

Greta Jans
KU Leuven University
Faculteit Letteren
Blijde Inkomststraat 21
3000 Leuven
Belgium

Akemi Kaneda
Faculty of Archaeology
Leiden University
P.O. Box 9514, 2300 RA Leiden
The Netherlands

David Kaniewski
EcoLab : Laboratoire d'Ecologie
 Fonctionnelle et Environnement
UMR 5245 CNRS UPS INPT
Université Paul Sabatier-Toulouse 3
Bâtiment 4R1
118 Route de Narbonne
31062 Toulouse cedex 9
France

Eva Kapteijn
Postdoctoral Researcher
Sagalassos Archaeological Research
 Project
Katholieke Universiteit Leuven
Belgium

Susanne Kerner
Associate Professor for Near Eastern
 Archaeology
Department of Cross-Cultural and
 Regional Studies

Karen Blixens Vej 4
2300 Copenhagen S
Denmark

Karel Van Lerberghe
Emeritus Professor
KU Leuven University
Faculteit Letteren
Blijde Inkomststraat 21
3000 Leuven
Belgium

Cheryl Makarewicz
Professorship for Archaeozoology and
    Isotope Research
Institut für Ur- und Frühgeschichte
Christian-Albrechts-Universität
Johanna-Mestorf-Str. 2–6
24098 Kiel
Germany

Richard H. Meadow
Senior Lecturer on Anthropology
Director, Zooarchaeology Laboratory
Department of Anthropology and
    Peabody Museum
Harvard University
11 Divinity Avenue
Cambridge, Massachusetts, 02138
    USA

Chris Meiklejohn
Professor Emeritus
Department of Anthropology
University of Winnipeg,
1244 Wolseley Ave.,
Winnipeg, MB
R3G 1H4, Canada

Deborah C. Merrett
Adjunct Faculty
Department of Archaeology
Simon Fraser University
8888 University Dr
Burnaby, BC
V5A 1S6, Canada

Olivier P. Nieuwenhuyse
Assistant Professor
Faculty of Archaeology
Leiden University
POB 9515
2300RA Leiden
The Netherlands

Johannes van der Plicht
Professor University of Groningen
Center for Isotope Research
Nijenborgh 4
9747AG Groningen
The Netherlands

Simone Riehl
Privatdozentin
Institute of Archaeological Sciences
    and Senckenberg Center of Human
    Evolution and Palaeoenvironment
    (HEP)
University of Tubingen
Rümelinstraße 23
72070 Tübingen
Germany

Neil Roberts
Professor of Physical Geography
School of Geography, Earth and
    Environmental Sciences
Plymouth University
Plymouth PL4 8AA
United Kingdom

Anna Russell
Faculty of Archaeology
Leiden University
POB 9515
2300RA Leiden
The Netherlands

Lasse Sørensen
Bornholms Museum
Sct. Mortensgade 29
3700 Rønne
Denmark

Jason Ur
Professor of Archaeology
Department of Anthropology
Harvard University
11 Divinity Avenue
Cambridge, Massachusetts, 02138
USA

Joshua Wright
Visiting Assistant Professor of EAS and Anthropology
Department of Anthropology at Oberlin College
Oberlin, OH, 44074
USA